INTERNATIONAL CENTRE FOR MECHANICAL SCIENCES

COURSES AND LECTURES - No. 272

IDENTIFICATION OF VIBRATING STRUCTURES

EDITED BY

H.G. NATKE

UNIVERSITY OF HANNOVER

SPRINGER-VERLAG WIEN GMBH

Originally published by Springer-Verlag Wien New York in 1982

ISBN 978-3-211-81651-6 ISBN 978-3-7091-2896-1 (eBook)
DOI 10.1007/978-3-7091-2896-1

PREFACE

The dynamic behaviour of novel and complicated structures often needs to be investigated by system analysis and system identification, since it usually has to meet certain requirements. A priori knowledge concerning the real system is gained by system analysis and/or previous tests, and it results in a non-parametric and/or in a parametric mathematical model. The identification of system parameters, i.e. experimental system analysis, is performed using measured quantities and taking into account deterministic and stochastic errors. If results of the identification have to be compared with the results of the system analysis, and if certain error bounds are exceeded, the model has to be improved. System identification has to take into account random aspects (errors, test signals), the real dynamic behaviour (damping coupling, non-linearities) and questions concerning practical handling (including large systems, economics). A broad understanding of system identification needs as its basis an extended theory of structural vibrations and estimation (stochastic processes), and must be coupled with practical aspects including experience and validated software.

The course on Identification of Vibrating Structures, the lecture notes of which are collected in this volume, deals with the topics mentioned above. First an introduction into the subject is given, and the theoretical background of vibrating structures and parameter estimation methods is dealt with. The following lectures deal with several identification methods including applications. The next papers discuss the indirect identification, that means the adjustment of theoretical models (results of system analysis) by the results of vibration tests (estimated values). These parts are supplemented by a presentation of an example of commercially available hard- and software including applications. For the practical application of system identification concerning large and complicated structures synthesizing techniques (substructure techniques) including error analysis are necessary, which are the subject of the following two papers, followed by identification of non-linear systems. The closing lectures deal with modern developments, firstly from the point of view of control theory, and secondly by coming from the theory of stochastic systems. The last three lectures deal with theoretical aspects, including examples of (simple) systems mainly with regard to non-deterministic system analysis. This seems to be a flexible and far-reaching tool worthwhile studying, because several applications with regard to early failure detection are possible, and it may give impulses towards the identification of linear and non-linear structures.

I wish to express my thanks to the coauthors, who made a very successful course possible at Udine, and I hope this volume will be equally well received by the reader.

H.G. Natke

CONTENTS

Page

Identification of vibrating structures: an introduction
by H.G. Natke 1

Multi-degree-of-freedom systems – a review
by H.G. Natke 15

Introduction to system identification using parameter estimation methods
by H. Unbehauen 53

Some identification methods using measured harmonic responses
by R. Fillod 121

Identification methods II – concepts and techniques
by E. Breitbach 147

Indirect identification methods I: adjustment of mathematical models by the results of vibrating tests: using eigensolutions
by G. Lallement 179

Correction of analytical models for damped linear systems using experimentally obtained forced vibration responses
by H.P. Felgenhauer 195

Indirect methods: correction of the results of system analysis by results of identification – a survey
by H.G. Natke 225

Application of system identification with commercially available hardware and software
by P. Catchpole 257

Modal synthesis – modal correction – modal coupling
by E. Breitbach 321

Linearized error analysis in synthesis techniques
by H.G. Natke 349

Identification of nonlinear systems
by E. Breitbach 373

Some modern developments in system identification using parameter estimation methods
by H. Unbehauen 405

Modern methods of covariance analysis and applications
by W. Wedig 445

Integration of nonlinear stochastic systems and applications to the crack identification
by W. Wedig 473

Stochastic systems with distributed parameters
by W. Wedig 493

IDENTIFICATION OF VIBRATING STRUCTURES: AN INTRODUCTION

H.G. Natke
Curt-Risch-Institut
Universität Hannover

The dynamic behaviour of novel and complicated structures has to be investigated by system analysis based on drawings (Fig. 1). Starting from a physical model, neglecting all physical effects which are not relevant to the problem to be investigated, one has to build up the mathematical model as a structured equation (parametric model). In certain cases it has to be simplified compared to the physical model. The parameter values of the model have to be calculated using the drawings. Taking into account the loadings (-assumptions) the dynamic system response can be predicted. System analysis results therefore in a parametric computational model. Its accuracy depends on the influence

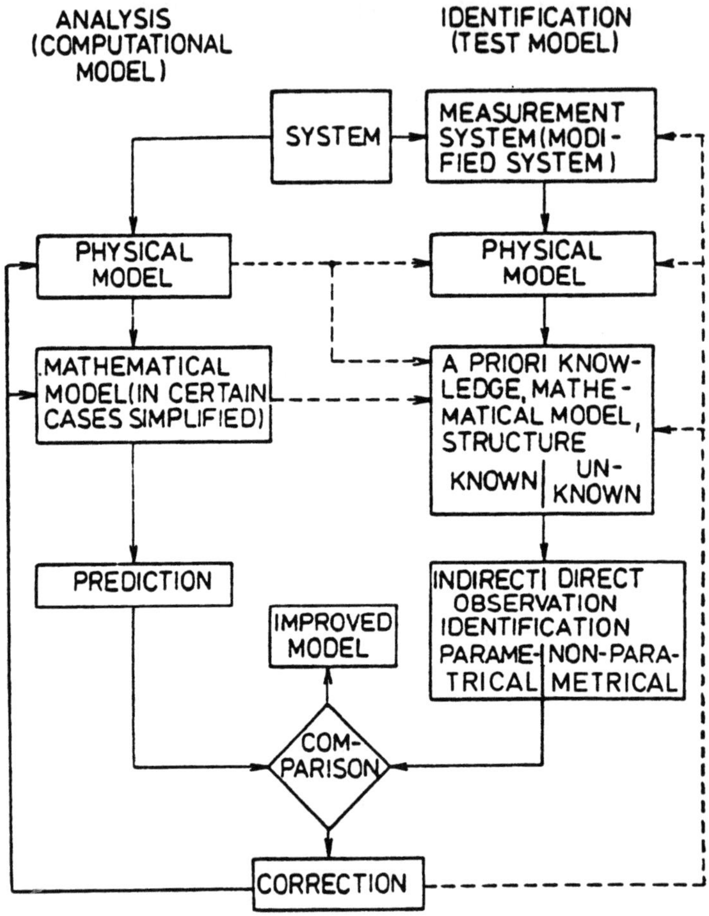

Fig. 1 Principal proceeding in system analysis and identification

of the introduced simplifications and assumptions. If the structure to be investigated is novel and experiences of comparable structures are not available the errors of the predicted results are unknown, the results may be useful for investigations but cannot prove the qualification of the structure.

Because of the unknown uncertainties of the computational model in the case mentioned above, tests have to be carried out additionally. These tests may concern the loading configurations (static, dynamic), but only a few loading configurations can be realized by testing for economic and testing limits e.g. Other more interesting tests concern the dynamic behaviour of the system itself: System identification (Fig. 1). The system under test is modified by the measuring and excitation equipment, it is called the measurement system (Fig. 1). The physical model of the measurement system and the a priori knowledge resulting from system analysis and/or previous tests (e.g. of subsystems) lead to a mathematical model which may be structured (parametric)or unstructured (non-parametric). The direct or indirect observed quantities enable an estimation of the model characteristics for describing the dynamic behaviour of the system under test. If the model is structured, it results in estimated parameter values. Because of the stochastic disturbances within the measured signals - besides a possible stochastic excitation - estimation methods have to be applied.

With the above briefly described identification process one yields the test model which can be compared with the computational model. The comparison may be done with the values of the system parameters or with the predicted or estimated results of the computational or the test model. In general the comparison is unsatisfactory regarding error bounds depending on the given problem, so that an improvement of the computational model is necessary. Sometimes the comparison results in an iterative procedure concerning an improved physical model of identification concerning e.g. the test equipment or environment. If the structured mathematical models of system analysis and identification are equal, the improvement of the computational model reduces in a parameter adjustment.

For better understanding Fig. 2 gives a classification of the identification problem to the structural problem. The direct problem is the known task of system analysis. The inverse problem is partitioned into the design, input, and identification problem as described briefly in Fig. 2.

The definition of identification by Zadek (1962) is: The determination on the basis of input and output (Fig. 3) of a system within a specified class of systems,to which the system under test is equivalent.

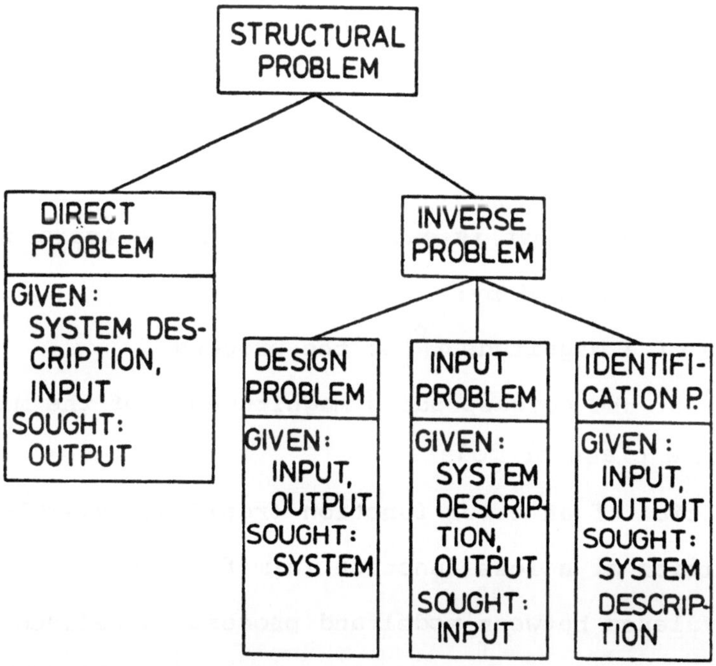

Fig. 2 Classification of the identification problem

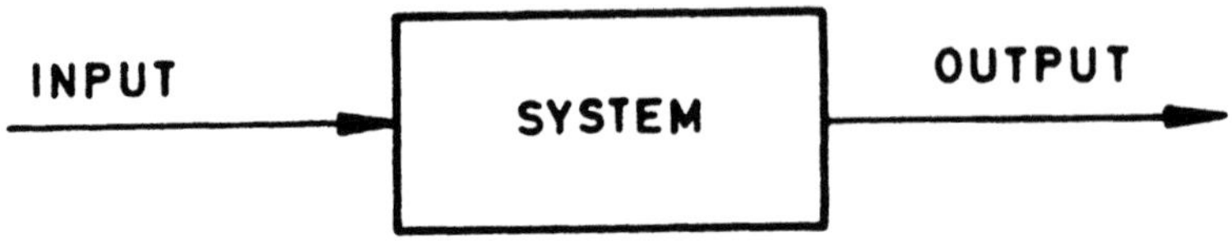

Fig. 3 Process: System under test

The system under test is called process. The approach may be briefly described by:

1. Choice of a magnitude α^P of the process
2. Choice of the corresponding magnitude α^M of the model within a class of models
3. Definition of an error function (residual) $v := \alpha^P - \alpha^M$
4. Definition of a loss function (aim functional) $J(v) = \text{Min.}$

The equivalence between model and process is defined by $J(v)$, the equivalence between two models is defined by $J_1 = J_2$.

Problems which arise with identification are e.g.:

- o the choice of models, model classes
- o choice of coordinates (discrete, continuous; physical, modal)
- o choice of input signals (operating conditions, test

signal)

o environmental conditions (see Fig. 4)

Indications to the identification problem are e.g.:

o Properties of dynamic operators

o Linearity investigations

An example for modelling and optimization by least squares: Given is a steel plate fixed on a rubber cube which itself is fixed on the rigid ground (Fig. 5a). Harmonic excitation in vertical direction using different force amplitudes in a specified frequency range leads to corresponding accelerations in the frequency domain (without Fig.). Possibilities for mathematical modelling are shown in Fig. 5b. The parameters to be identified are assembled in the vector $\underline{a}$ which is defined in the case of Kelvin's model by $\underline{a}^T := (m, b_1, k_1)$. Choosing the force residual $v(t) := p^P(t) - p^M(t) = p^P(t) - m u^P(t) + \ldots$,the aim functional $J(\underline{a}) := \int_0^T v^2(t)dt = \mathrm{Min}(\underline{a})$, T - measuring time, the following results are achieved, Merrit:[1]

$$0.077\, \ddot{u}^M(t) + 3.57\, \dot{u}^M(t) + 1662\, u^M(t) = p^M(t)$$

$$-0.55\cdot 10^{-5}\, \dddot{u}^M(t) + 0.1\, \ddot{u}^M(t) + 0.55\, \dot{u}^M(t) + 1310\, u^M(t) = p^M(t) + 0.7\cdot 10^{-3}\, \dot{p}^M(t)$$

$$0.0085\, \ddot{u}^M(t) + 3.93\, \dot{u}^M(t) + 3980\, u^M(t) + -0.012\, \dot{u}^{M3}(t) - 2.1\cdot 10^6 u^{M3}(t) = p^M(t)$$

The identification yields results for each model. The question arises: Which model represents the dynamic be-

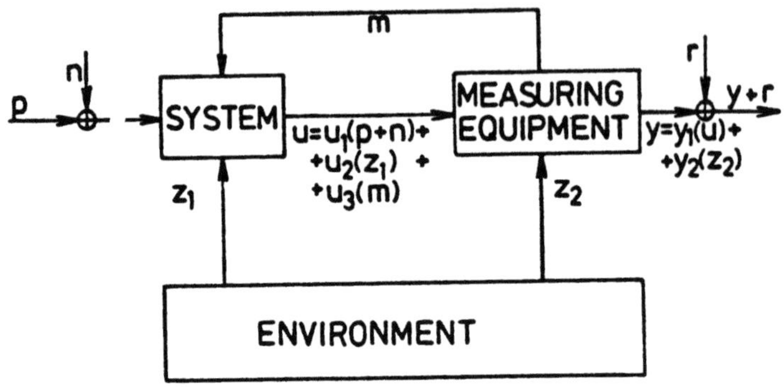

Fig. 4 Possible disturbances

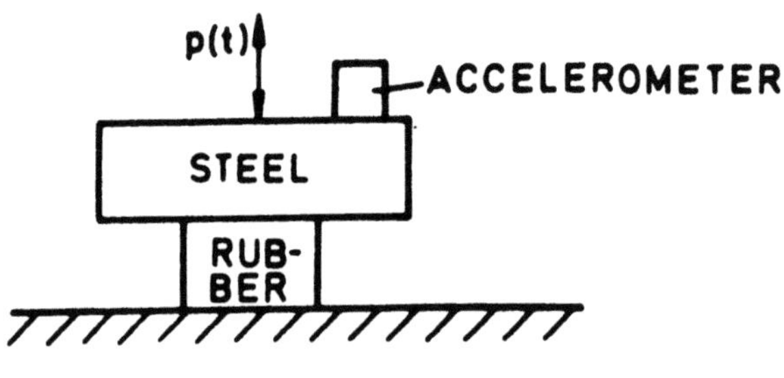

Fig. 5a An example for modelling and optimization

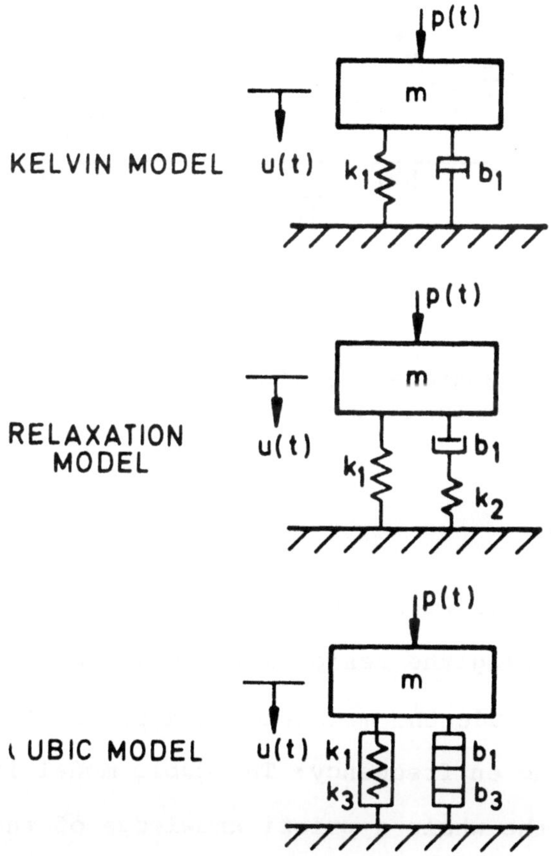

Fig. 5b Examples of feasible mathematical models

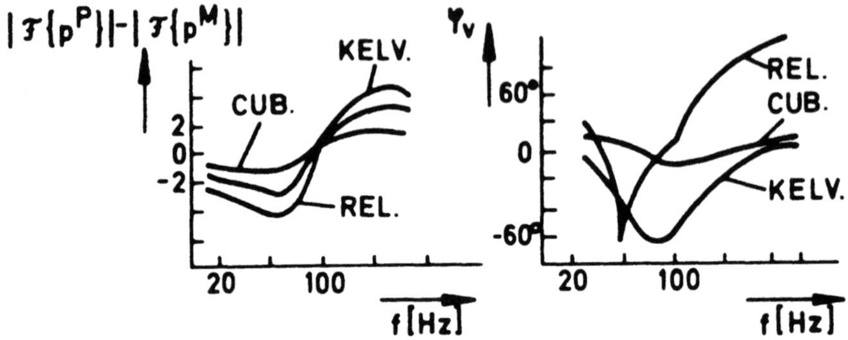

Fig. 5c Residual values versus frequency

haviour of the system in the best way? The decision may be get by considering the residual values. Fig. 5c contains for one force amplitude the magnitude and phase of the residuals depending on frequency: The cubic model is the best one. But not the whole a priori knowledge of the system is taken into account, e.g. the possibility of mass measurements.

Depending on the choice of residuals combined with the mathematical model three different error models can be established (Fig. 6). Using the output signal of the process and of the model for error definition, the output error, the resulting model is called the forward model, which is

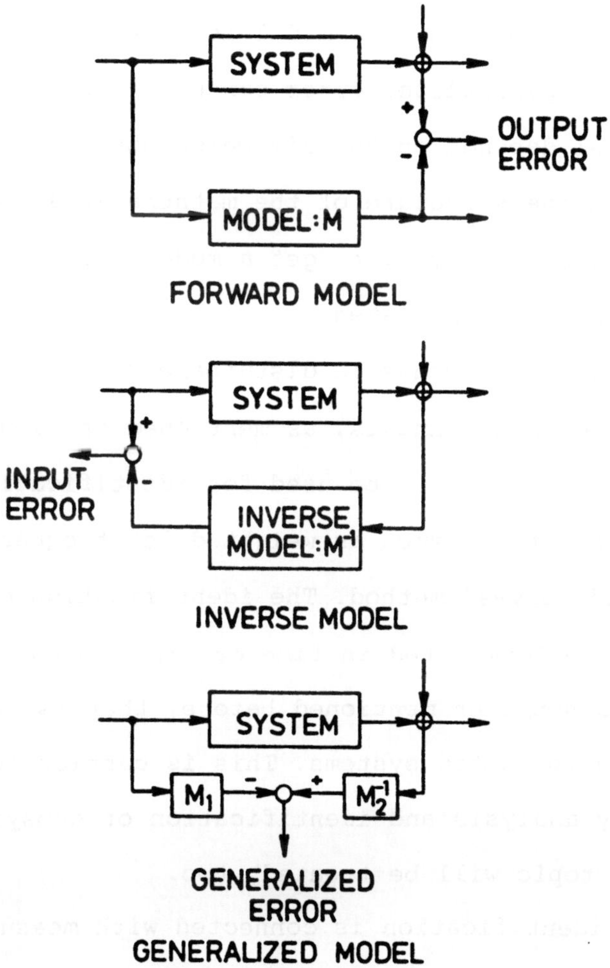

Fig. 6 Model descriptions

often used. If the inverse formulation of the model exists, the inputs may be used for the input error of the inverse model. If one part of the model is invertible then a mixed model and a generalized error can be defined. The choice of error model depends on the aim which should be achieved, it depends on the structure of the mathematical model and last but not least on trying to get a model that is linear in the parameters to be estimated.

Because of stochastic disturbances in the test model and applied random inputs, as mentioned previously, statistical methods have to be used for identification, i.e. estimation methods such as weighted least squares, maximum likelihood, Bayes' method. The identification methods ref. (2-4) may be formulated in time or frequency domain. One aspect has not been mentioned before, that is the handling of large complicated systems. This is carried out advantageously by analysis and identification of subsystems. This important topic will be treated, too.

The identification is connected with measurement and test techniques, with time series analysis as sampling and estimation of spectral densities etc. These topics will only be touched; in this connection the ref. (5-7) may be mentioned.

The engagement in identification of vibrating structures requires as basis two different things: The theory of vibrating structures and the theory of estimation. The first one may be discussed by multi-degree-of-freedom systems because numerical handling and measurement in single points and sampling the signals means discretization. With respect to estimation combined with the input - output problem in general the structural engineer can profit from the control theory. This is the reason why this topic will be presented by an authority of control theory. We are interested in classical, approved, and new identification methods, in handling and application including experience. Identification of non-linear systems is a very sophisticated task, which has to be distinguished from error models non-linear in the parameters. Finally we are interested in modern methods and developments. These topics and themes will be discussed in this course.

References

1 Merrit, P.H., W.E. Baker, A Method of System Identification with an Experimental Investigation, Shock and Vibration Bull. 47, No. 4, 1977

2 Eykhoff, P., System Identification: Parameter and State Estimation, John Wiley and Sons, 1974

3 Strobel, H., Experimentelle Systemanalyse, Akademie-Verlag, 1975

4 Isermann, R., Prozeßidentifikation - Identifikation und Parameterschätzung dynamischer Prozesse mit diskreten Signalen, Springer-Verlag, 1974

5 Bendat, J.S., A.G. Piersol, Random Data: Analysis and Measurement Procedures, Wiley-Interscience, 1971

6 Jenkins, G.M., D.G. Watts, Spectral Analysis and its Applications, Holden-Day, 1968

7 Otnes, R.K., L. Enochson, Applied Time Series Analysis, Vol. I: Basic Techniques, John Wiley and Sons, 1978.

MULTI-DEGREE-OF-FREEDOM SYSTEMS- A REVIEW

H.G. Natke
Curt-Risch-Institut
Universität Hannover

1. Introduction

Investigations and qualifications of the dynamic behaviour of complex vibrating structures require system analysis (based on drawings) and system identification if experience with comparable structures is not available. System identification may be understood as a three-stage process:

1. Summarizing the a-priori-knowledge of the system to be investigated
2. Carrying out pre-tests: Non-parametric identification (investigations concerning linearity, damping behaviour etc.)

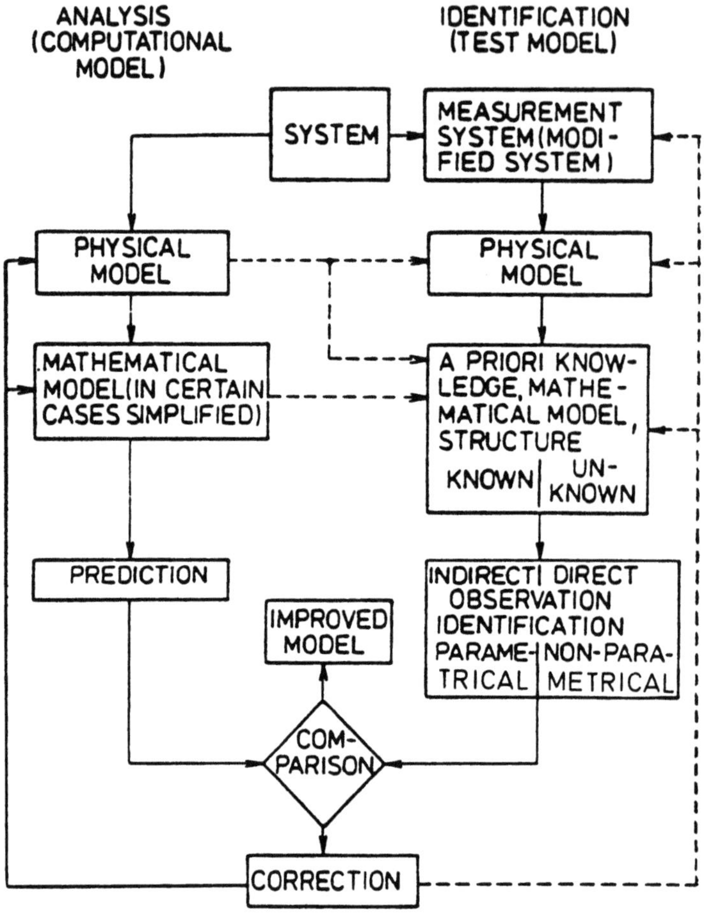

Fig. 1. Principal proceeding in system analysis and identification.

3. Parameter estimation: Parametric identification.

Depending on what is to be achieved the problem is sometimes solved by stage 2, and sometimes one has to go beyond stage 3, i.e. to correct the computational model of system analysis. Figure 1 shows the principal steps.

In addition to test problems concerning excitation and measurement, and in addition to data acquisition and data reduction, when one considers the estimation procedures urgently required, the general knowledge of system properties and system relations are necessary to identify vibrating systems, Natke.[1] The non-parametric input-output relation is the basis of non-parametric identification (Fig. 2), and the equations of motion of the system under test are the fundamental equations of parametric identification (Fig. 3). A review of important equations and relations used for identification of vibrating structures is therefore given as an introduction to this course.

The structures under test are continuous systems which vibrate about their equilibrium positions. Their displacements can be specified by continuous coordinates. However, powerful methods, such as the Rayleigh-Ritz technique (Galerkin), the finite element method, and in general the numerical handling of the system, enable us to approximate the descriptions of the dynamic behaviour of continuous systems by means of a system with a finite number of coordi-

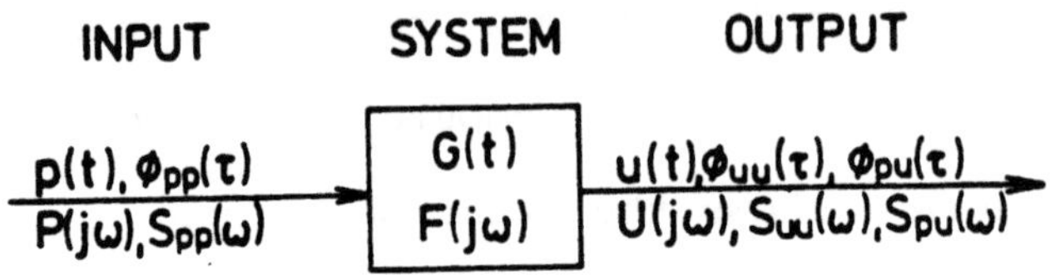

Fig. 2. Input - output relationship of a linear single degree-of-freedom system

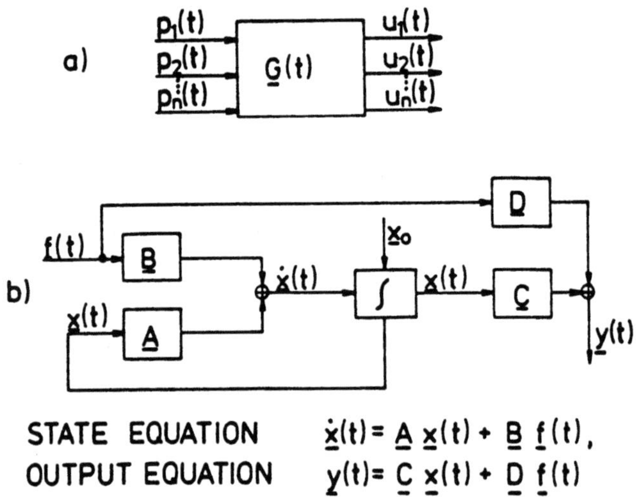

Fig. 3. Input - output relationships of a linear multi-degree-of-freedom system

nates. Therefore we will agree on the assumption that the state of our mechanical systems can be described by a finite number of coordinates, which means that we shall be involved in multi-degree-of-freedom systems.

2. Energy Expressions, Dissipation Functions, Equations of Motion

The chosen coordinates may be $u_1(t),\dots,u_n(t)$, assuming that these coordinates can be varied arbitrarily and independently of each other without violating any constraints which may act on the system (set of minimal coordinates, principal, generalized coordinates, holonomic system). Therefore we can formulate equations of motion for the system by means of the classical Lagrangean method. In order to do this, we first have to set up the energy expressions and to take into account damping effects as well as external forces. As is usually the case, we will denote time derivatives by dot notation and the transpose of a matrix by a superscript T.

2.1 Energy Expressions, Dissipation Functions

If an equilibrium position $\underline{u} = \underline{o}$, $\underline{u} = \underline{o}$ exists, and the system is skleronomic, and if the applied forces depend only on the displacements $\underline{u}$ or their derivatives $\dot{\underline{u}}$, then - within the classical linear theory of elasticity - the kinetic energy may be written in the form

$$E_{kin}(t) := \frac{1}{2}\, \dot{\underline{u}}^T(t)\; \underline{M}\; \dot{\underline{u}}(t) \tag{1}$$

and the potential energy of the elastic restoring forces, assumed as linear and conservative, may be written in the form

$$E_{pot}(t) := \frac{1}{2}\, \underline{u}^T(t)\; \underline{K}\; \underline{u}(t). \tag{2}$$

$\underline{u}(t)$ is the vector with the generalized velocities as components, and $\underline{M}$ is the real symmetric inertia matrix. Because $E_{kin}(t)$ is always positive, it follows that $\underline{M}$ is positive definite. Dependent on $\underline{u} = \underline{o}$, $\dot{\underline{u}} = \underline{o}$ is a stable or a neutrally stable equilibrium position,and then it holds true for the potential energy that $E_{pot}(t) \geq 0$, and the stiffness matrix $\underline{K}$ follows as non-negative definite. A viscous damped system with the internal damping force

$$\underline{f}_d(t) := -\, \underline{B}\; \dot{\underline{u}}(t) \tag{3}$$

with the real quadratic matrix $\underline{B}$ of the order n leads from the rate of energy dissipation $\underline{f}_d^T(t)\ \dot{\underline{u}}(t)$ to the dissipation function (Rayleigh)

$$E_{dis}(t) := \frac{1}{2}\ \dot{\underline{u}}^T(t)\ \underline{B}\ \dot{\underline{u}}(t)\ . \tag{4}$$

Non-conservative damping forces yield $\underline{B} = \underline{B}^T$ with non-negative definite $\underline{B}$.

Forces giving rise to a rate of energy dissipation which is independent of the frequency will be denoted as hysteretic damping forces

$$\underline{f}_D(t) := -\ j\ \underline{D}\ \underline{u}(t),\ j := \sqrt{-1} \tag{5}$$

which are restricted to sinusoidal motions. They can be derived from the dissipation function

$$E_{Dis}(t) := \frac{1}{2}\ j\ \underline{u}^T(t)\ \underline{D}\ \underline{u}(t). \tag{6}$$

The damping matrix $\underline{D}$ contains real constant elements. It is symmetrical and non-negative definite.

2.2 Equations of Motion

When describing the system with viscous and hysteretic damping forces, the Lagrangean equations for sinusoidal motions are given by $(i = 1(1)n)$

$$\frac{d}{dt}\left(\frac{\partial E_{kin}}{\partial \dot{u}_i}\right) + \frac{\partial E_{dis}}{\partial \dot{u}_i} + \frac{\partial E_{Dis}}{\partial u_i} + \frac{\partial E_{pot}}{\partial u_i} = p_i \ , \tag{7}$$

where $p_i(t)$ denotes the generalized external forces. The resulting equations of motion can be written in the matrix form

$$\underline{M}\ \underline{\ddot{u}}(t) + \underline{B}\ \underline{\dot{u}}(t) + (j\ \underline{D} + \underline{K})\ \underline{u}(t) = \underline{p}(t) = \hat{\underline{p}} e^{j\Omega t} \tag{8}$$

with the frequency of sinusoidal motion Ω. It follows that

$$(-\Omega^2 \underline{M} + j\Omega \underline{B} + j\ \underline{D} + \underline{K})\ \hat{\underline{u}} = \hat{\underline{p}}, \tag{9}$$

where $\hat{\underline{u}}$ is the time independent displacement amplitude vector.

Arbitrary motions of the system - $\underline{D}$ cannot appear - will be described by the matrix equation

$$\underline{M}\ \underline{\ddot{u}}(t) + \underline{B}\ \underline{\dot{u}}(t) + \underline{K}\ \underline{u}(t) = \underline{p}(t). \tag{10}$$

The parametric model (10) will be used for identification, if it is not known a-priori that the motion is sinusoidal.

We will define systems as passive systems (non-active) if the equations of motion contain symmetrical and non-negative definite matrices (the associated undamped system is conservative and without gyroscopic effects; and the damping matrix does not result from potential functions), otherwise we will describe it as an active system. The equations of motion for active systems (with gyroscopic effects and non-conservative restoring forces) may be written in the form

$$\underline{M}\,\ddot{\underline{u}}(t) + (\underline{B} + \underline{B}^{a})\,\dot{\underline{u}}(t) + (\underline{K} + \underline{K}^{a})\,\underline{u}(t) = \underline{p}(t). \qquad (11)$$

$\underline{B}^{a} = -\,\underline{B}^{aT}$ is the real and skew-symmetric gyroscopic matrix, $\underline{K} = -\,\underline{K}^{aT}$, is the matrix resulting from circular forces (non-conservative restoring forces).

The equations of motion (8) - (11) are the classical system descriptions written in the form of differential equations of second order. Through the introduction of the state vector

$$\underline{x}(t) := \begin{Bmatrix} \underline{u}(t) \\ \dot{\underline{u}}(t) \end{Bmatrix} \qquad (12)$$

and the abbreviations

$$\underline{A}_V := \begin{bmatrix} \underline{B} & \underline{M} \\ \underline{M} & \underline{O} \end{bmatrix}, \quad \underline{B}_V := \begin{bmatrix} \underline{K} & \underline{O} \\ \underline{O} & -\underline{M} \end{bmatrix}, \quad \underline{f}_V(t) := \begin{Bmatrix} \underline{p}(t) \\ \underline{O} \end{Bmatrix} \tag{13}$$

or

$$\underline{A} := \begin{bmatrix} \underline{O} & \underline{I} \\ -\underline{M}^{-1}\underline{K} & -\underline{M}^{-1}\underline{B} \end{bmatrix}, \quad \underline{f}(t) := \begin{Bmatrix} \underline{O} \\ \underline{M}^{-1}\underline{p}(t) \end{Bmatrix} \tag{14}$$

$\underline{I}$: identity matrix

the equations of motion (10) and (11) pass into the state equations

$$\underline{A}_V \, \dot{\underline{x}}(t) + \underline{B}_V \, \underline{x}(t) = \underline{f}_V(t) \tag{15}$$

or, in a more general form, the state equation is given by

$$\dot{\underline{x}}(t) = \underline{A} \, \underline{x}(t) + \underline{B} \, \underline{f}(t) \tag{16}$$

instead of (15), which is combined with the output (measuring) equation

$$\underline{y}(t) = \underline{C} \, \underline{x}(t) + \underline{D} \, \underline{f}(t). \tag{17}$$

Here

$\underline{A}$ is the system matrix

$\underline{B}$ the input or distribution matrix (unequal to the damping matrix in spite of the same letter; confusion seems to be excluded)

$\underline{C}$ the output or measurement matrix

$\underline{D}$ the input-output matrix.

The measuring equation is needed in identification, e.g. if the measuring points are not equal to the system points by which the displacement components are defined. The input-output relationship of the equations (17) is shown in Fig.3.

3. Eigenvibrations

3.1 The Passive Undamped System

The eigenvibrations of the passive undamped system follow from equation (10) with $\underline{p}(t) = \underline{0}$ and

$$\underline{u}(t) = \underline{\hat{u}}_o \, e^{j\omega_o t} \tag{18}$$

leading to the matrix eigenvalue problem

$$(-\omega_o^2 \underline{M} + \underline{K}) \, \underline{\hat{u}}_o = \underline{0}. \tag{19}$$

The eigensolutions of equation (19) hold the following properties (without proofs), Wilkinson:[2]

- o The eigenvalues $\lambda_o := \omega_o^2$ are real (symmetry of matrices and $\underline{M}$ positive definite)

- The eigenvalues are , moreover, non-negative
- There are n eigenvalues λ_{oi}, i = 1(1)n, counted according to their multiplicities (det $\underline{M} \neq 0$ is the coefficient of λ^n of the characteristic polynominal)
- Right and left eigenvectors are equal (symmetry of matrices)
- For each eigenvalue λ_{oi} one eigenvector $\hat{\underline{u}}_{oi} \neq \underline{0}$ exists, which can be normalized in such a manner that $\hat{\underline{u}}_{oi}$ is real (homogeneous system of equations with real coefficients and singular system matrix)
- The eigenvectors fulfil the generalized orthogonality relations

$$\hat{\underline{u}}_{oi}^T \underline{M}\, \hat{\underline{u}}_{ok} = m_{gi}\, \delta_{ik},\ m_{gi} > 0$$
$$\hat{\underline{u}}_{oi}^T \underline{K}\, \hat{\underline{u}}_{ok} = k_{gi}\, \delta_{ik},\ k_{gi} = \lambda_{oi}\, m_{gi} \qquad (20)$$

δ_{ik} is the Kronecker delta.

When defining the modal matrix $\hat{\underline{U}}_o := (\hat{\underline{u}}_{o1}, \ldots\ldots, \hat{\underline{u}}_{on})$ and the diagonal matrix of the eigenvalues $\underline{\Lambda}_o := \text{diag}\,(\lambda_{oi})$, the matrix eigenvalue problem (19) and the orthogonality relations can be rewritten in the form

$$- \underline{M}\hat{\underline{U}}_o \underline{\Lambda}_o + \underline{K}\, \hat{\underline{U}}_o = \underline{0}$$
$$\underline{M}_g := \hat{\underline{U}}_o^T \underline{M}\, \hat{\underline{U}}_o = \text{diag}\,(m_{gi}), \qquad (21)$$
$$\underline{K}_g := \hat{\underline{U}}_o^T \underline{K}\, \hat{\underline{U}}_o = \text{diag}\,(k_{gi}) = \underline{M}_g\, \underline{\Lambda}_o .$$

The matrix $\underline{M}_g$ of generalized masses m_{gi} may be normalized to $\underline{M}_g = \underline{I}$ or otherwise (e.g. in such a manner that the maximum absolute value of the components of each eigenvector is equal to one).

3.2 The Passive Hysteretic Damped System

The matrix eigenvalue problem follows from equation (9) with $\underline{B} = \underline{O}$ and $\hat{\underline{p}} = \underline{O}$:

$$\left[-\omega_D^2 \underline{M} + (\underline{K} + j\underline{D})\right] \hat{\underline{u}}_D = \underline{O}, \ \underline{u}(t) = \hat{\underline{u}}_D \, e^{j\omega_D t}. \tag{22}$$

The properties of the eigensolutions are similar to those of the passive undamped system, with the difference that the eigenvalues $\lambda_{Di} := \omega_{Di}^2 = \lambda_{Di}^{re} + j\lambda_{Di}^{im}$ are complex with non-negative real and imaginary parts, and therefore in general the eigenvectors are also complex, and that the modal matrix $\hat{\underline{U}}_D := (\hat{\underline{u}}_{D1}, \ldots\ldots, \hat{\underline{u}}_{Dn})$ does not need to be regular. But $\hat{\underline{U}}_D$ is regular in the case of distinct eigenvalues. The associated matrix equations are:

$$\begin{aligned} &- \underline{M}\, \hat{\underline{U}}_D \underline{\Lambda}_D + (\underline{K} + j\,\underline{D})\, \hat{\underline{U}}_D = \underline{O} \\ &\underline{M}_D := \hat{\underline{U}}_D^T \underline{M}\, \hat{\underline{U}}_D = \text{diag}\,(m_{Di})\ (\text{e.g.} = \underline{I}) \\ &\underline{S}_D := \hat{\underline{U}}_D^T (\underline{K} + j\,\underline{D})\, \hat{\underline{U}}_D = \underline{M}_D \underline{\Lambda}_D,\ \underline{\Lambda}_D := \text{diag}\,(\lambda_{Di}). \end{aligned} \tag{23}$$

The frequency of vibration yields the effective damping

$$\gamma_i := \lambda_{Di}^{im} / \lambda_{Di}^{re} \tag{24}$$

from

$$\omega_{Di} = \pm \sqrt{\lambda_{Di}} = \omega_{Di}^{re} + j\omega_{Di}^{im} = \pm\sqrt{\lambda_{Di}^{re}} \sqrt{1 + j\gamma_i}$$

to

$$\omega_{Di}^{re} = \pm \frac{\lambda_{Di}^{re}}{2} (1 + |\sqrt{1 + \gamma_i^2}|). \tag{25}$$

The real part of λ_{Di},

$$\lambda_{Di}^{re} := \omega_{Gi}^2 = \frac{\hat{\underline{u}}_{Di}^{*T} \underline{K} \hat{\underline{u}}_{Di}}{\hat{\underline{u}}_{Di}^{*T} \underline{M} \hat{\underline{u}}_{Di}}, \tag{26}$$

gives the eigenfrequency of the i-th associated generalized undamped single-degree-of-freedom-system.

In general the matrix of generalized dampings

$$\underline{B}_g := \hat{\underline{U}}_o^T \underline{D} \hat{\underline{U}}_o, \tag{27}$$

- generalized with regard to a congruence transformation with the modal matrix $\hat{\underline{U}}_o$ of the associated undamped system - is non-diagonal.

If

$$\underline{B}_g = \underline{B}_E := \text{diag}\,(d_{Ei}),\ d_{Ei} := \hat{\underline{u}}_{oi}^T \underline{D} \hat{\underline{u}}_{oi}, \tag{28}$$

then the eigenvectors are real; they are equal to those of

the associated passive undamped system, $\underline{\hat{u}}_{Di} = \underline{\hat{u}}_{oi}$, and the damping of each degree-of-freedom is modal

$$g_i := \lambda_{Di}^{im} / \lambda_{Di}^{re} = \frac{d_{Ei}}{\omega_{oi}^2 m_{gi}} \quad . \tag{29}$$

This is true for commutative matrices $\underline{M}$, $\underline{D}$, $\underline{K}$, Caughey et al.[3,4]:

$$\underline{K}\ \underline{M}^{-1}\ \underline{D} = \underline{D}\ \underline{M}^{-1}\ \underline{K}. \tag{30}$$

A special case of this "ease-hypothesis" is given if real constants a, b exist such that

$$\underline{D} = a\ \underline{M} + b\ \underline{K} \qquad \text{(proportional damping)}. \tag{31}$$

For small effective damping values $g_i \ll 1$, the g_i may be interpreted as loss angles:

$$k_{gi} + j\ d_{EI} = k_{gi}(1 + j\ g_i) \doteq k_{gi} e^{jg_i}. \tag{32}$$

3.3 The Passive Viscous Damped System

The statement

$$\underline{u}(t) = \hat{\underline{u}}_B \, e^{\lambda_B t}$$

leads with $\underline{p}(t) = \underline{O}$ from equation (10) to the quadratic matrix eigenvalue problem of the order n

$$(\lambda_B^2 \underline{M} + \lambda_B \underline{B} + \underline{K}) \hat{\underline{u}}_B = \underline{O}. \tag{34}$$

The properties of the eigensolutions are, Lancaster:[5]

- The eigenvalues λ_B are real or complex, the complex values appear as complex conjugate pairs.
- There are 2n eigenvalues λ_{Bl}, $l = 1(1)2n$.
- Right and left eigenvectors are equal.
- The eigenvectors associated with real eigenvalues can be normalized to real vectors, and the eigenvectors of complex conjugate pairs of eigenvalues are also complex conjugate.
- The rank of the n x 2n modal matrix $\hat{\underline{U}}_B$ is n for distinct eigenvalues.
- The matrix eigenvalues problem and the orthogonality equations with the normalization (second line) can be

written in the form:

$$\begin{gathered}\underline{M} \hat{\underline{U}}_B \underline{\Lambda}_B^2 + \underline{B} \hat{\underline{U}}_B \underline{\Lambda}_B + \underline{K} \hat{\underline{U}}_B = \underline{O}, \\ \underline{\Lambda}_B \hat{\underline{U}}_B^T \underline{M} \hat{\underline{U}}_B + \hat{\underline{U}}_B^T \underline{M} \hat{\underline{U}}_B \underline{\Lambda}_B + \hat{\underline{U}}_B^T \underline{B} \hat{\underline{U}}_B = \underline{I},\end{gathered} \tag{35}$$

$$\underline{\Lambda}_B \, \hat{\underline{U}}_B^T \, \underline{M} \, \hat{\underline{U}}_B \, \underline{\Lambda}_B - \hat{\underline{U}}_B^T \, \underline{K} \, \hat{\underline{U}}_B = \underline{\Lambda}_B \tag{35}$$

$$\underline{\Lambda}_B : = \text{diag } (\lambda_{Bl}).$$

In the state-space formulation the equivalent orthogonality expressions are easier to describe:

$$\begin{aligned} \hat{\underline{X}}_V^T \, \underline{A}_V \, \hat{\underline{X}}_V &= \underline{I} \\ \hat{\underline{X}}_V^T \, \underline{B}_V \, \hat{\underline{X}}_V &= - \underline{\Lambda}_B \end{aligned} \tag{36}$$

where

$$\hat{\underline{X}}_V^T : = (\hat{\underline{U}}_B^T, \underline{\Lambda}_B \, \hat{\underline{U}}_B^T).$$

Weakly damped systems vibrate with the eigenfrequency

$$\omega_{Bl} : = \lambda_{Bl}^{im} \tag{37a}$$

and possess effective damping

$$\begin{aligned} \alpha_l : &= - \lambda_{Bl}^{re} / \lambda_{Bl}^{im} = \delta_{Bl} / \omega_{Bl}, \\ \delta_{Bl} : &= - \lambda_{Bl}^{re} \geq 0. \end{aligned} \tag{37b}$$

With the "ease - hypothesis" (30) with $\underline{B}$ instead of $\underline{D}$, the eigenvectors are those of the associated undamped system $\hat{\underline{u}}_{Bi} = \hat{\underline{u}}_{oi}$, $i = 1(1)n$, and the eigenvalues for weakly damped systems hold

$$\lambda_{Bi} = -\frac{b_{Ei}}{2m_{gi}} \pm j\omega_o \sqrt{1 - \frac{b_{Ei}^2}{4\omega_{oi}^2 m_{gi}^2}}\,, \tag{38}$$

$$b_{Ei} := \hat{\underline{u}}_{oi}^T \, \underline{B} \, \hat{\underline{u}}_{oi}.$$

It follows that the vibration frequency is equal to

$$\omega_{Bi} = \lambda_{Bi}^{im} = \omega_{oi} \sqrt{1 - \alpha_i^2} \tag{39}$$

with the effective damping (modal damping)

$$\alpha_i := \frac{b_{Ei}}{2\omega_{oi} m_{gi}} = \frac{\delta_{Ei}}{\omega_{oi}}\,. \tag{40}$$

3.4 Active Systems

The equations of motion (11) for active systems may be written as

$$\underline{A}_2 \, \ddot{\underline{u}}(t) + \underline{A}_1 \, \dot{\underline{u}}(t) + \underline{A}_o \, \underline{u}(t) = \underline{p}(t) \tag{41}$$

with real quadratic matrices $\underline{A}_2$, $\underline{A}_1$, $\underline{A}_o$ of the order n, which are not necessarily symmetric. The homogeneous equation of (41) describes, for example, the flutter phenomenon. The principal difference from the matrix eigenvalue problem of the passive viscous damped system follows from the non-symmetrical matrices:

$$\begin{aligned} (\lambda_B^2 \underline{A}_2 + \lambda_B \underline{A}_1 + \underline{A}_o)\, \hat{\underline{u}} &= \underline{O} \\ \hat{\underline{v}}^T (\lambda_B^2 \underline{A}_2 + \lambda_B \underline{A}_1 + \underline{A}_o) &= \underline{O}. \end{aligned} \tag{42}$$

By defining the matrix of left eigenvectors, $\hat{\underline{V}}_B := (\hat{\underline{v}}_{B1}, \ldots, \hat{\underline{v}}_{B2n})$ one obtains the ortho-normalization

$$\begin{aligned} \underline{\Lambda}_B \hat{\underline{V}}_B^T \underline{A}\, \hat{\underline{U}}_B + \hat{\underline{V}}_B^T \underline{A}_2 \hat{\underline{U}}_B \underline{\Lambda}_B + \hat{\underline{V}}_B^T \underline{A}_1 \hat{\underline{U}}_B &= \underline{I} \\ \underline{\Lambda}_B \hat{\underline{V}}_B^T \underline{A}_2 \hat{\underline{U}}_B \underline{\Lambda}_B - \hat{\underline{V}}_B^T \underline{A}_o \hat{\underline{U}}_B &= \underline{\Lambda}_B. \end{aligned} \tag{43}$$

Annotation: In the theory of eigenmagnitude identification (resonance testing) the following question leads to a latent root problem (eigenvalue problem): can a real force amplitude vector be found such that the components of the response vector are in phase with one another, but are not necessarily in phase with the components of the force vector? The solutions are "proper" or characteristic solutions consisting of dependent linearly independent proper (characteristic) vectors with real elements. (Fraeijs des Veubeke;[6,7] characteristic phase lag theory, method of stationary phase).

4. Relations between System Parameters and Eigenmagnitudes

The magnitudes to be identified may be the system parameters directly, or the eigenmagnitudes. Both pieces of information are equivalent because of their definite relationships, which are derived for the passive undamped system.

The orthonormalization (21) in the form $\underline{\hat{U}}_o^T \underline{M} \underline{\hat{U}}_o = \underline{I}$ leads with the regular modal matrix $\underline{\hat{U}}_o$ to the inertia matrix

$$\underline{M} = (\underline{\hat{U}}_o \underline{\hat{U}}_o^T)^{-1} \tag{44}$$

or to the inverse inertia matrix

$$\underline{M}^{-1} = \underline{\hat{U}}_o \underline{\hat{U}}_o^T = \sum_{i=1}^{n} \underline{\hat{u}}_{oi} \underline{\hat{u}}_{oi}^T . \tag{45}$$

The equation of the generalized stiffness matrix permits the representation

$$\underline{K} = (\underline{\hat{U}}_o^T)^{-1} \underline{\Lambda}_o \underline{\hat{U}}_o^{-1}, \tag{46a}$$

in which the inverse modal matrix from (44) can be substituted:

$$\underline{K} = \underline{M} \underline{\hat{U}}_o \underline{\Lambda}_o \underline{\hat{U}}_o^T \underline{M} = \sum_{i=1}^{n} \lambda_{oi} \underline{M} \underline{\hat{u}}_{oi} \underline{\hat{u}}_{oi}^T \underline{M}. \tag{46b}$$

The flexibility matrix $\underline{G} := \underline{K}^{-1}$, if it exists, follows from equation (46a) as

$$\underline{G} = \hat{\underline{U}}_o \underline{\Lambda}_o^{-1} \hat{\underline{U}}_o^T = \sum_{i=1}^{n} \frac{1}{\lambda_{oi}} \hat{\underline{u}}_{oi} \hat{\underline{u}}_{oi}^T . \tag{47}$$

The relationships (46b) und (47) show that in the representation for $\underline{K}$ (the terms of the sum are proportional to the eigenvalues) the degrees-of-freedom with the higher eigenfrequencies are emphasized , and in that for $\underline{G}$ (the terms of the sum are proportional to the reciprocal eigenvalues) the degrees-of-freedom with the smaller eigenfrequencies are emphasized.

The derivation from the generalized orthogonality relations of the corresponding equations of the viscous and hysteretic damped system are similar. Tables 1.1 - 1.3 contain a combination of the results. The last line provides the frequency response matrix (receptance, resolvent matrix) of the non-parametric input-output relationship, e.g. with the Fourier transformed equations of motion:

$$\begin{aligned} \underline{U}(j\omega) &= \underline{F}(j\omega)\, \underline{P}(j\omega), \\ \underline{U}(j\omega) :&= \mathcal{F}\{\underline{u}(t)\} , \\ \underline{P}(j\omega) :&= \mathcal{F}\{\underline{p}(t)\} . \end{aligned} \tag{48}$$

SYSTEM / SPECIFICATION	CONSERVATIVE SYSTEM
MATRIX EIGENVALUE PROBLEM	$(-\lambda_{0i}\underline{M}+\underline{K})\hat{\underline{u}}_{0i}=\underline{0}$ $i=1(1)n,\ \underline{M}_g=\underline{I}$
INERTIA MATRIX	$\underline{M}=(\hat{\underline{U}}_0\hat{\underline{U}}_0^T)^{-1}$
INVERSE INERTIA MATRIX	$\underline{M}^{-1}=\hat{\underline{U}}_0\hat{\underline{U}}_0^T=\sum_{i=1}^{n}\hat{\underline{u}}_{0i}\hat{\underline{u}}_{0i}^T$
DAMPING MATRIX	——
STIFFNESS MATRIX	$\underline{K}=(\hat{\underline{U}}_0^T)^{-1}\underline{\Delta}_0\hat{\underline{U}}_0^{-1}$ $=\underline{M}\hat{\underline{U}}_0\underline{\Delta}_0\hat{\underline{U}}_0^T\underline{M}$ $=\sum_{i=1}^{n}\lambda_{0i}\underline{M}\hat{\underline{u}}_{0i}\hat{\underline{u}}_{0i}^T\underline{M}$
FLEXIBILITY MATRIX	$G=K^{-1}=\hat{\underline{U}}_0\underline{\Delta}_0^{-1}\hat{\underline{U}}_0^T$ $=\sum_{i=1}^{n}\frac{1}{\lambda_{0i}}\hat{\underline{u}}_{0i}\hat{\underline{u}}_{0i}^T$
FREQUENCY RESPONSE MATRIX	$\underline{F}(j\omega)=(-\omega^2\underline{M}+\underline{K})^{-1}$ $=\hat{\underline{U}}_0(-\omega^2\underline{I}+\underline{\Delta}_0)^{-1}\hat{\underline{U}}_0^T$ $=\sum_{i=1}^{n}\frac{\hat{\underline{u}}_{0i}\hat{\underline{u}}_{0i}^T}{\omega_{0i}^2-\omega^2}$

Table 1.1 Relations between system parameter matrices and modal magnitudes

SYSTEM / SPECI-FICATION	VISCOUS DAMPED SYSTEM
MATRIX EIGEN-VALUE PRO-BLEM	$(\lambda_{Bl}^2 \underline{M} + \lambda_{Bl} \underline{B} + \underline{K}) \hat{\underline{u}}_{Bl} = \underline{0}$ $l = 1(1)2n, \; \underline{\Delta}_B \hat{\underline{U}}_B^T \underline{M} \hat{\underline{U}}_B + \hat{\underline{U}}_B^T \underline{M} \hat{\underline{U}}_B \underline{\Delta}_B +$ $+ \hat{\underline{U}}_B^T \underline{B} \hat{\underline{U}}_B = \underline{I}$
INERTIA MATRIX	$\underline{M} = (\hat{\underline{U}}_B \underline{\Delta}_B \hat{\underline{U}}_B^T)^{-1}$
INVERSE INER-TIA MATRIX	$\underline{M}^{-1} = \hat{\underline{U}}_B \underline{\Delta}_B \hat{\underline{U}}_B^T = \sum_{l=1}^{2n} \lambda_{Bl} \hat{\underline{u}}_{Bl} \hat{\underline{u}}_{Bl}^T$
DAMPING MATRIX	$\underline{B} = -\underline{M} \hat{\underline{U}}_B \underline{\Delta}_B^2 \hat{\underline{U}}_B^T \underline{M}$ $= -\sum_{l=1}^{2n} \lambda_{Bl}^2 \underline{M} \hat{\underline{u}}_{Bl} \hat{\underline{u}}_{Bl}^T \underline{M}$
STIFFNESS MATRIX	$\underline{K} = -\underline{M} \hat{\underline{U}}_B \underline{\Delta}_B^2 (\underline{I} - \hat{\underline{U}}_B^T \underline{M} \hat{\underline{U}}_B \underline{\Delta}_B) \cdot$ $\cdot \underline{\Delta}_B \hat{\underline{U}}_B^T \underline{M}$
FLEXIBILITY MATRIX	$\underline{G} = \underline{K}^{-1} = -\hat{\underline{U}}_B \underline{\Delta}_B^{-1} \hat{\underline{U}}_B^T$ $= -\sum_{l=1}^{2n} \frac{1}{\lambda_{Bl}} \hat{\underline{u}}_{Bl} \hat{\underline{u}}_{Bl}^T$
FREQUENCY RESPONSE MATRIX	$\underline{F}(j\omega) = (-\omega^2 \underline{M} + j\omega \underline{B} + \underline{K})^{-1}$ $= \hat{\underline{U}}_B (j\omega \underline{I} - \underline{\Delta}_B)^{-1} \hat{\underline{U}}_B^T$ $= \sum_{l=1}^{2n} \frac{\hat{\underline{u}}_{Bl} \hat{\underline{u}}_{Bl}^T}{j\omega - \lambda_{Bl}}$

Table 1.2 Relations between system parameter matrices and modal magnitudes

SYSTEM / SPECIFICATION	HYSTERETIC DAMPED SYSTEM
MATRIX EIGENVALUE PROBLEM	$(-\lambda_{Di}^2 \underline{M} + \underline{K} + j\underline{D})\hat{\underline{u}}_{Di} = \underline{0}$ $i = 1(1)n,\ \underline{M}_D = \underline{I}$
INERTIA MATRIX	$\underline{M} = (\hat{\underline{U}}_D \hat{\underline{U}}_D^T)^{-1}$
INVERSE INERTIA MATRIX	$\underline{M}^{-1} = \hat{\underline{U}}_D \hat{\underline{U}}_D^T = \sum_{i=1}^{n} \hat{\underline{u}}_{Di} \hat{\underline{u}}_{Di}^T$
DAMPING MATRIX	$\underline{B} = Im[(\hat{\underline{U}}_D^T)^{-1} \underline{\Delta}_D \hat{\underline{U}}_D^{-1}]$ $= Im(\underline{M} \hat{\underline{U}}_D \underline{\Delta}_D \hat{\underline{U}}_D^T \underline{M})$
STIFFNESS MATRIX	$\underline{K} = Re[(\hat{\underline{U}}_D^T)^{-1} \underline{\Delta}_D \hat{\underline{U}}_D^{-1}]$ $= Re(\underline{M} \hat{\underline{U}}_D \underline{\Delta}_D \hat{\underline{U}}_D^T \underline{M})$
FLEXIBILITY MATRIX	$(\underline{K} + j\underline{D})^{-1} = \hat{\underline{U}}_D \underline{\Delta}_D^{-1} \hat{\underline{U}}_D^T$ $= \sum_{i=1}^{n} \frac{1}{\lambda_{Di}} \hat{\underline{u}}_{Di} \hat{\underline{u}}_{Di}^T$
FREQUENCY RESPONSE MATRIX	$\underline{F}(j\omega) = (-\Omega^2 \underline{M} + \underline{K} + j\underline{D})^{-1}$ $= \hat{\underline{U}}_D(-\Omega^2 \underline{I} + \underline{\Delta}_D)^{-1} \hat{\underline{U}}_D^T$ $= \sum_{i=1}^{n} \frac{\hat{\underline{u}}_{Di} \hat{\underline{u}}_{Di}^T}{\lambda_{Di} - \Omega^2}$

Table 1.3 Relations between system parameter matrices and modal magnitudes

It can be seen that the relationships for active systems can be derived in the same way, Lancaster:[5]

$$\begin{aligned} \underline{A}_2^{-1} &= \hat{\underline{U}}_B \underline{\Lambda}_B \hat{\underline{V}}_B^T , \\ \underline{A}_1 &= -\underline{A}_2 \hat{\underline{U}}_B \underline{\Lambda}_B^2 \hat{\underline{V}}_B^T \underline{A}_2 , \\ \underline{A}_o^{-1} &= -\hat{\underline{U}}_B \underline{\Lambda}_B^{-1} \hat{\underline{V}}_B^T . \end{aligned} \tag{49}$$

The above-mentioned relationships are not useful for calculating the system parameter matrices from the identified eigenmagnitudes, because the latter are generally incomplete and inaccurate, and all errors will enter into the computation. But the knowledge of these equations is helpful for checking tests, and may be used in connection with estimation methods.

5. Spectral Decomposition of Dynamic Responses

The spectral decomposition in the ω-space is given, for example, by (48), and the given expressions in the last line of Tables 1.1 - 1.3. But because the decoupling of the equations of motion is shown by modal transformation, the spectral decomposition should be derived by using the Laplace transformation

$$\underline{U}(s) := \mathcal{L}\{\underline{u}(t)\} ,$$
$$\underline{P}(s) := \mathcal{L}\{\underline{p}(t)\} , \qquad (50)$$

with the initial condition equal to zero.

The spectral decomposition of the dynamic response is the basis of most programmes used in digital Fourier analysis in order to identify modal magnitudes, Natke.[8,9,10]

5.1 The Passive Undamped System

The congruence transformation with the regular modal matrix

$$\underline{u}(t) = \underline{\hat{U}}_o \, \underline{q}(t) \qquad (51)$$

applied to the passive system (10) without damping effects, $\underline{B} = \underline{0}$, when the orthogonality conditions (21) are taken into account, results in

$$\underline{M}_g \left[\underline{\ddot{q}}(t) + \underline{\Lambda}_o \, \underline{q}(t)\right] = \underline{\hat{U}}_o^T \, \underline{p}(t) . \qquad (52)$$

Here the system matrices are diagonal and the equations of motions are decoupled:

$$m_{gi}\left[\ddot{q}_i(t) + \lambda_{oi}\, q_i(t)\right] = \underline{\hat{u}}^T_{oi}\, \underline{p}(t),\; i = 1(1)n. \tag{53}$$

The $q_i(t)$ are the principle coordinates of the system.
Eq. (52) Laplace transformed,

$$\underline{M}_g\left[s^2\underline{I} + \underline{\Lambda}_o\right]\underline{Q}(s) = \underline{\hat{U}}^T_o\,\underline{P}(s),\; \underline{Q}_s{:} = \mathcal{L}\{\underline{q}(t)\}\,, \tag{54}$$

leads to

$$\underline{Q}_s(s) = \left[s^2\,\underline{I} + \underline{\Lambda}_o\right]^{-1}\underline{M}_g^{-1}\,\underline{\hat{U}}^T_o\,\underline{P}(s)$$

and with (51) in the s-domain to the spectral decomposition

$$\underline{U}(s) = \underline{\hat{U}}_o\left[s^2\,\underline{I} + \underline{\Lambda}_o\right]^{-1}\underline{M}_g^{-1}\,\underline{\hat{U}}^T_o\,\underline{P}(s)$$

$$= \sum_{i=1}^{n}\frac{\underline{\hat{u}}^T_{oi}\,\underline{P}(s)}{m_{gi}(\lambda_{oi}+s^2)}\,\underline{\hat{u}}_{oi}. \tag{55}$$

This equation contains all the well-known information about the dynamic behaviour of the systems, e.g. the resonance phenomenon, and the excitation condition in order to excite only one natural mode.

5.2 The Passive Hysteretic Damped System

The harmonic excited system is described by eq. (9) with $\underline{B} = \underline{0}$. With the congruence transformation applied to eq. (9) with the regular modal matrix

$$\hat{\underline{u}} = \hat{\underline{U}}_D \hat{\underline{q}}, \tag{56}$$

and when one bears in mind the orthogonality equations (23) in the normalization $\underline{M}_D = \underline{I}$ of the eigenvectors $\hat{\underline{u}}_{Di}$ this results in

$$(-\Omega^2 \underline{I} + \underline{\Lambda}_D) \hat{\underline{q}} = \hat{\underline{U}}_D^T \hat{\underline{p}}. \tag{57}$$

Since the matrices on the left side of eq. (57) are diagonal, the equations of motion are decoupled, and we get

$$(-\Omega^2 + \lambda_{Di}) \hat{q}_i = \hat{\underline{u}}_{Di}^T \hat{\underline{p}}, \tag{58}$$

and gain the response amplitude vector through

$$\begin{aligned} \hat{\underline{u}} &= \hat{\underline{U}}_D (-\Omega^2 \underline{I} + \underline{\Lambda}_D)^{-1} \hat{\underline{U}}_D^T \hat{\underline{p}} \\ &= \sum_{i=1}^{n} \frac{\hat{\underline{u}}_{Di}^T \hat{\underline{p}}}{\lambda_{Di} - \Omega^2} \hat{\underline{u}}_{Di}. \end{aligned} \tag{59}$$

Eq. (59) represents the spectral decomposition of the dynamic response to a harmonic excitation with the force amplitude vector $\hat{\underline{p}}$ and the real excitation frequency Ω using the eigenmodes of the damped system. The physical interpretation can be done easily, and the information needed for resonance testing etc. is included.

In general, the decomposition is possible only if the eigenvectors of the damped system are used. If the ease-hypothesis (30) is valid, then $\hat{\underline{u}}_D$ is equal to $\hat{\underline{u}}_o$, and the decomposition into the real natural modes of the associated undamped system is possible.

5.3 The Passive Viscous Damped System

The spectral decomposition of the passive viscous damped system can be derived in a corresponding manner (similar to that of the active system using the left eigenvectors). Congruence transformation of eq. (15) with

$$\underline{x}(t) = \hat{\underline{X}}_V\, \underline{q}_x(t) \tag{60}$$

leads to

$$\hat{\underline{X}}_V^T\, \underline{A}_V\, \hat{\underline{X}}_V\, \dot{\underline{q}}_x(t) + \hat{\underline{X}}_V^T\, \underline{B}_V\, \hat{\underline{X}}_V\, \underline{q}_x(t) = \hat{\underline{X}}_V^T\, \underline{f}_V(t)\,. \tag{61a}$$

When the orthogonalization (36) is taken into account, (61a) converts into

$$\dot{\underline{q}}_x(t) - \underline{\Lambda}_B \underline{q}_x(t) = \hat{\underline{X}}_v^T \underline{f}_v(t). \tag{61b}$$

The Laplace transformation applied to (60) and (61a) produces

$$\begin{aligned} &\underline{Q}_x(s) = (s \underline{I} - \underline{\Lambda}_B)^{-1} \hat{\underline{X}}_v^T \underline{F}_v(s), \\ &\underline{Q}_x(s) := \mathcal{L}\{\underline{q}_x(t)\}, \ \underline{F}_v(s) := \mathcal{L}\{\underline{f}_v(t)\}, \end{aligned} \tag{62}$$

and finally

$$\begin{aligned} \underline{X}(s) &:= \mathcal{L}\{\underline{x}(t)\} = \\ &= \hat{\underline{X}}_v \underline{Q}_x(s) = \\ &= \hat{\underline{X}}_v (s \underline{I} - \underline{\Lambda}_B)^{-1} \hat{\underline{X}}_v^T \underline{F}_v(s) \\ &= \sum_{l=1}^{2n} \frac{\hat{\underline{x}}_{vl}^T \underline{F}_v(s)}{s - \lambda_{Bl}} \hat{\underline{x}}_{vl}. \end{aligned} \tag{63}$$

Taking into account the definition of the state space vector, and assuming 2s real eigensolutions, one obtains the spectral decomposition

$$\underline{U}(s) = \sum_{r=1}^{2s} \frac{\hat{\underline{u}}_{Br}^T \underline{P}(s)}{s - \lambda_{Br}} \hat{\underline{u}}_{Br} + \sum_{k=1}^{n-s} \left[\frac{\hat{\underline{u}}_{B2s+k}^T \underline{P}(s)}{s - \lambda_{B2s+k}} \hat{\underline{u}}_{B2s+k} + \frac{\hat{\underline{u}}_{B2s+k}^{*T} \underline{P}(s)}{s - \lambda_{B2s+k}^*} \hat{\underline{u}}_{B2s+k}^* \right]. \tag{64}$$

The statement made in section 5.3 concerning the physical interpretation etc. and the possibility of a decomposition into $\hat{\underline{u}}_{oi}$ also holds true in this case.

Because of its importance in identification, the transient excitation may be considered i.e. $\underline{p}(t) \neq \underline{0}$ only in the time interval $0 \leq t \leq T$. The free decaying vibrations in the s-domain (that means the dynamic response for $t \geq T$) follows from equation (64) for active weakly damped systems to

$$\underline{U}(s) = \sum_{k=1}^{2n} \frac{\underline{a}_k}{s-\lambda_{Bk}} = \sum_{k=1}^{n} \left[\frac{\underline{a}_k}{s-\lambda_{Bk}} + \frac{\underline{a}_k^*}{s-\lambda_{Bk}^*} \right],$$
$$\underline{a}_k := \hat{\underline{u}}_{Bk} \left[\hat{\underline{v}}_{Bk}^T \int_0^T \exp(-\lambda_{Bk} t)\, \underline{p}(t)dt \right]. \tag{65}$$

6. Sensibility, Observability, Controllability, Identifiability

With regard to the point in the heading of this section, only a few remarks are to be given, and details may be quoted from the references, e.g. Müller,[11] and Eykhoff.[12]

Generally it can be observed that the estimation of eigenfrequencies is more accurate than, for instance, that of the effective dampings. Or one may estimate modal magnitudes of the system when looking for the dynamic response or the system parameters. We will learn in the next lectures how to calculate the standard deviations of these magnitudes, and these deviations are also based on a deterministic reflection. This means sensitivity (influence) investigations by a linear differential error calculation. The sensitivity functions are then defined by the corresponding partial derivatives. In a simplified manner we consider the linearized, approximate influence of a small additive inertia variation $\Delta \underline{M}$ on the eigenfrequencies of the passive undamped system neglecting the eigenvector changes:

$$\left[- (\omega_o + \Delta\omega_o)^2 (\underline{M} + \Delta\underline{M}) + \underline{K} \right] \hat{\underline{u}}_o \doteq \underline{0}$$

$$(-\omega_o^2 \underline{M} - \omega_o^2 \Delta\underline{M} - 2\Delta\omega_o \omega_o \underline{M} + \underline{K}) \hat{\underline{u}}_o \doteq \underline{0}$$

$$-(\omega_o \Delta\underline{M} + 2\Delta\omega_o \underline{M}) \hat{\underline{u}}_o \doteq \underline{0}$$

$$\Delta\omega_o \doteq - \frac{\omega_o}{2} \frac{\Delta m_g}{m_g} ,$$

$$\Delta m_g := \hat{\underline{u}}_o^T \Delta \underline{M} \hat{\underline{u}}_o .$$

An illustration of the concepts of observability and controllability is shown in Fig. 4. Time-invariant, linear, deterministic, and asymptotic stable systems, as described by the state equation (16) and the output equation (17) may be investigated regarding observability and controllability by means of Kalman's criteria, using the rank of specified matrices composed of the matrices of the state and output equation or by the associated eigenvectors of the transposed system matrix using Hautus' criteria. That means, for example with regard to the controllability, that the eigenvectors are not orthogonal to the force vector. With regard to the classical phase resonance method, simple relationships can be derived from the properties of the matrices of the state and output equation; Kortüm, Niedbal.[13] The process is called identifiable if it is possible from the measurements of the state variables to determine the system matrix $\underline{A}$ by a unique solution. Rank criteria exist for investigations.

7. Remarks on Weak Non-linearities

In parameter and state estimation one has to distinguish linear and non-linear models in the unknowns (parameters), independent of whether the input-output relation of the system under test is linear or not. This statement is true with regard to estimation, but the distinction between

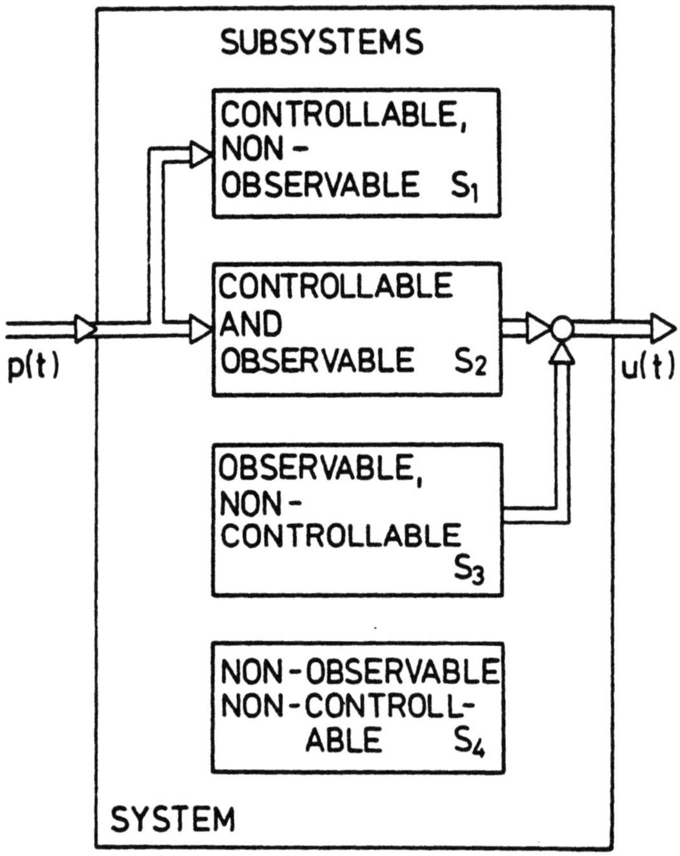

Fig. 4 Division of a system into (non-) controllable and (non-) observable subsystems

linear and non-linear dynamic behaviour is essential, too. There are mainly two reasons for this. When testing a real system one should meet at least weak non-linearities, as may be caused by dry friction, which in certain cases are not interesting, but which sometimes produce large effects on near-resonance testing when one is working with the assumption of a linear system,Dat.[14] This is the first reason for thinking about non-linearities in identification: modified dynamic behaviour concerning the assumed linear system. The other reason is, for example, the dependence of the dynamic behaviour on the inital conditions. Therefore the identified model describes only the system in the considered (measured) hyperspace, but not in general.

In non-linear vibrating systems the existence of orthogonal eigenmodes is not certain. If they exist, e.g. in conservative non-linear systems, their properties are different from those of linear conservative systems. The dynamic response of free non-linear systems can be described only to a less extent. In contrast to linear systems, the development of the dynamic response with time (stable, unstable) depends greatly on the initial conditions given. In addition to the damping coupling of linear systems, one has to consider the symmetry properties of the matrices of discretized non-linear systems. It can be shown that systems with non-linear damping and restoring forces lead to nonsymmetric damping and stiffness matrices, Chandivert.[15]

The effects of weak non-linear damping and stiffness characteristics lead to large distortions in the frequency response in the neighbourhood of the resonance frequency, when compared with the frequency response of the associated linear system. If one is interested in the parameters of the associated linear system, then near-resonance testing ought to be avoided. Further aid consists in the introduction of an additional arbitrary linear damping force; Dat.[16] If dry friction is to be overcome, an additional high-frequency force superimposed on the low frequency excitation has to be applied. Identification methods for non-linear systems are presented in another lecture (for a single dof system with cubic damping and stiffness term s. Natke.[17]).

References:

1 Natke, H.G., Probleme zur Modellierung und Kenndatenermittlung dynamischer Systeme am Beispiel alternativer Fragestellungen; Fachseminar spurgeführter Fernverkehr, Rad/Schiene - Technik - Zusammenwirken von Fahrzeug und Fahrweg, Augsburg - Mai 1979

2 Wilkinson, J.H., The algebraic eigenvalue problem, Clarendon Press, 1965

3 Caughey, T.K., Classical normal modes in damped linear dynamic systems, J. Appl. Mech. 27, 269, 1960

4 Caughey, T.K., M.E.J. O'Kelly, Classical normal modes in damped linear dynamic systems, J. Appl. Mech. 32, 583, 1965

5 Lancaster, P., Lambda-matrices and vibrating systems, Pergamon Press, 1966

6 Veubeke de Fraejis, B.M., A variational approach to pure mode excitation based on characteristic phase lag theory, AGARD Rep. 39, 1956

7 Veubeke de Fraejis, B.M., Les déphasages caractéristiques en présence des modes rigides et de modes non amortis, Academic royal de Belgique, Bulletin de la Classe des Sciences, (5) 51, 1965

8 Natke, H.G., A method for computing natural oscillation magnitudes from the results of vibration testing in one exciter configuration, NASA-TT-F-12446, 1969

9 Natke, H.G., Die Berechnung der Eigenschwingungsgrößen eines gedämpften Systems aus den Ergebnissen eines Schwingungsversuches in einer Erregerkonfiguration, Jahrbuch der DGLR, 98, 1971

10 Natke, H.G. et al., Zeitreihen- und Modalanalyse, Identifikation technischer Konstruktionen, VDI-BW 32-22-03, Düsseldorf, 1981

11 Müller, P.C., W.O. Schiehlen, Lineare Schwingungen, Akademische Verlagsanstalt, 1976

12 Eykhoff, P., System identification - Parameter and state estimation, John Wiley and Sons, 1974

13 Kortüm, W., N. Niedbal, Application of modern control theory to modal survey techniques, Lecture held at the EUROMECH 131, Besançon, 1980

14 Dat, R., L'essai de vibration d'une structure imperfaitement lineaire, La Rech. Aerospatiale, No. 4, 223, 1975

15 Chandivert, M.G., Considerations relevant to the vibration testing of aeronautical structures having non-ideal characteristics, Dep. of Supply, Australian Defence Scientific Service, ARL/SM. 355, 1970

16 Dat, R., R. Otiayon, Structural vibration testing methods and the correction of the theoretical models, Lecture held at the EUROMECH 131, Besançon, 1980

17 Natke, H.G., Fehlerbetrachtungen zur parametrischen Identifikation eines Systems mit kubischem Steifigkeits- und Dämpfungsterm; Czerwenka-Festschrift, TU München, 1979

INTRODUCTION TO SYSTEM IDENTIFICATION USING PARAMETER ESTIMATION METHODS

H. Unbehauen
Ruhr-Universität
Bochum

Contribution to the book
"Identification of Vibrating Structures"
(Ed.: H.G. Natke) Springer Verlag, 1982.

CONTENTS

1. Introduction

2. Model design and parameter estimation methods

2.1. Model design for identification methods

2.2. Parameter estimation methods

2.2.1. Least-Squares and extended Least-Squares methods

2.2.2. Instrumental Variable method

2.2.3. Uniform presentation of the recursive estimation equations

2.3. Determination of structural coefficients

2.3.1. Determination of model order

2.3.2. Determination of a system deadtime

2.3.3. Stationary signal values superposed to the system signals

3. Conditions for optimal identification and practical aspects

3.1. Determination of sampling interval

3.2. Input signals

3.3. Setting the initial values

3.4. Selection of starting time and length of measurement

3.5. Observation of the progress of the estimation results

3.6. Steps for improving the convergence (Identification guidance)

3.7. Interpretation of the results and a comparison of the efficiency of the methods

3.8. Interpretation of the results by the identification of real systems

3.9. Identification program packages

4. Conclusions

Appendix A: Recursive version of the unweighted estimation algorithm

Appendix B: Recursive version of the weighted estimation algorithm

References

1. INTRODUCTION

The rapid progress in the field of computer technology during recent years has led to an increasing use of process computers to analyse, supervise and control technical processes. Process computers provide the opportunity to sample and store measuring data in a short time, and to process them immediately for better process operation (e.g. with optimum efficiency, best product quality, optimum time behaviour etc.). An important condition for optimum process operation is the knowledge of the process behaviour in the past, at present and in the future. The use of computers allows a permanent analysation or identification of the process dynamics by evaluating the input and output signals of the system. The result of such a process identification is usually a mathematical model, by which the static and dynamic behaviour can be estimated or predicted.

There are numerous reasons for setting up such models for technical processes. The most important are listed in the following:

- the design of modern optimal control algorithms can only be realized on the basis of a precise process model;
- a model provides the opportunity to understand the internal process behaviour;
- models can be used for the simulation of the actual process behaviour in a special situation, for instance the simulation of malfunctions in a nuclear power plant;

- in some cases the operators of complex technical plants are trained by simulators, which, as an essential component, contain a process model (e.g. simulators for aeroplanes, nuclear plants, cars, etc.);
- in modern central control room technology, process models are frequently used by the operators, e.g. for the prediction of the process behaviour.

Dependent on the various types of processes there are different possibilities for the mathematical description of models (e.g. differential equations, state space description, transfer functions, non-parametric models, etc.). For the choice of the optimum form of the mathematical description it must be noted that all essential characteristics of the process have to be described by the model.

In many cases the optimum development of a mathematical model can be performed by an interpretation of technical and physical laws underlying the process. But those models based only on a *theoretical analysis* are often either connected with uncertain assumptions, or they are useless for further processing because of their special form. Therefore, in most practical applications an *experimental analysis* is more appropriate for process identification. Experimental system identification evaluates how the observable process output signals react upon the controllable process input signals and unmeasurable disturbances.

If well-defined test signals are used as input signals, for instance step or impulse functions, *deterministic methods* are appropriate for system identification. In many processes, however, no additional deterministic test signal can be superposed to the input signal. The natural process signals mostly have to be used for excitation of the process. Because of the stochastic character of these signals, only stochastic procedures can then be applied for process identification. Today *statistical parameter estimation methods* are mostly used besides the classical *nonparametric correlation methods*. In the following only parameter estimation methods will be discussed.

In modern control theory two forms of parametric models become important: the description of *input-output behaviour*, e.g. by transfer functions, differential - or difference equations, and the *state space representation*. While the description by state space representation has become dominant in the field of system synthesis during the last ten years, this is not the case in the field of system analysis, where the input-output description form has remained dominant. This in the first instance hardly understandable separate evolution has its origin in the use of state variables, which contain important information for a regulator design, but which are considered as unneccessary additional estimation variables during the analysis.

The *historical development* of statistical parameter estimation methods using input-output description forms started around 1964. It can be divided into three phases:

- basic theoretical studies (1964 - 1972),
- practical applications and comparative studies for linear single-input single-output systems (1970 - 1976),
- identification of special processes, e.g. multi-input multi-output systems, non-linear systems, systems operating in closed loops, etc. (since 1975).

The now almost historical first phase of theoretical research, development and application led to a great variety of statistical estimation algorithms. The most important parameter estimation methods are summarized in Table 1.1.

During recent years different modifications and some new estimation methods have been suggested. However, they will never achieve the importance of the above-mentioned methods, because these are simple and robust in application.

Method	Abbre-viations	References
Least Squares	LS	Eykhoff (1967), Aström (1968, 1970), Göhring (1973), Unbehauen (1973a)
Generalized LS	GLS	Clarke (1967), Hasting-James (1969), Talmon (1971)
Instrumental Variable	IV	Wong-Polak (1967), Young (1970)
Maximum-Likelihood	ML	Bohlin (1968, 1970), Aström (1970), Niemann (1971), Gustavsson (1972)

Table 1.1. Summary of the most important parameter estimation methods

In the second phase of development the first experiments in practical application of the estimation methods were conducted, e.g. in navigation (Aström (1972)). The broad scope of application spectrum is contained in the proceedings of the IFAC symposia on 'Identification and System parameter Estimation' edited by Eykhoff (1973), Rajbman (1976), and Isermann (1979).

Another research direction in this phase of development was the comparison of different estimation methods in order to select an optimal identification method for the type of application (Saridis (1974), Göhring (1973), Unbehauen (1973b), Isermann (1973)).

During the third phase the development of identification methods for processes with special properties can be observed. These are especially multi-input / multi-output systems, nonlinear systems, time-variant systems, systems with distributed parameters and closed loop systems. This development will be discussed in a later chapter in this volume, entitled 'Some modern Developments in System Identification Using Parameter Estimation Methods'.

In this chapter the model design, the estimation procedure and the conditions for optimum identification of linear, time-invariant single-input / single-output systems will be discussed. Because it is impossible to describe the different properties of various systems with one unique model, first the possibilities of model structure are presented in section 2. The Least-Squares, the extended Least-Squares and the Instrumental Variable method are parameter estimation methods which use the model structures presented here. They are developed in a uniform presentation for recursive algorithms. When there is no a priori information about the system it is a task of the identification to determine the model order and a system deadtime and to detect stationary values superposed to the system signals.

In section 3 the conditions for an optimum identification are given, and practical aspects for application will be discussed. It is very important to choose an appropriate sampling interval and to determine input signals which excite the system such that all dynamical modes can be detected. The choice of initial values and the selection of an appropriate starting time are necessary conditions for consistent estimation results for recursive algorithms. During the identification procedure the progress of the results must be observed, and perhaps the convergence can be improved by identification guidance. After the identification the estimation results must be interpreted and examined for their validity.

This chapter represents an introduction to system analysis by parameter estimation methods. The presentation is summarized in the context of the topics of this book. The special references must be consulted for further details.

2. MODEL STRUCTURE AND PARAMETER ESTIMATION METHODS

The purpose of identification is to develop a mathematical model which describes the static and dynamic behaviour of the system in a sufficiently accurate manner. A critical examination of the quality of the model is obtained by a comparison of the system output y and the model output y_M, where the system and the model are both excited by the same input signal u. The measurable system output y consists of the unmeasurable output signal x and the noise signal r. In Fig. 2.1 the model is arranged parallel to the system.

When the output signal error

$$e = y - y_M \tag{2.1}$$

is small enough, it is assumed that an optimal model for the unknown

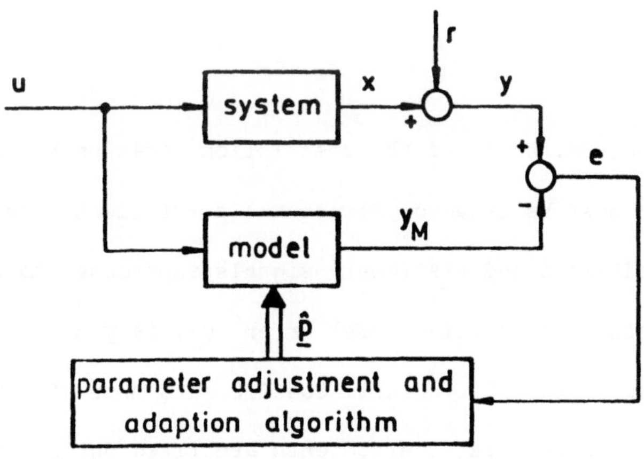

Fig. 2.1. Principal identification procedure using parameter estimation methods

system is obtained. When this is not the case, the variable parameters of the model must be corrected by a parameter adjustment and adaption algorithm. This correction is made until the error becomes minimal.

When there is no a priori information about the system, the identification consists of the following three steps:

- determination of the description equations,
- determination of the structural coefficients, and
- estimation of the equation parameters.

The choice of the description equation depends on the physical characteristics of the system and is normally fixed during the identification procedure. In the following only linear continuous, time-invariant systems are considered, which can be described by a discrete difference equation. Various forms of difference equations are presented in the following.

After the determination of the description equation their structural coefficients must be assumed. Structural coefficients are the model order n, the deadtime d and stationary signals superposed to the input and output signals. When a noise model or an auxiliary model are used during identification, their structural coefficients must also be determined. Usually the structural coefficients are fixed during the identification procedure. They are only changed when no sufficient error minimization can be obtained.

The parameters of the description equation can be estimated by the parameter estimation methods presented in section 2.2.

2.1. Model structures for identification methods

Most processes to be analysed are time continuous systems. Therefore it would be appropriate to examine the time continuous output signal error

$$e(t) = y(t) - y_M(t). \tag{2.2}$$

Then every deviation of the estimated output signal $y_M(t)$ could be detected. But this is impossible, because the output signal of the model is time discrete when the model is estimated by a digital computer. The model output signal $y_M(t)$ is only available for the time instants $t = kT$, for $k = 0,1,2,\ldots,N$, where T is the sampling interval. Usually the time interval T is not directly used, so that the output signal error can be written in the discrete form by

$$e(k) = y(k) - y_M(k) \tag{2.3}$$

for the sampling instant k.

According to Fig. 2.1 the measurable output signal consists of the undisturbed output signal x(k) and a measuring and system noise r(k):

$$y(k) = x(k) + r(k). \tag{2.4}$$

Because the model estimates the undisturbed system output x(k), the error e(k) always contains the noise r(k).

The model output signal $y_M(k)$ can be calculated in discrete form by the difference equation

$$y_M(k) = -\sum_{\nu=1}^{n} a_\nu y_M(k-\nu) + \sum_{\nu=0}^{n} b_\nu u(k-\nu), \tag{2.5}$$

where the coefficients a_ν and b_ν represent the parameters of the mathematical model which have to be estimated. Eq.(2.5) shows that the output signal $y_M(k)$ is calculated using the measured values of the input signal $u(\mu)$ for $\mu = k,\dots,k-n$ and the previously calculated values of the output signal $y_M(\mu)$ for $\mu = k-1,\dots,k-n$. Applying the shift operation of the z-transformation to Eq.(2.5) the form

$$\mathfrak{Z}\{y_M(k)\} = Y_M(z) = -Y_M(z)(a_1 z^{-1} + \dots + a_n z^{-n}) + U(z)(b_o + b_1 z^{-1} + \dots + b_n z^{-n}) \tag{2.6}$$

is obtained. Using the abbreviations

$$A(z^{-1}) = 1 + a_1 z^{-1} + a_2 z^{-2} + \dots + a_n z^{-n} = 1 + A^*(z^{-1}) \tag{2.7}$$

$$B(z^{-1}) = b_o + b_1 z^{-1} + \dots + b_n z^{-n} \tag{2.8}$$

the discrete model transfer function $G_M(z)$ is defined as

$$G_M(z) = \frac{b_o+b_1z^{-1}+\ldots+b_nz^{-n}}{1+a_1z^{-1}+\ldots+a_nz^{-n}} = \frac{B(z^{-1})}{1+A^*(z^{-1})} \; . \tag{2.9}$$

The output signal error of Eq.(2.3) can now be calculated by

$$E(z) = Y(z) + A^*(z^{-1}) \; Y_M(z) - B(z^{-1}) \; U(z) \; . \tag{2.10}$$

This equation corresponds to the block diagram of Fig. 2.2.

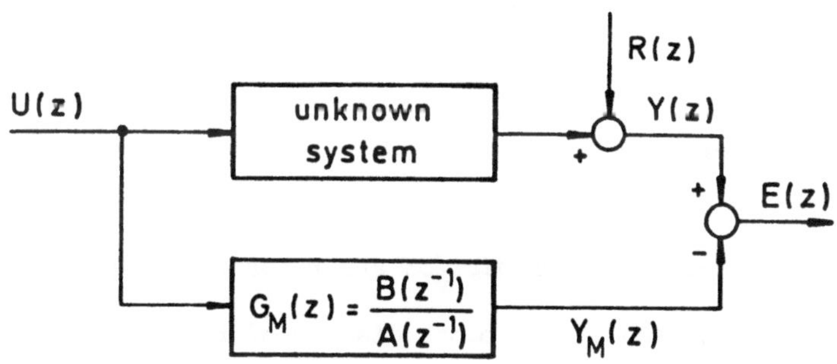

Fig. 2.2. Model for defining the output signal error

In Eq.(2.5) the output signal of the model $y_M(k)$ is based on the calculation of the previous output signals $y_M(k-\nu)$, i.e.

$$y_M(k) = y_M(k/k-1) \; . \tag{2.11}$$

When these previous values, especially in recursive parameter estimation methods, are already calculated with errors, the momentary output signal $y_M(k)$ will also contain these errors. This disadvantage can be circumvented by using the measured output signals $y(k-\nu)$ instead of the calculated signals $y_M(k-\nu)$ in Eq.(2.5). Then $y_M(k)$ can be obtained by

$$y_M(k) = -\sum_{\nu=1}^{n} a_\nu y(k-\nu) + \sum_{\nu=0}^{n} b_\nu u(k-\nu) = y(k) \qquad (2.12)$$

and the error E(z) according to Eq.(2.10) is given by

$$E(z) = A(z^{-1})\, Y(z) - B(z^{-1})\, U(z) . \qquad (2.13)$$

Since the model output signal $y_M(k)$ is now based on the previously measured values of the system output signal $y(k-\nu)$, i.e.

$$y_M(k) = y(k/k-1) , \qquad (2.14)$$

therefore E(z) is called the *prediction error* (or *equation error*). Eq. (2.13) can be interpreted by the block diagram of Fig. 2.3.

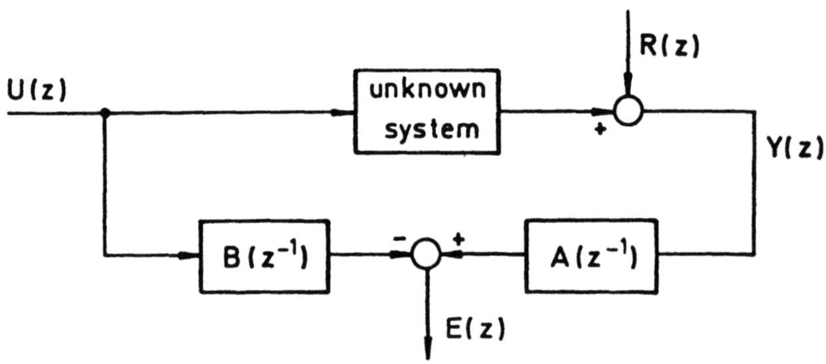

Fig. 2.3. Model for defining the equation error

Another advantage of the equation error is the linear connection between the variation of the parameters a_i and b_i and the error E(z). This causes a simple mathematical algorithm.

When the measured output signal, however, is used to calculate the model output signal $y_M(k)$ in Eq.(2.12), the superposed noise r(k) leads to an autocorrelated error in the parameter estimation. This systematical error is called a *bias*. As shown later, the Least-Squares method is based on the equation error E(z) of Eq.(2.13). That means that the Least-Squares method always provides a biased parameter estimation, if the residual E(z) does not behave like white noise.

To avoid this disadvantage two possibilities can be considered:

- using the estimated undisturbed system output signal x(k) for the calculation of $y_M(k)$ or
- transforming the error e(k) into white noise using a filter with an appropriate transfer function.

In order to make use of the first possibility, the instrumental variable method (IV-method) was developed, which - starting with a model structure according to Eq.(2.13) - produces instrumental variables by filtering the input signal, which can be used for the estimation of the parameters. These instrumental variables are strongly correlated with the signals of the system but not with the superposed noise. Therefore it is possible to get consistent estimates of the parameters a_i and b_i for (i=1,2,...,n). The IV-parameter estimation algorithm is presented in section 2.2.2.

In most cases, however, an attempt is made to transform the error e(k) into white noise by an additional disturbance model $G_R(z)$. This procedure is based on the fact that every stochastic noise signal can be reduced to white noise by appropriate filtering. With this additional filtering the *extended general equation error* follows from Eq.(2.13)

$$E(z) = G_R^{-1}(z)\{A(z^{-1})\ Y(z) - B(z^{-1})\ U(z)\} \tag{2.15}$$

and

$$G_R(z)\ E(z) = A(z^{-1})\ Y(z) - B(z^{-1})\ U(z) = V(z) \tag{2.16}$$

which also includes for $G_R(z) = 1$ the model structure used for the LS- and IV-method. The extended model is presented in Fig. 2.4.

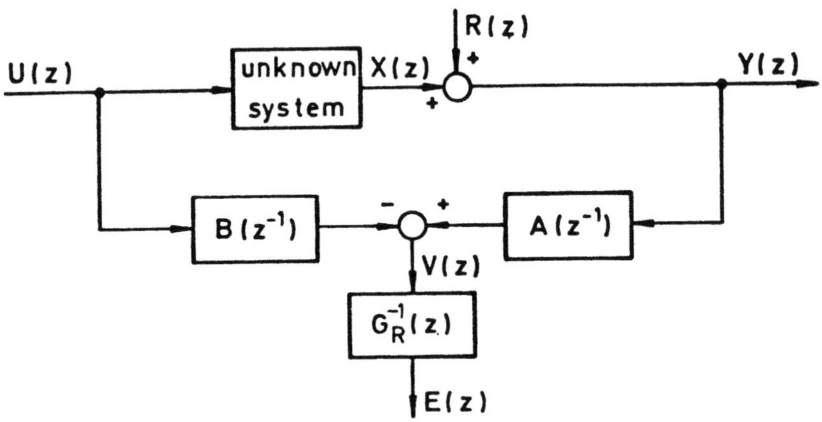

Fig. 2.4. The extended model

All the models presented in the following can be described by the extended general equation error, Eq.(2.15). This leads to the possibility of developing the parameter estimation methods in a unique form.

The auto-regressive model (AR)

For the AR-model it is assumed that there is no excitation of the system by an input signal or by an additive noise. The output signal is only dependent on previous values $y(k-\nu)$ (for $\nu=1,\ldots,n$) and on the parameters a_i (for $i=1,\ldots,n$):

$$y_M(k) = -\sum_{\nu=1}^{n} a_\nu y(k-\nu) . \tag{2.17}$$

Then the extended general equation error reduces to

$$E(z) = A(z^{-1})\ Y(z) . \tag{2.18}$$

<u>The moving average model (MA)</u>

If the output signal of the system only depends on a superposed noise signal $r(k)$, the moving average model is appropriate. The output signal is the "average" of a "moving" noise signal. For the polynomials $A(z^{-1}) = 1$ and $B(z^{-1}) = 0$ the extended general equation error is reduced to

$$E(z) = Y(z) - C^*(z^{-1})\ E(z) \tag{2.19}$$

where the polynomial

$$C^*(z^{-1}) = c_1 z^{-1} + \ldots + c_n z^{-n} . \tag{2.20}$$

is contained in the transfer function

$$G_R(z) = 1 + C^*(z^{-1}) = C(z^{-1}) . \tag{2.21}$$

Thus it is also possible to write Eq.(2.19) in the form

$$Y(z) = [1 + C^*(z^{-1})]\ E(z) .$$

The auto-regressive, moving average model (ARMA)

The ARMA-model is the combination of the AR- and the MA-models with $A(z^{-1}) \neq 1$. Consequently the extended general equation error is defined as

$$E(z) = A(z^{-1})\ Y(z) - C^*(z^{-1})\ E(z). \qquad (2.22)$$

For simplification of the mathematical treatment the order of $C^*(z^{-1})$ is set equal to the order of $A(z^{-1})$. Contrary to the AR- and the MA-model the ARMA-model is used for practical application. The ARMA-model was suggested by Aström (1970).

The extended auto-regressive, moving average model (ARMAX)

The extended ARMA-model corresponds to the extended model in Fig. 2.4. The system is excited by an input signal u(k), and the output signal x(k) is superposed by a noise signal. The ARMAX-model is the most important model in practical application.

The extended general equation error is defined as

$$G_R(z)\ E(z) = A(z^{-1})\ Y(z) - B(z^{-1})\ U(z) . \qquad (2.23)$$

For the MA- and the ARMA-model the special form of a polynomial $C(z^{-1})$

was defined for the transfer function $G_R(z)$. A general presentation of $G_R(z)$, however, is given by the fractional, rational transfer function

$$G_R(z) = \frac{C(z^{-1})}{D(z^{-1})} = \frac{1+c_1 z^{-1}+\ldots+c_n z^{-n}}{1+d_1 z^{-1}+\ldots+d_n z^{-n}}, \tag{2.24}$$

whereby the polynomial $C(z^{-1})$ is newly defined. Again the order of the polynomials C and D is set equal to the order of the model. This is permissible, because usually the best estimation results are obtained in this case.

The general error equation follows from Eqs.(2.23) and (2.24)

$$A(z^{-1})\ Y(z) - B(z^{-1})\ U(z) = \frac{C(z^{-1})}{D(z^{-1})}\ E(z) \tag{2.25}$$

with

$$\frac{C(z^{-1})}{D(z^{-1})}\ E(z) = V(z)\ . \tag{2.26}$$

Dependent on the form of the transfer function $G_R(z)$ three different model structures can be obtained for the ARMAX-model.

1st extended model (1. EM)

The polynomial $C(z^{-1})$ is set equal to one so that the equation error of Eq.(2.25) becomes

$$A(z^{-1})\ Y(z) - B(z^{-1})\ U(z) = \frac{1}{D(z^{-1})}\ E(z)\ . \tag{2.27}$$

2nd extended Model (2. EM)

It is assumed that the polynomial $D(z^{-1}) = 1$ so that the equation error is

$$A(z^{-1})\ Y(z) - B(z^{-1})\ U(z) = C(z^{-1})\ E(z)\ . \tag{2.28}$$

3rd extended model (3. EM)

For the 3rd EM the transfer function $G_R(z)$ is

$$G_R(z) = \frac{C(z^{-1})}{D(z^{-1})} \tag{2.29}$$

so that the general equation error according to Eq.(2.25) can be obtained.

Another possibility is that the transfer function $G_R(z)$ becomes one. Then the equation error is

$$A(z^{-1})\ Y(z) - B(z^{-1})\ U(z) = E(z)\ . \tag{2.30}$$

This model structure - as already mentioned above - is used for the Least Squares method (LS) and the Instrumental Variable method (IV). The different possibilities for the extended ARMAX model are summarized in Table 2.1.

parameter estimation method	equation error	$G_R(z)$
LS and IV	$AY - BU = E$	$G_R = 1$
1. EM	$AY - BU = \frac{1}{D} E$	$G_R = \frac{1}{D}$
2. EM	$AY - BU = CE$	$G_R = C$
3. EM	$AY - BU = \frac{C}{D} E$	$G_R = \frac{C}{D}$

Table 2.1. Model structures for the ARMAX model

The recursive Maximum-Likelihood (ML) method uses the model structure of the 2nd extended model (2. EM). The corresponding difference equation for the 3rd extended model for any sampling instant $k \geq (k\,n+1)$ is

$$y(k) + \sum_{\nu=1}^{n} a_\nu y(k-\nu) - \sum_{\nu=0}^{n} b_\nu u(k-\nu)$$

$$= - \sum_{\nu=1}^{n} d_\nu v(k-\nu) + \sum_{\nu=1}^{n} c_\nu e(k-\nu) + e(k) \tag{2.31}$$

where

$$v(k) = - \sum_{\nu=1}^{n} d_\nu v(k-\nu) + \sum_{\nu=1}^{n} c_\nu e(k-\nu) + e(k) \, . \tag{2.32}$$

The difference equations for the other model structures can be obtained in a similar way.

The structural coefficents must be determined for all these model structures. The structural coefficients are

- the order n,
- the deadtime d, and
- stationary values superposed to the input and output signals.

The problems of determination are discussed in section 2.3.

2.2. Parameter estimation methods

After the choice of the model structure and of the structural coefficients the parameters a_ν, b_ν, c_ν and d_ν (=1,2,...,n) of the model must be determined, whereby for the simplification of the numerical treatment the parameter b_o is set to zero. This is done by the minimization of a quadratic function of the error e(k). The procedure of the numerical treatment of this optimization problem partly differs with the various methods. Therefore a separate presentation of the different methods is necessary for the further discussion. To give a uniform presentation of the methods, the LS-, the 1. EM-, the 2. EM-, and the 3. EM-methods are treated together. The IV-method is presented afterwards.

2.2.1. Least-Squares and extended Least-Squares methods

Starting with the 3rd extended model structure according to Fig. 2.4 and Eq.(2.25) the parameter vector

$$\underline{p}^T = \left[a_1 \ldots a_n \mid b_1 \ldots b_n \mid c_1 \ldots c_n \mid d_1 \ldots d_n\right]$$

$$= \left[\underline{p}_a^T \mid \underline{p}_b^T \mid \underline{p}_c^T \mid \underline{p}_d^T\right] \tag{2.33}$$

and the data vector

$$\underline{m}^T(k) = \left[-y(k-1) \ldots -y(k-n) \mid u(k-1) \ldots u(k-n) \mid e(k-1) \ldots e(k-n) \mid -v(k-1) \ldots -v(k-n)\right]$$

$$= \left[\underline{m}_y^T(k) \mid \underline{m}_u^T(k) \mid \underline{m}_e^T(k) \mid \underline{m}_v^T(k)\right] \tag{2.34}$$

are introduced. Thus Eq. (2.31) leads to

$$y(k) = \underline{m}^T(k)\, \underline{p} + e(k) \; . \tag{2.35}$$

For N subsequent pairs of samples $\{u(k),\ y(k)\}$ with $k = n+1, \ldots, n+N$, Eq. (2.35) can be written in the vector form

$$\underline{y} = \underline{M}\,\underline{p} + \underline{e} \tag{2.36}$$

with the abbreviations

$$\underline{y}^T = \left[y(n+1) \;\ldots\; y(n+N)\right] , \tag{2.37}$$

$$\underline{e}^T = \left[e(n+1) \;\ldots\; e(n+N)\right] \tag{2.38}$$

and the data matrix

$$\underline{M} = \begin{bmatrix} \underline{m}^T(n+1) \\ \underline{m}^T(n+2) \\ \vdots \\ \underline{m}^T(n+N) \end{bmatrix} = \left[\begin{array}{cc|cc|cc|cc} -y(n) & \ldots -y(1) & u(n) & \ldots u(1) & e(n) & \ldots e(1) & -v(n) & \ldots -v(1) \\ -y(n+1) & \ldots -y(2) & u(n+1) & \ldots u(2) & e(n+1) & \ldots e(2) & -v(n+1) & \ldots -v(2) \\ \vdots & & \vdots & & \vdots & & \vdots & \\ -y(n+N-1) & \ldots -y(N) & u(n+N-1) & \ldots u(N) & e(n+N-1) & \ldots e(N) & -v(n+N-1) & \ldots -v(N) \end{array}\right]. \tag{2.39}$$

The generation of the vectors $\underline{p}$ and $\underline{m}$ and the matrix $\underline{M}$ is different for the LS, 1st, 2nd, and 3rd extended model and therefore summarized in Table 2.2.

To obtain the model parameters which correspond in an optimum way to the system parameters an optimization criterion in form of a square error function

$$I_1(\underline{p}) = \frac{1}{2}\,\underline{e}^T\,\underline{e} = \frac{1}{2}\sum_{k=n+1}^{n+N} e^2(k) \overset{!}{=} \text{Min.} \tag{2.40}$$

is defined on the basis of the error vector $\underline{e}$ after N+n samples. Taking Eq.(2.36) into consideration, the minimum of $I_1(\underline{p})$ is obtained by setting

$$\left.\frac{\partial I_1(\underline{p})}{\partial \underline{p}}\right|_{\underline{p}=\hat{\underline{p}}} = -\underline{M}^T\underline{y} + \underline{M}^T\underline{M}\,\hat{\underline{p}} = \underline{0}\,. \tag{2.41}$$

By solving this equation for the desired estimated parameter vector $\hat{\underline{p}}$, the equation of the LS-method

$$\boxed{\hat{\underline{p}} = [\underline{M}^T\,\underline{M}]^{-1}\,\underline{M}^T\underline{y}} \tag{2.42}$$

method	structure for $G_r(z)$	$\hat{\underline{p}}$	$\underline{w}(k)$	$\underline{m}(k)$	$\hat{e}(k)$	$\hat{v}(k)$	initial values
RLS	1	$[\hat{a}_1 \ldots \hat{a}_n \; \hat{b}_1 \ldots \hat{b}_n]^T$	$\underline{w}(k)=\underline{m}(k)$	$\underline{m}(k) = [-y(k-1) \ldots -y(k-n) \; u(k-1) \ldots u(k-n)]^T$	$\hat{e}(k) = y(k) - \underline{m}^T()\hat{\underline{p}}(k-1)$	—	$\hat{\underline{p}}$, $\underline{P}$
1. EM	$\frac{1}{D(z^{-1})}$	$[\hat{a}_1 \ldots \hat{a}_n \; \hat{b}_1 \ldots \hat{b}_n \; \hat{d}_1 \ldots \hat{d}_n]^T$	$\underline{w}(k)=\underline{m}(k)$	$\underline{m}(k) = [-y(k-1) \ldots -y(k-n) \; u(k-1) \ldots u(k-n) \; -\hat{v}(k-1) \ldots -\hat{v}(k-n)]^T$		$\hat{v}(k) = y(k) - \underline{m}_{yu}^T(k)\hat{\underline{p}}_{ab}(k-1)$	$\hat{\underline{p}}$, $\underline{P}$, $\hat{v}$
2. EM	$C(z^{-1})$	$[\hat{a}_1 \ldots \hat{a}_n \; \hat{b}_1 \ldots \hat{b}_n \; \hat{c}_1 \ldots \hat{c}_n]^T$	$\underline{w}(k)=\underline{m}(k)$	$\underline{m}(k) = [-y(k-1) \ldots -y(k-n) \; u(k-1) \ldots u(k-n) \; \hat{e}(k-1) \ldots \hat{e}(k-n)]^T$		—	$\hat{\underline{p}}$, $\underline{P}$, $\hat{e}$
3. EM	$\frac{C(z^{-1})}{D(z^{-1})}$	$[\hat{a}_1 \ldots \hat{a}_n \; \hat{b}_1 \ldots \hat{b}_n \; \hat{c}_1 \ldots \hat{c}_n \; \hat{d}_1 \ldots \hat{d}_n]^T$	$\underline{w}(k)=\underline{m}(k)$	$\underline{m}(k) = [-y(k-1) \ldots -y(k-n) \; u(k-1) \ldots u(k-n) \; \hat{e}(k-1) \ldots \hat{e}(k-n) \; -\hat{v}(k-1) \ldots -\hat{v}(k-n)]^T$		$\hat{v}(k) = y(k) - \underline{m}_{yu}^T(k)\hat{\underline{p}}_{ab}(k-1)$	$\hat{\underline{p}}$, $\underline{P}$, $\hat{e}$, $\hat{v}$
IV	1	$[\hat{a}_1 \ldots \hat{a}_n \; \hat{b}_1 \ldots \hat{b}_n]^T$	$\underline{w}(k)=[-y_H(k-1) \ldots -y_H(k-n) \; u(k-1) \ldots u(k-n)]^T$ $y_H(k) = \underline{w}^T(k)\underline{p}_H(k)$ $\underline{p}_H(k)$ acc. Eq.(2.65)	$\underline{m}(k) = [-y(k-1) \ldots -y(k-n) \; u(k-1) \ldots u(k-n)]^T$		—	$\hat{\underline{p}}$, $\underline{P}$, y_H

Table 2.2. Definition of the variables in a unified presentation of the recursive algorithms

is directly obtained, if the error signal e(k) is an $N(o,\sigma^2)$-distributed white noise and u(k) is a sufficient and persistent excitation of the process. Then the parameters estimated in this way with the LS-method are consistent in the quadratic average.

As the model structure of the LS-method according to Fig. 2.4 with $G_R(z) = 1$ is practically not sufficient because the noise signal normally is not white noise, an autocorrelated signal will appear applying this method, and thus the estimates will have a bias. This bias is avoided for the other *extended LS-methods* (using the extended model structures according to Table 2.1) by an additional filtering of the correlated error, which thus is reduced to white noise.

For the immediate estimation of the parameters according to Eq.(2.42) all samples have to be stored in a matrix $\underline{M}$. Then the parameters can be calculated at once. If this direct estimation is to be used additionally in on-line operation, then an inversion of the matrix $\underline{M}^T\underline{M}$ has always to be performed for newly measured data after this matrix has been updated with the new measurements. This considerable numerical expense and the high storage requirements can be avoided if the estimation equations are derived in a recursive form.

For the derivation of an algorithm for the *recursive parameter estimation* by the LS-method, we start with Eq.(2.42) or with an equivalent summation notation for k samples

$$\underline{\hat{p}}(k) = \underline{P}(k) \cdot \sum_{i=n+1}^{k} \underline{m}(i)\, y(i) \tag{2.43}$$

where

$$\underline{P}(k) = \left[\sum_{i=n+1}^{k} \underline{m}(i)\, \underline{m}^T(i) \right]^{-1} . \tag{2.44}$$

After some lengthy transformations, described in appendix A, the following recursive algorithm for the LS-method is obtained:

$$\underline{\hat{p}}(k+1) = \underline{\hat{p}}(k) + \underline{q}(k+1)\, \hat{e}(k+1) , \tag{2.45a}$$

$$\underline{q}(k+1) = \underline{P}(k)\, \underline{m}(k+1) \left[1 + \underline{m}^T(k+1)\, \underline{P}(k)\, \underline{m}(k+1) \right]^{-1} , \tag{2.45b}$$

$$\underline{P}(k+1) = \underline{P}(k) - \underline{q}(k+1)\, \underline{m}^T(k+1)\, \underline{P}(k) , \tag{2.45c}$$

$$\hat{e}(k+1) = y(k+1) - \underline{m}^T(k+1)\, \underline{\hat{p}}(k) . \tag{2.45d}$$

Eqs.(2.45a) and (2.45d) show the basic structure of all recursive estimation methods:

$$\begin{matrix}\text{new}\\ \text{estimated}\\ \text{value}\end{matrix} = \begin{matrix}\text{old}\\ \text{estimated}\\ \text{value}\end{matrix} + \begin{matrix}\text{correction}\\ \text{vector}\end{matrix} \left[\begin{matrix}\text{new}\\ \text{sample}\end{matrix} - \begin{matrix}\text{predicted value}\\ \text{using the parameters}\\ \text{of the last step}\end{matrix} \right] .$$

$\hat{e}(k+1)$ is, according to the expression in brackets, a prediction error, because of its calculation by the difference between the sample $y(k+1)$ and its predicted value based on the parameters of the last step and the measured system output signal $y(k-\nu)$ for $(\nu=1,\ldots,n)$ (see also section 2.1):

$$\hat{y}_m(k+1) = \hat{y}(k+1/k) = \underline{m}^T(k+1)\ \hat{\underline{p}}(k)\ . \tag{2.46}$$

Since the expression in the brackets of Eq.(2.45b) is a scalar, there is no need for an inversion of the matrix in the recursive algorithm. Storage of the new measured data is also not necessary, because of the immediate processing. Besides these considerable advantages, there are some disadvantages common to all recursive algorithms concerning the starting phase of these algorithms, which will be dealt with in section 3.3.

The derived direct or recursive equations are immediately applicable for the LS-method. If one looks at the extended LS-methods, the data vector $\underline{m}$ according to Eq.(2.34) and therefore the data matrix $\underline{M}$ also contain values of the errors e(k) and v(k). But these are unknown in reality, and we have to find suitable approximations. An easy method is, instead of e(k) and v(k), to use their estimation values

$$\hat{e}(k+1) = y(k+1) - \underline{m}^T(k+1)\ \hat{\underline{p}}(k) \tag{2.47}$$

and

$$\hat{v}(k+1) = y(k+1) - \underline{m}^T_{yu}(k+1)\ \hat{\underline{p}}_{ab}(k) \tag{2.48}$$

according to Eq.(2.31) and Eq.(2.45d), for the modified data vector

$$\underline{m}^T(k) = \left[-y(k-1)\ldots-y(k-n) \;\vdots\; u(k-1)\ldots u(k-n) \;\vdots\; \hat{e}(k-1)\ldots\hat{e}(k-n) \;\vdots\; -\hat{v}(k-1)\ldots-\hat{v}(k-n)\right]\ . \tag{2.49}$$

Thus the complete recursive algorithm of the extended LS-method is composed from Eq.(2.45), where $\underline{m}^T(k)$ has to be used from Eq.(2.49). The exact meaning of the values used here in particular is listed in Table 2.2, where once again a uniform and summarizing presentation of the investigated algorithms can be found.

2.2.2. Instrumental-Variable method (IV-method)

Whereas with the extended LS-methods the problem of a correlated residual is solved by a reduction to white noise using a filter transfer function $G_R(z)$, the IV-method starts with the model structure of the LS-method according to Eq.(2.30) and Fig. 2.4 for $G_R(z) = 1$ which leads to the expression

$$A(z^{-1})\ Y(z) - B(z^{-1})\ U(z) = E(z) \tag{2.50}$$

or

$$y(k) + \sum_{i=1}^{n} a_i\ y(k-i) - \sum_{i=1}^{n} b_i\ u(k-i) = e(k) \tag{2.51}$$

where no special assumptions for the residual e(k) have to be made. In order to determine the model parameters, Eq.(2.51) is again written in vector notation according to Eq.(2.36)

$$\underline{y} = \underline{M}\underline{p} + \underline{e}\,, \tag{2.52}$$

where the meaning of the values $\underline{p}$ and $\underline{M}$ is the same as with the LS-method (see Table 2.2). Using this error vector $\underline{e}$ a quadratic error function

$$I_2(\underline{p}) = \frac{1}{2}\,\underline{e}^T \underline{W}\,\underline{W}^T \underline{e} \tag{2.53}$$

is now defined, where

$$\underline{W} = \begin{bmatrix} \underline{w}^T(n+1) \\ \vdots \\ \underline{w}^T(n+N) \end{bmatrix} \tag{2.54}$$

denotes the instrumental variable matrix, when the following relations are valid:

$$\underset{N\to\infty}{p\,\lim} \frac{1}{N}\,\underline{W}^T\,\underline{M} \quad \text{regular} \tag{2.55}$$

and

$$\underset{N\to\infty}{p\,\lim} \frac{1}{N}\,\underline{W}^T\,\underline{e} = \underline{0}\,. \tag{2.56}$$

Using Eq.(2.52), the minimum of $I_2(p)$ is finally obtained by setting

$$\left.\frac{\partial I_2(\underline{p})}{\partial \underline{p}}\right|_{\underline{p}=\hat{\underline{p}}} = -\,\underline{M}^T\,\underline{W}\,\underline{W}^T\,\underline{y} + \underline{M}^T\,\underline{W}\,\underline{W}^T\,\underline{M}\,\hat{\underline{p}} = \underline{0}\,. \tag{2.57}$$

The solution of this expression for the desired estimated parameter vector $\hat{\underline{p}}$ leads to the expression

$$\hat{\underline{p}} = \left[\underline{M}^T\,\underline{W}\,\underline{W}^T\,\underline{M}\right]^{-1}\,\underline{M}^T\,\underline{W}\,\underline{W}^T\,\underline{y}\,, \tag{2.58}$$

from which, by consideration of Eq.(2.55), the direct equation of the IV-method follows for large values of N

$$\boxed{\hat{\underline{p}} = \left[\underline{W}^T \underline{M}\right]^{-1} \underline{W}^T \underline{y}} \; . \qquad (2.59)$$

As long as the conditions in Eqs.(2.55) and (2.56) are satisfied, the parameters estimated by the IV-method are consistent.

Analog to section 2.2.1, the recursive version of the estimation equation for the IV-method is obtained by starting with the direct equation, Eq.(2.59), or the equivalent summation notation for k samples

$$\hat{\underline{p}}(k) = \underline{P}(k) \sum_{i=n+1}^{k} \underline{w}(i)\, y(i) \qquad (2.60)$$

where

$$\underline{P}(k) = \left[\sum_{i=n+1}^{k} \underline{w}(i)\, \underline{m}^T(k)\right]^{-1} \qquad (2.61)$$

and $\underline{m}$ are defined according to Table 2.2. The corresponding procedure shown in appendix B, with $\rho = 1$, supplies the following recursive algorithm for the IV-method

$$\hat{\underline{p}}(k+1) = \hat{\underline{p}}(k) + \underline{q}(k+1)\, \hat{e}(k+1) \qquad (2.62a)$$

with

$$\underline{q}(k+1) = \underline{P}(k)\, \underline{w}(k+1) \left[1 + \underline{m}^T(k+1)\, \underline{P}(k)\, \underline{w}(k+1)\right]^{-1} , \qquad (2.62b)$$

$$\underline{P}(k+1) = \underline{P}(k) - \underline{q}(k+1)\, \underline{m}^T(k+1)\, \underline{P}(k) , \qquad (2.62c)$$

$$\hat{e}(k+1) = y(k+1) - \underline{m}^T(k+1)\, \hat{\underline{p}}(k) . \qquad (2.62d)$$

The speed of convergence as well as the efficiency of the estimation depend essentially on the selection of the IV-matrix $\underline{W}$. To satisfy the condition of Eqs.(2.55) and (2.56), the elements of $\underline{W}$ have to be selected so that they are strongly correlated to the signals of the matrix $\underline{M}$, but not with the noise. The best constellation exists if $\underline{W}$ contains the undisturbed signals of the system. Since the output signal x is not measurable, estimation values of this signal are used as instrumental variables y_H. These instrumental variables ordered in the matrix $\underline{W}$ as

$$\underline{w}^T(k) = [-y_H(k-1)\ldots -y_H(k-n) \mid u(k-1)\ldots u(k-n)] \quad (2.63)$$

are obtained from the recursive version of the IV-method with the expression

$$y_H(k) = \underline{w}^T(k)\ \underline{p}_H(k) \quad (2.64)$$

as the output of an auxiliary model with a transfer function $B_H(z^{-1})/A_H(z^{-1})$, which is excited with the input signal u(k) according to Fig. 2.5.

Here $\underline{p}_H$ contains the parameters of the auxiliary model, which are calculated from the estimated parameters $\hat{\underline{p}}$. Since the noise r(k) influences the determination of $\hat{\underline{p}}(k)$, the instrumental variables are correlated to the noise, and therefore also to the residual e(k). This leads to a systematical estimation error, because of Eq.(2.56), therefore Wong/Polak (1967) suggested introducing a dead time d between the estimated parameters $\hat{\underline{p}}$ and those parameters $\underline{p}_H$ adjusted in the auxiliary model, where d

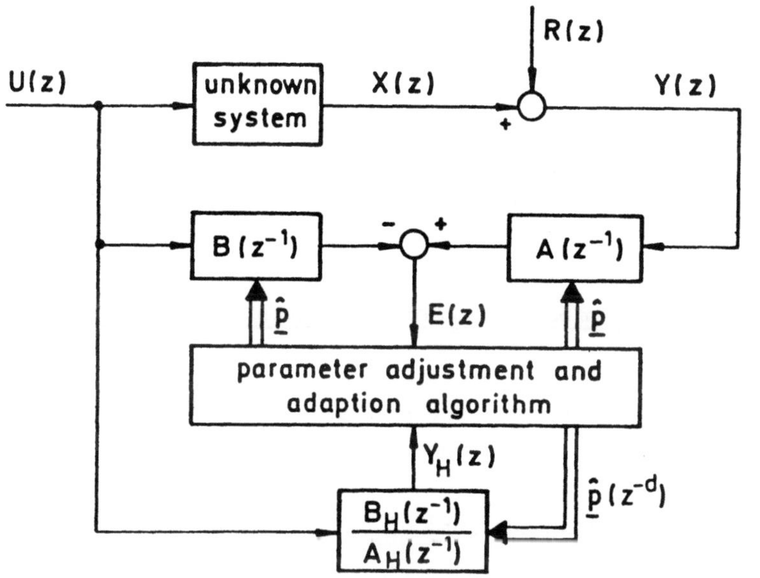

Fig. 2.5. Model for the recursive version of the IV-method

is selected in such a way that e(k+d) is independent of e(k). If, in addition to this, a discrete low pass filter

$$\underline{p}_H(k) = (1-\gamma)\ \underline{p}_H(k-1) + \gamma\hat{\underline{p}}(k-d) \qquad (2.65)$$

is used as described by Young et al (1970), then the selection of d does not need to be so accurate and the estimated parameter values will be smoothed, so that fast changes of the parameters of the auxiliary model will be avoided. γ is selected according to Baur (1976).

The efficiency of the estimation with the recursive version of the IV-method depends on a suitable auxiliary model, the parameters of which are adjusted such that a correlation between the instrumental variable $y_H(k)$ and the residual e(k) is avoided. The complete recursive algorithm of the IV-method is described by Eqs.(2.62), (2.64), and (2.65), where $\underline{w}^T(k)$ has to be used according to Eq.(2.63). The exact meaning of all variables used here in particular, is once more listed in Table 2.2.

2.2.3. Uniform presentation of the recursive estimation equations

The strong similarity of the recursive estimation equations derived in the previous sections gives rise to a uniform presentation (as already used in Eq.(2.62a-d)) in the form of

$$\underline{\hat{p}}(k+1) = \underline{\hat{p}}(k) + \underline{q}(k+1)\ \hat{e}(k+1) \tag{2.66a}$$

with

$$\underline{q}(k+1) = \underline{P}(k)\ \underline{w}(k+1)\ \left[1 + \underline{m}^T(k+1)\ \underline{P}(k)\ \underline{w}(k+1)\right]^{-1}, \tag{2.66b}$$

$$\underline{P}(k+1) = \underline{P}(k) - \underline{q}(k+1)\ \underline{m}^T(k+1)\ \underline{P}(k), \tag{2.66c}$$

$$\hat{e}(k+1) = y(k+1) - \underline{m}^T(k+1)\ \underline{\hat{p}}(k), \tag{2.66d}$$

which makes the formal analogy of the recursive algorithms clear. The respective meanings as well as the calculation procedure of the variables used are noted in Table 2.2, particularly for the 5 parameter estimation methods discussed above.

A problem occurring with all recursive parameter estimation methods indicated by the description of the extended LS-method, is connected with the start of these recursive algorithms. Resulting from Eq.(2.66) and Table 2.2 it is necessary to define initial values especially for $\hat{\underline{p}}$ and $\underline{P}$ as well as for $\hat{e}$ and $\hat{v}$ within the 1st, 2nd, and 3rd EM-methods and for y_H within the IV-method, which, however, do not have a strong influence on the convergence rate.

2.3. Determination of the structural coefficients

The following structural coefficients must be determined for every model, independent of the different structures

- the order n of the polynomials,
- the deadtime d of the systems, and
- stationary signal values superposed to the input and output signals.

If there is no a priori information about these coefficients they must be determined prior to the identification procedure by various testing methods, by a test identification or by a comparison of the identification results obtained for different coefficients. These possibilities are discussed globally in the following. However, for details reference must be made to the special literature.

2.3.1. Determination of model order

For the calculation of the general equation error Eq.(2.25) it was assumed that the order n of the polynomials A, B, C and D are known. In most practical applications there is no a priori information about the order n, therefore it must be determined by various testing methods. The difficulty in the determination of the model order arises from the fact that, on the one hand, the model should contain as many poles as are necessary for a precise description of the dynamic system behaviour, and on the other hand that the model order should be low with regard to the realization purposes, such as, for example, simulation studies or controller design. The important question is therefore, how accurately the model has to describe the real system behaviour, because for many applications the model does not have to contain all system modes. In other applications it may also happen that special system modes are not measurable, and therefore cannot be identified. The model order is determined practically by the desired model accuracy in the related field of application.

In order to avoid this subjective judgement in choosing the order of a model, several procedures for model-structure testing had been proposed during recent years. These methods, which are listed in Table 2.3 with the corresponding valuation and references, can be divided into three groups:

A priori test methods

These methods are applied before starting the parameter estimation using the input and output signals of a test measurement in order to estimate a rough model order.

test method	abbreviation	efficiency without disturbances	efficiency with disturbances	references
Determinant ratio	DR	good	bad	Woodside, 1971
Extended determinant ratio	EDR	good	fair	Göhring, 1973
Instrumental determinant ratio	IDR	good	good	Wellstead, 1978; Young, 1980
Independence of output signal	IO	good	fair	Chow, 1972
Loss function	LF	good	good	Van den Boom, 1974
Signal error	SE	good	good	Unbehauen, 1973b; Göhring, 1973
F-test	FT	good	good	Söderström, 1977
Final prediction error	FPE	good	good	Akaike, 1969, 1970; Söderström, 1977
Akaike information criterion	AIC	good	good	Akaike, 1978
Polynomial-test	PT	good	good	Unbehauen, 1973b

Table 2.3. Test methods for model order determination

Determinant ratio test: Data matrices are constructed for different model orders. If the ratio of the determinants of the matrices for the order $\hat{n}$ and $\hat{n}+1$ differs only little from the previous determinant ratios for smaller orders, then $\hat{n}$ is the most appropriate order of the model.

Extended determinant ratio test: This test has the same examination criteria as the determinant ratio test. It is possible, however, to examine systems where noise is superposed to the input-output signals.

Instrumental determinant ratio test: It is assumed that noise is superposed only to the output signals. The undisturbed system output signals are estimated by an instrumental variable method, and with these signals the determinant ratio test is performed.

Independence of output signals: The correlation of the output signals is calculated and then it will be checked for which order $\hat{n}$ a statistical independence occurs.

Test methods for the accuracy of signal estimation

The output signals of the unknown system and of the model are compared either during the estimation procedure or afterwards. It is assumed that for an optimum model of the system the equation error becomes minimal. Either the error or a function of the error is used for the test methods.

Loss function test: During the identification procedure the squared equation error is summed up, and after identification this loss function value is compared with the loss function values for other model orders. When the loss function value is very small compared with the loss function values for lower orders it is assumed that the according order is the optimum order of the model. This test is very simple but also very good.

Signal error test: The parameters for models with different orders are estimated. After identification, time responses of the estimated and simulated system are compared with the time response of the system.

Statistical F-test: It is examined whether the loss function for the order $\hat{n}+1$ is significantly better than the loss function for the order $\hat{n}$. If this is not the case, the order $\hat{n}$ is an appropriate order for the model.

Final prediction error (FPE) test: Again the loss functions of models with different orders are compared. If the FPE-value calculated for every order with the according loss function becomes minimal, the optimal order for the model is obtained.

Akaike information criterion: This test is based on the investigation of the likelihood function. The F-test, the Final Prediction Error test and the Akaike Information criterion have asymptotically same results as shown by Söderström (1977).

Test of pole configuration of the model

The parameters of models with different orders are estimated. After a decomposition of the estimated transfer functions to partial fractions it is tested whether poles and zeros can be compensated or neglected. The optimum order can be determined by the uncompensated poles. This test is the only one which is not based on the investigation of the signal estimation, but on the estimated parameters.

A comparison of the efficiency is listed in Table 2.3, where references of the various methods are also given. The efficiency has also been compared by Unbehauen (1974b) and Söderström (1977).

2.3.2. Determination of a system deadtime

For the development of different models a system deadtime was neglected in order to simplify the numerical treatment. If the system, however, has a deadtime it must be recognized for the calculation of the model output signal. Since the models have discrete transfer functions, only a discrete deadtime can be chosen for the model. This means that the deadtime is a multiple of the sample interval T

$$d = \nu T \quad \text{for} \quad \nu = 1,2,\ldots \; . \tag{2.67}$$

This limitation is sufficient for usual practical application, and represents only a disadvantage if the sample interval is very large.

For systems with deadtime the discrete model transfer function of Eq.(2.9) is changed to

$$G_M(z) = \frac{B(z^{-1})}{A(z^{-1})} z^{-d} = \frac{b_o+b_1z^{-1}+\ldots+b_nz^{-n}}{1+a_1z^{-1}+\ldots+a_nz^{-n}} z^{-d} . \quad (2.68)$$

By multiplying the terms of the polynomial $B(z^{-1})$ by the deadtime the transfer function

$$G_M(z) = \frac{b_i z^{-(d+1)}+\ldots+b_n z^{-(d+n)}}{1+a_1z^{-1}+\ldots+a_nz^{-n}} \quad (2.69)$$

is obtained. Then the model output signal y_M for the time instant k can be calculated according to Eq.(2.12) by

$$y_M(k) = -a_1y(k-1) - a_2y(k-2) - \ldots - a_ny(k-n)$$

$$+b_1u(k-d-1) + \ldots + b_nu(k-d-n) . \quad (2.70)$$

Comparing Eqs.(2.12) and (2.70) there is only a shifting of the input signals $u(k-\nu)$ to the previous values $u(k-d-\nu)$. Therefore in the parameter estimation algorithms only the data vector Eq.(2.34) must be varied:

$$\underline{m}^T(k) = [-y(k-1)\ldots-y(k-n) \mid u(k-d-1)\ldots u(k-d-n)$$

$$e(k-1)\ldots e(k-n) \mid -v(k-1)\ldots-v(k-n)] . \quad (2.71)$$

Then the numerical treatment of systems with deadtime is very simple.

Contrary to the model order, the deadtime is usually known from a priori information. In most practical applications a deadtime arises from a transportation operation of material, which can be measured exactly. In chemical processes, however, the deadtime is often unknown because it is not possible to measure the important state variables. Then the dead time must be determined experimentally. This can be performed by one of the following possibilities:

- investigation of the cross correlation function between input and output signal,
- examination of the step responses and
- comparison of the loss function for different deadtimes.

While the 1st and the 3rd method are always possible, the second test presumes that deterministic input signals are allowed.

In most practical applications of parameter estimation methods the deadtime is determined by a comparison of the loss function for different deadtimes, because this method can also be applied for recursive algorithms. Often the order $\hat{n}$ and the deadtime are determined in a combined calculation. It is assumed that the loss function becomes minimal for the optimum estimated deadtime and the most appropriate order of the model. This method, however, may have a great disadvantage: Instead of a small deadtime a higher order $\hat{n}$ can be estimated, because the output signal reacts in a similar way if the system has a small deadtime or a higher order n. This error can only be detected through a critical examination.

Concluding the discussion about the deadtime it must be mentioned that in most cases the determination of the deadtime is no problem for system identification. Simple detection and determination of the dead-time, and also simple numerical treatment of the identification is always possible.

2.3.3. Stationary signal values superposed to the system signals

If stationary signal values are superposed to the input and output signals of the system this must be noted for identification. The numerical treatment for such a disturbance is very simple. For off-line identification the measurement period is examined and the mean value is calculated. Then every signal is corrected by the mean value. Therefore it is not necessary to change the identification algorithm.

For on-line identification the mean value must also be calculated by a test measurement. Then the signals can be corrected before they are used for identification. This is much simpler than a correction of the model difference equation.

3. CONDITIONS FOR OPTIMUM IDENTIFICATION AND PRACTICAL ASPECTS

In the following section those conditions for the model and the measurements are to be considered which lead to optimum identification results. Usually a model identified from a special series of measurements contains only the system information from these measurements of perhaps only one special operating condition. Therefore, in general the measurements must be performed so that as many as possible of the operating conditions of the investigated system are contained in the series of measurements which is used for identification. Only if this condition is fulfilled can the identified model describe the real system behaviour exactly. Therefore it is very important to choose an appropriate sampling time to determine suitable input signals, to find good initial values for the recursive algorithms, and to select the starting time and the length of measurement.

Those who have already handled identification problems know that the problems do not consist only in programming the equations of the parameter estimations method, but in applying these to practical systems. The operator must have had some experience in order to observe the progress of the estimation results, to improve the convergence by identification guidance, and to interpret the results. Some useful notes are given in this section.

3.1. Determination of sampling interval

In the application of parameter estimation methods, input and output signals are always processed in discrete form. The selection of the sampling time needed for the discrete signal processing usually depends on several facts:

- The sampling time should be at least so small that the fastest modes of the system can be processed.
- The sampling time may, however, not be essentially smaller than the minimum time constant of the investigated system. If this were the case, constant sampling values would be obtained over several sampling intervals, and this would cause numerical difficulties in the estimation algorithm, as the matrix $\underline{P}$ of Eq.(2.44) would obtain too small coefficients during a very short time.
- For on-line parameter estimation methods the smallest possible sampling time is determined in real-time operation by the computing-time for one estimation of all model parameters.

Two methods have proven to be most appropriate for the determination of the sampling time T:

1) The spectral power densities of the measured input and output signals are calculated. Then a sampling time can be determined by $T=\pi/\omega_m$, where ω_m is a positive frequency so that $|G(j\omega)|\approx 0$ for $|\omega|>\omega_m$ according to *Shannon's sampling theorem*.

2) If a rough knowledge of the step response of the system is available, e.g. from previous test identification or from direct (not too precise) measurements, then the sampling time can be determined by $T = (1/6 \ldots 1/10)\ T_{63}$, where T_{63} represents the time when the step

response has reached 63 % of its assymptotic final value.

Obviously the sampling time has to fulfil the condition of being within a limited region in order to obtain an optimum estimation.

Finally, it should be mentioned here that the *measuring time* of the input and output signals is also of great importance for the optimum result of parameter estimation methods. The upper limit for the measuring time is usually given by the limited stationarity of a real process, e.g. caused by drift in the signals. The necessary measuring time is determined on the one hand by the form and magnitude of the noise in the output signal, and on the other hand by the desired accuracy of the mathematical model. With increasing signal/noise-ratio the measuring time increases. This also holds in the case of multi-input/multi-output systems (MIMO-systems) when the number of input and/or output signals is increased.

3.2. Input signals

In the identification of a single-input/single-output system (SISO-system) by analysing the input signal u and the output signal y, two cases can be distinguished according to Fig. 3.1 and Table 3.1. Here the measurable input and output signals u and y respectively are comprised of different signals.

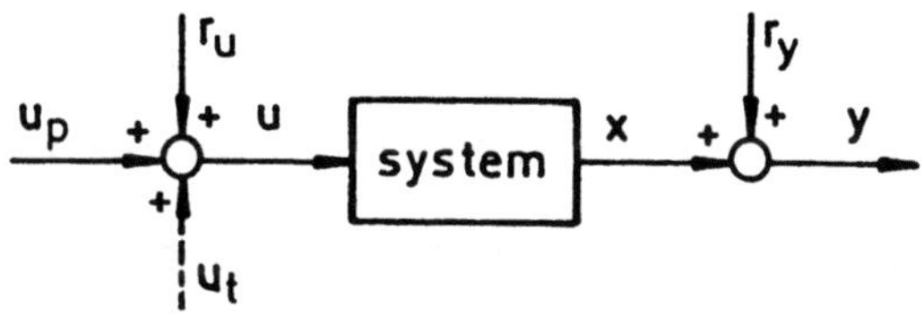

Fig. 3.1. Signals of a SISO-system
u_p process input signal
r_u input disturbance signal
u_t test input signal
x process output signal
r_y output disturbance signal

Case 1: The process signal u_p is of deterministic form. If there is no noise superposed to the input and output signals, the measurable signals u and y are of deterministic form too. If a stochastic disturbance r_u or r_y occurs, the measurable output signal y is described by a stochastic signal.

Case 2: If the input signal u_p is a stochastic signal, then the measur-

case	u_p proc. sign.	r_u dis-turb.	u meas-urable	x proc. sign.	r_y dis-turb.	y meas-urable
1	d	- --- s	d ----- s	d --- s	- --- s --- - --- s	d ---- s ---- s
2	s	- --- s	s	s	- --- s	s

Table 3.1. Possible forms of input-output signals
(s=stochastic signal, d=deterministic signal)

If there is no noise superposed to the process signal, the parameter estimation can be performed without any difficulty if a sufficiently broad frequency range of the system is excited by the input signals. If non-measurable stochastic signals (noise) occur, then a detailed judgement of this disturbance signal has to be conducted, which comprises of

- type of the disturbance signal (e.g. Gaussian process, drift etc.) and its frequency spectrum;
- magnitude of the disturbance (e.g. signal power, maximum, effective and mean value);
- statistic independence of the disturbance signal (i.e. not correlated with itself and with the input and output signal of the investigated system);
- stationarity of the disturbance signal.

As the disturbance signal is included into the model structure by the filter transfer function $G_R(z)$, all a priori information about it should be used for the above-mentioned judgement.

If the analysis of the signals proves that the power or the amplitude spectrum of the natural process input signal, is not sufficient for a parameter estimation, then the system must be excited by a *test signal* u_t. Such a test signal should be limited as little as possible. However, it must excite permanently the whole frequency range of the investigated system. Furthermore, the mean value and variance of the test signal should remain constant. These conditions are fulfilled by pseudo-random binary sequences (PRBS). By an appropriate selection of the test signals, the computational expense of the parameter estimation can also be reduced, see Goodwin et al (1976). The selection of the magnitude of a PRBS-signal is dependent on the application. The magnitude must not be selected as being too big, in order not to disturb the normal operation of the investigated system, as well as to remain within the linear system behaviour.

3.3. Setting the initial values

The problems of setting initial values arise from the application of recursive estimation methods. The recursion algorithm computes the newly estimated parameter vector $\underline{\hat{p}}$ and the matrix $\underline{P}$ from Eqs.(2.45a) to (2.45c), where the new value is composed of the old value and a correction term. At the beginning of the calculation initial values have to be selected both for $\underline{\hat{p}}$ and $\underline{P}$.

If a priori estimations of the parameters are available, they can be used as initial values. If not, it can be set $\underline{\hat{p}}(0) = \underline{0}$. The 1st, 2nd and 3rd EM-method as well as the IV-method have to be started by the LS-estimation, and in about 30-50 steps appropriate values have to be found. These are the initial values for the actual estimation, which would have led to instability of the estimation algorithm with an initial value of zero.

Because the matrix $\underline{P}$ is mainly responsible for the speed of convergence, the elements of its main diagonal are set to large values. Fig. 3.2 shows a typical curvature, and how the values are reduced in the following steps. $\underline{P} = \alpha \underline{I}$ with $\alpha = 10^4$ are recommended as initial values for fast convergence, where $\underline{I}$ represents the unity matrix. The correction of the matrix $\underline{P}$ during the computation is treated in connection with the improvements for convergence discussed later.

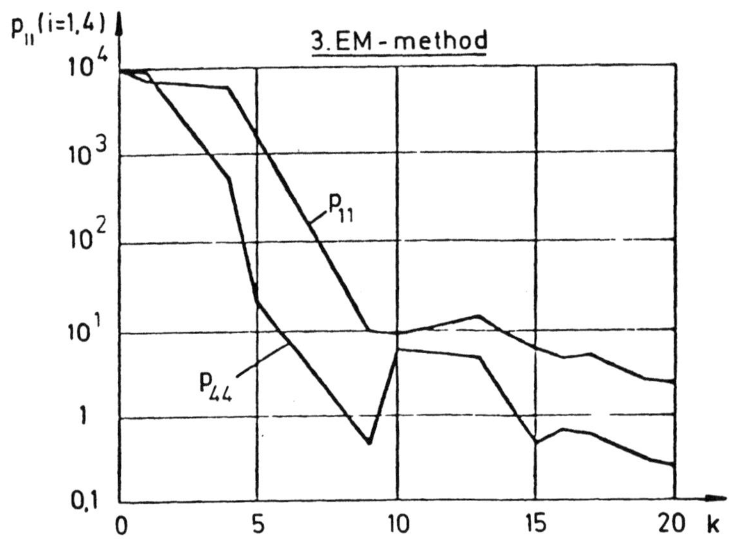

Fig. 3.2. Typical decrease of the matrix elements p_{ii} depending on the sampling steps k

3.4. Selection of starting time and length of measurement

The algorithm has to start at such a moment when stationary signals occur. While this is simple to realize by a digital or analog simulation, an examination has to be made in practical application as to whether the system is in steady state. For the starting time a moment should be selected when the normalized output signal is moving close to zero. There can be an improvement of the convergence if after the start of the estimation procedure the input signal excites the low frequencies of the system, because, in particular for the first estimations after filling up the vectors and matrices during (n+1)steps, the disturbances caused by the fast modes of the system have considerable disadvantages for the estimation.

3.5. Observation of the progress of the estimation results

While the non-recursive estimation algorithms only compute the final result, the recursive methods calculate a new actual solution after each sample. Of course, the user is very interested in the progress of the estimation results, so that he can start the manipulation of the identification procedure if necessary (as described in section 3.6) or stop the procedure because of insufficient convergence or instability of the estimation results. Computer implementation with two independently running programs according to Fig. 3.3 is suitable for this, but also for manipulating the identification. These programs are a dialog and a calculation program:

- *Dialog program:* This program reads all data and writes all results. There are no calculations for the identification. The dialog program

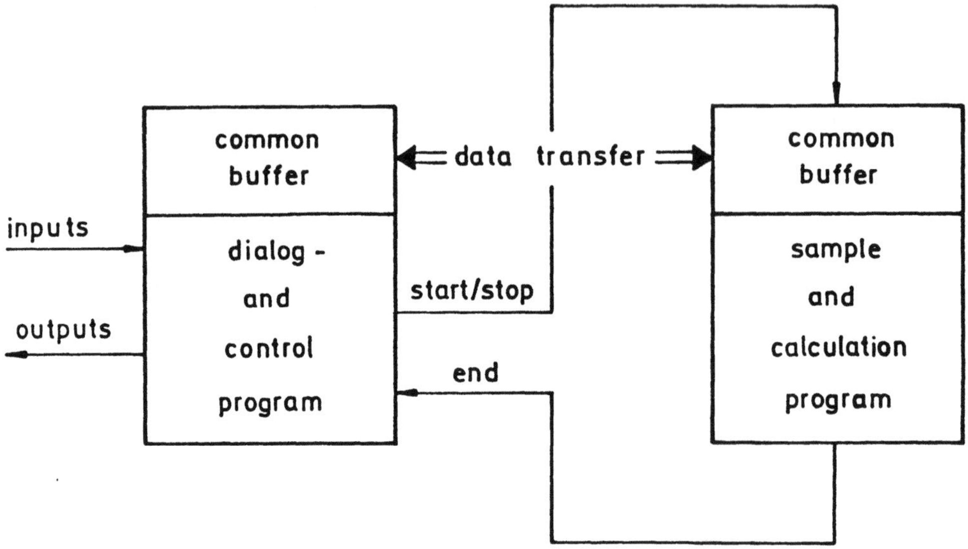

Fig. 3.3. Structure of the computer programs

can start and stop the calculation program.

- *Calculation program:* The estimation results are computed only in this program, but there are no dialog input or output facilities.

Both programs can exchange data by using a COMMON-block. The data needed for calculation are read by the dialog program and the calculation program is started. This first takes the input data from the COMMON-block and returns all the interesting results to the common part of the memory. From there, the dialog program can obtain intermediate results at any time, which then are also available to the user.

Of special importance is the observation of the progress of the parameter estimation, the estimation error of the gain and the correction values. It is mostly sufficient to observe four of five characteristic values, according to their graphical curvature. These are mainly the estimation error and the estimated parameters $\hat{a}_1$ up to $\hat{a}_n$. Fig. 3.4 shows as a typical result the curvature of the parameter $\hat{a}_n$ for the identification of the test-system.

3.6. Steps for improving the convergence (Identification guidance)

The calculation of the correction term is essentially dependent on the matrix $\underline{P}$ of Eq.(2.45c). Because of the relatively fast decrease of the elements of the main diagonal, shown in Fig. 3.2, although they have not yet reached the true parameter values, there can be a convergence improvement if the elements of the diagonal are weighted. For this several methods have been developed, which are applied on a large scale in the identification of time-variant systems, because there also has to be a parameter adaptation after a greater number of measurements. In the following, five weighting methods are to be discussed. They are presented in Table 3.2.

The "setting-back" technique sets back the matrix $\underline{P}$ once or repeatedly to the initial values, while the other methods reach a continuous alteration. Here the weighting factor may be constant (G1, M3) or time-variant (G2, G3). A typical progress of the estimated parameter $\hat{a}_1$, without weighting, with weighting and with too strong weighting is shown in Fig.3.5.

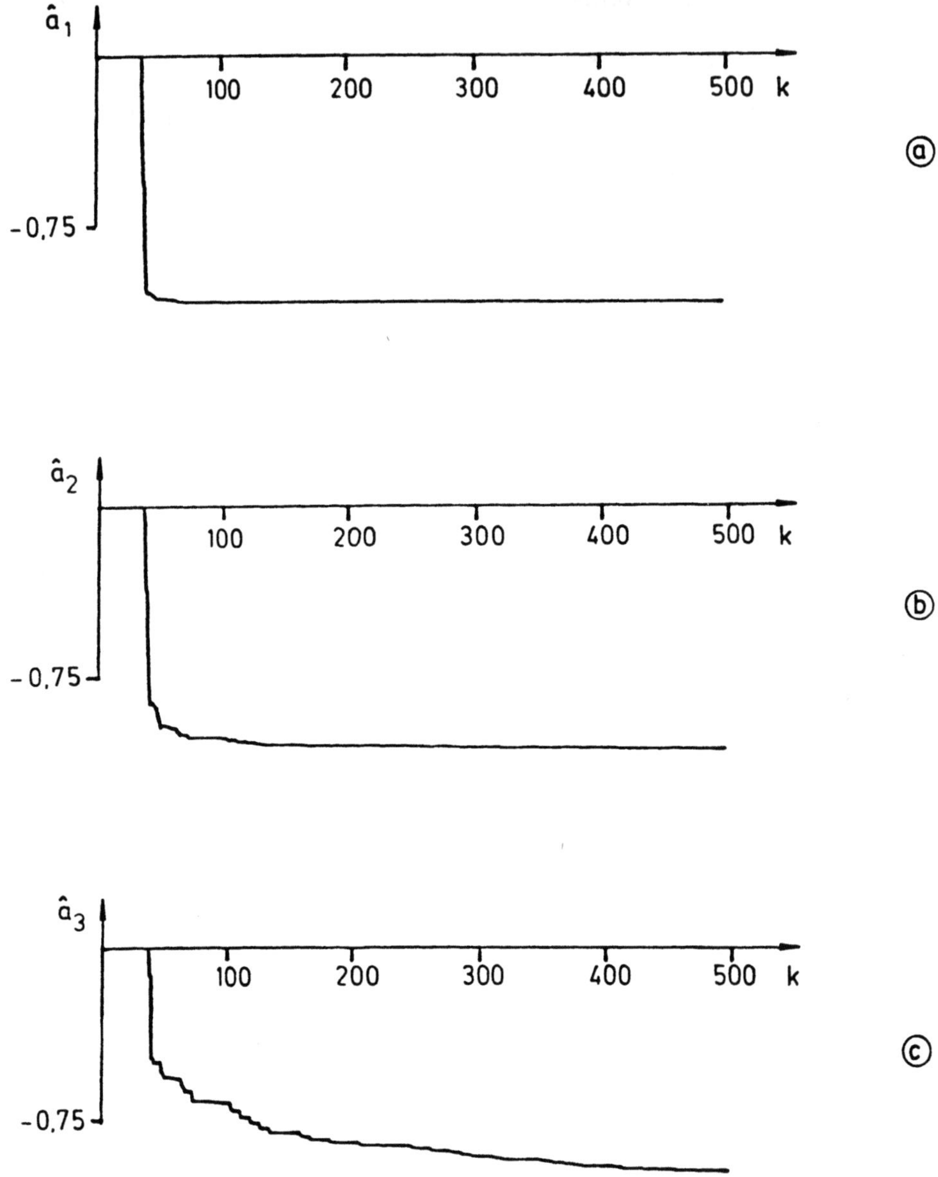

Fig. 3.4. Typical curvature for the estimated system parameters $\hat{a}_n$ for a system with (a) order n=1, (b) order n=2 and (c) order n=3

weighting method	abbre-viation	equation
setting-back of $\underline{P}$	R	setting of $\underline{P}$ back to $\underline{P}(0)$ after k_R samples
constant weighting	G1	$\underline{P}'(k+1)=\frac{1}{\rho(k+1)}\underline{P}(k+1)$ with $\rho(k+1)=\rho_o < 1$
variable weighting	G2	$\underline{P}'(k+1)=\frac{1}{\rho(k+1)}\underline{P}(k+1)$ with $\rho(k+1)=\rho(k)+[1-\rho(k)]\Delta\rho$
variable weighting	G3	$\underline{P}'(k+1)=\frac{1}{\rho(k+1)}\underline{P}(k+1)$ with $\rho(k+1)=1-\frac{\rho}{1-(1-\rho)^{k+1}}$
constant increase	M3	$\underline{P}'(k+1) = \underline{P}(k+1) + \underline{G}^*$ where $\underline{G}^*$ has elements g_{ii}

Table 3.2. Weighting methods for recursive estimation algorithms

It is particularly the setting-back technique that is suitable for manipulations during the parameter estimation (identification guidance). If the convergence rate is too slow, as can be recognized during the recursive estimation from the curvature of the parameters, the matrix $\underline{P}$ is

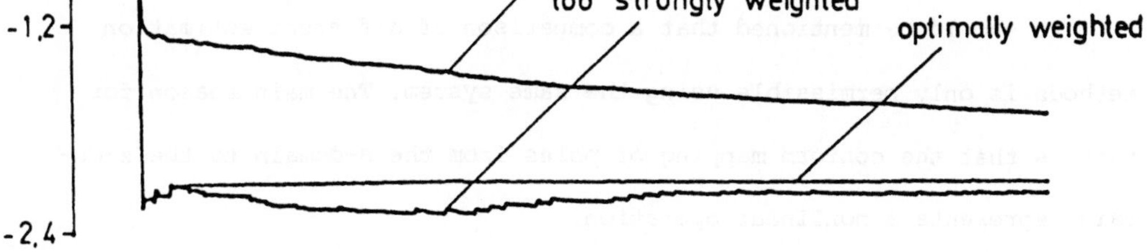

Fig. 3.5. Progress of the estimated parameter $\hat{a}_1$

set back to the initial values by the dialog program. The same results with regard to the improvement of the convergence rate can also be achieved by the application of the other weighting methods.

3.7. Interpretation of the results and a comparison of the efficiency of the methods

The valuation of the estimation results and the comparison of the efficiency of the different estimation methods is possible without difficulties by using digital and analog simulated systems. The estimated parameter $\hat{a}_n$, the gain $\hat{K}_S$ and the simulated step response of the estimated model are suitable values for valuation. A comparison of the 3rd EM-method and the IV-method has been made to serve as an example. Fig. 3.6 a,b shows the influence of the number k of measured samples on the estimated parameter $\hat{a}_2$ and the gain factor $\hat{K}_S$ in the case of an undisturbed output signal. Fig. 3.6 c,d contains the step responses of the estimated models for a noise level of $\lambda_S = 0$ and $\lambda_S = 0.1$, where λ_S is defined as the ratio of the effective values of the disturbance and the input signal.

It should be mentioned that a comparison of different estimation methods is only permissible using the same system. The main reason for this is that the conform mapping of poles from the s-domain to the z-domain represents a nonlinear operation.

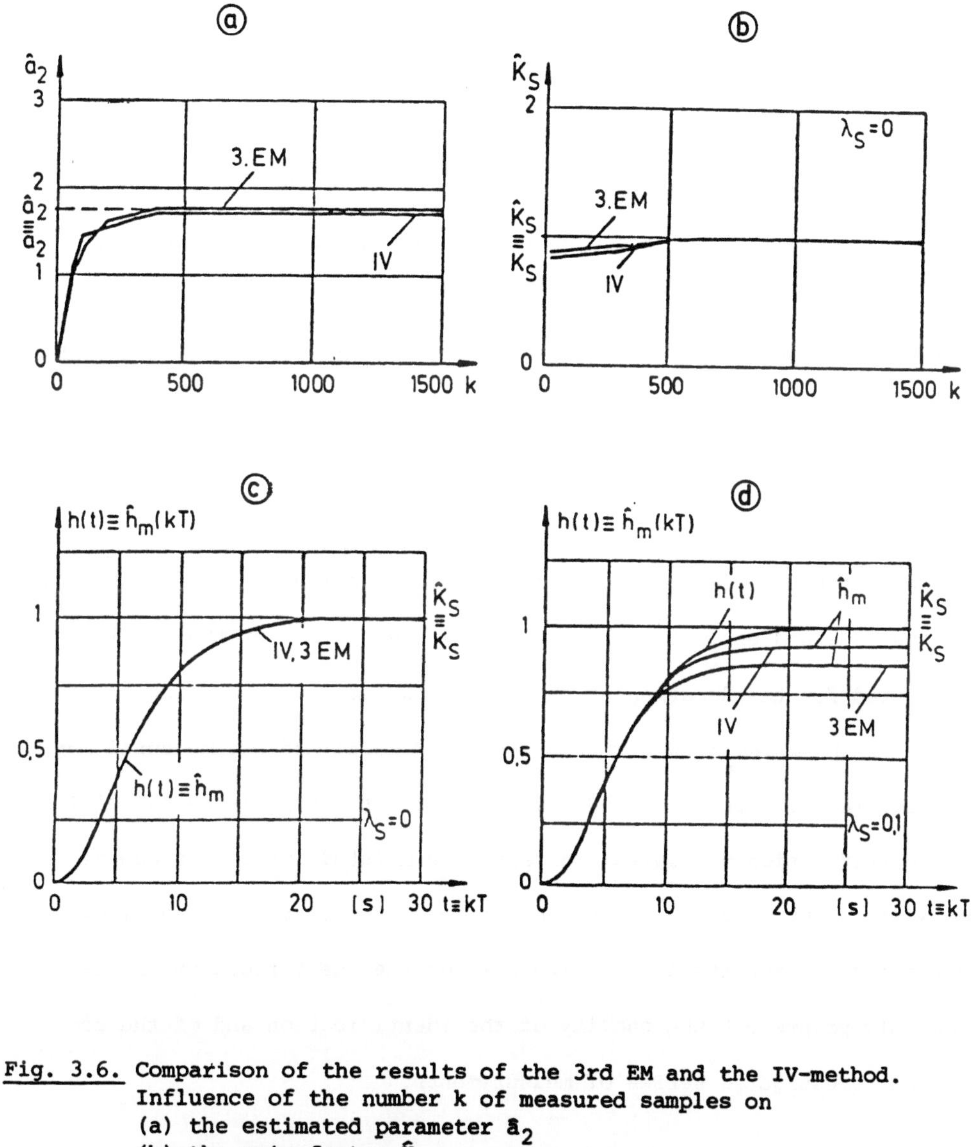

Fig. 3.6. Comparison of the results of the 3rd EM and the IV-method.
Influence of the number k of measured samples on
(a) the estimated parameter $\hat{a}_2$
(b) the gain factor $\hat{K}_S$
Step responses of the estimated models for
(c) the undisturbed output signal (λ_S=0)
(d) the disturbed output signal (λ_S=0.1)

3.8. Interpretation of the results by the identification of real systems

The objective of the identification of a real system is the realization of a model which is sufficiently accurate for the applications by the user. The valuation of the quality of a model is made only by the user, by applying the model, for instance, to simulations or to controller design. The user has to decide whether and how far the model is suitable for his purposes.

The user of the identification programs can determine the probability of identity between model and system from the identification procedure and the obtained estimation results only. The quality of the model is only assured for the duration of the measuring time, during which the output y_M can be calculated with the estimated model and the measured input and compared with the measured values y. From the experience gained in the identification of digital and analog simulated systems it can be decided whether the progress during the parameter estimation has already led to the correct results or not. The user of the identification program thus only points out the quality of the identification and of the model for the investigated series of measurements.

3.9. Identification program packages

In the past many program packages for identification of unknown systems have been developed by various institutions, see Table 3.3. They differ considerably in size and quality. When introducing one of these program packages, care should be taken that the experience gained with

Name	Editor	Contents
PNE 1	Mc Corkell, Jordanstown, N. Ireland (1979)	Strejc's method; Levy's method pulse technique
L-A-S	Bingulac, Campinas, Brasil (1976)	various method
TRIP	V.D. Bosch, Delft, Netherlands (1977)	identification and signal analysis methods
SATER	V.D. Boom, Eindhoven, Netherlands (1977)	various identification methods
OLID	Isermann, Darmstadt, F.R.G. (1976)	on-line identification of SISO and 2 input, 2 output processes using different methods
IDPAC	Wieslander, Lund, Schweden (1976)	various identification methods
LISPID	Källström, Lund, Schweden (1976)	ML-method for MIMO systems
KEDDC	Unbehauen, Bochum, F.R.G. Schmid, Bochum, F.R.G. (1979)	on- and off-line identification of SISO and MIMO systems with max. 5 inputs including controller design
CLA	Edmunds, Cambridge, England (1979)	frequency response analysis
IDCOM	Richalet, Adersa, France (1979)	on-line identification and indirect adaptive control
MERCEDES	Keviczky, Budapest, Hungary Haber, Budapest, Hungary (1976)	various identification methods

Table 3.3. Various identification packages

these program packages by the developing institution are well documented, because this technical know-how is an important aspect in identification.

4. CONCLUSIONS

This contribution represents an introduction to parameter estimation methods for statistical system identification. As was written by a control engineer, the system theoretical aspects of discribing dynamical systems has been put in the foreground. Various identification methods have been introduced, and recursive estimation methods have been specially dealt with. Besides the description of different methods, conditions for optimum identification and practical aspects are broadly discussed. The detailed mathematical derivation of unweighted and weighted recursive estimation algorithms is contained in the appendix.

APPENDIX A: Recursive version of the unweighted estimation algorithm

In order to derive a recursive version of the direct estimation equation, the expression obtained from Eq. (2.44)

$$\underline{P}^{-1}(k) = \underline{P}^{-1}(k-1) + \underline{m}(k)\ \underline{m}^{T}(k) \tag{A.1}$$

is multiplied by $\underline{P}(k)$ from the left-hand side and by $\underline{P}(k-1)$ from the right-hand side, so that

$$\underline{P}(k-1) = \underline{P}(k) + \underline{P}(k)\ \underline{m}(k)\ \underline{m}^{T}(k)\ \underline{P}(k-1) \tag{A.2}$$

is obtained. Multiplication from the right-hand side by $\underline{m}(k)$ yields to

$$\underline{P}(k-1)\underline{m}(k) = \underline{P}(k)\ \underline{m}(k)\left[1 + \underline{m}^{T}(k)\ \underline{P}(k-1)\ \underline{m}(k)\right]. \tag{A.3}$$

Multiplication from the right-hand side by

$$\left[1 + \underline{m}^{T}(k)\ \underline{P}(k-1)\ \underline{m}(k)\right]^{-1}\ \underline{m}^{T}(k)\ \underline{P}(k-1)$$

supplies

$$\underline{P}(k-1)\ \underline{m}(k)\left[1 + \underline{m}^{T}(k)\ \underline{P}(k-1)\ \underline{m}(k)\right]^{-1}\ \underline{m}^{T}(k)\ \underline{P}(k-1) =$$
$$= \underline{P}(k)\ \underline{m}(k)\ \underline{m}^{T}(k)\ \underline{P}(k-1)$$

and with Eq. (A.2) follows

$$\underline{P}(k-1)\ \underline{m}(k)\left[1 + \underline{m}^{T}(k)\ \underline{P}(k-1)\ \underline{m}(k)\right]^{-1}\underline{m}^{T}(k)\ \underline{P}(k-1) = \underline{P}(k-1) - \underline{P}(k)$$

or

$$\underline{P}(k) = \underline{P}(k-1) - \underline{P}(k-1)\ \underline{m}(k)\left[1 + \underline{m}^{T}(k)\ \underline{P}(k-1)\ \underline{m}(k)\right]^{-1}\ \underline{m}^{T}(k)\ \underline{P}(k-1).$$

Using the correction vector

$$\underline{q}(k+1) = \underline{P}(k)\,\underline{m}(k-1)\left[1 + \underline{m}^T(k+1)\,\underline{P}(k)\,\underline{m}(k+1)\right]^{-1}, \tag{A.4}$$

results in the recursive equation

$$\underline{P}(k+1) = \underline{P}(k) - \underline{q}(k+1)\,\underline{m}^T(k+1)\,\underline{P}(k)\ . \tag{A.5}$$

For the desired parameter vector $\hat{\underline{p}}$ finally follows from Eqs. (2.43), (A.4) and (A.5)

$$\begin{aligned}
\hat{\underline{p}}(k+1) &= \underline{P}(k+1)\left[\sum_{i=n+1}^{k} \underline{m}(i)\,y(i) + \underline{m}(k+1)\,y(k+1)\right] \\
&= \left[\underline{P}(k) - \underline{q}(k+1)\,\underline{m}^T(k+1)\,\underline{P}(k)\right]\left[\sum_{i=n+1}^{k} \underline{m}(i)y(i) + \right. \\
&\qquad \left. + \underline{m}(k+1)\,y(k+1)\right] \\
&= \hat{\underline{p}}(k) + \underline{P}(k)\,\underline{m}(k+1)\,y(k+1) - \underline{q}(k+1)\,\underline{m}^T(k+1)\,\hat{\underline{p}}(k) - \\
&\quad - \underline{q}(k+1)\,\underline{m}^T(k+1)\,\underline{P}(k)\,\underline{m}(k+1)\,y(k+1) \\
&= \hat{\underline{p}}(k) - \underline{q}(k+1)\,\underline{m}^T(k+1)\,\hat{\underline{p}}(k) + \\
&\quad + \left[\underline{P}(k)\,\underline{m}(k+1) - \underline{q}(k+1)\,\underline{m}^T(k+1)\,\underline{P}(k)\,\underline{m}(k+1)\right]y(k+1) \\
&= \hat{\underline{p}}(k) - \underline{q}(k+1)\,\underline{m}^T(k+1)\hat{\underline{p}}(k) + \left[\underline{P}(k)\,\underline{m}(k+1) - \right. \\
&\quad \left. - \underline{P}(k)\,\underline{m}(k+1) + \underline{q}(k+1)\right]\,y(k+1) \\
&= \hat{\underline{p}}(k) + \underline{q}(k+1)\,\left[y(k+1) - \underline{m}^T(k+1)\,\hat{\underline{p}}(k)\right]
\end{aligned}$$

and thus

$$\hat{\underline{p}}(k+1) = \hat{\underline{p}}(k) + \underline{q}(k+1)\,\hat{e}(k+1) \tag{A.6}$$

where

$$\hat{e}(k+1) = y(k+1) - \underline{m}^T(k+1)\,\hat{\underline{p}}(k)\ . \tag{A.7}$$

Eqs. (A.4) to (A.7) thus describe the recursive version of the estimation algorithm.

APPENDIX B: Recursive version of the weighted estimation algorithm

A recursive version of the direct estimation equation of the weighted IV-method is obtained from the following expression derived from Eq. (2.61) using the weighting factor $\rho(k)$

$$\underline{P}^{-1}(k) = \rho(k)\left[\underline{P}^{-1}(k-1) + \underline{w}(k)\,\underline{m}^{T}(k)\right] \tag{B.1}$$

by multiplying this equation from the left-hand side by $\underline{P}(k)$ and from the right-hand side by $\underline{P}(k-1)$. Thus follows

$$\underline{P}(k-1) = \rho(k)\left[\underline{P}(k) + \underline{P}(k)\,\underline{w}(k)\,\underline{m}^{T}(k)\,\underline{P}(k-1)\right] \tag{B.2}$$

and after the right-hand side multiplication by

$$\underline{P}(k-1)\,\underline{w}(k) = \rho(k)\,\underline{P}(k)\,\underline{w}(k)\left[1 + \underline{m}^{T}(k)\,\underline{P}(k-1)\,\underline{w}(k)\right]. \tag{B.3}$$

The right-hand side multiplication by

$$\left[1 + \underline{m}^{T}(k)\,\underline{P}(k-1)\,\underline{w}(k)\right]^{-1}\underline{m}^{T}(k)\,\underline{P}(k-1)$$

leads to

$$\underline{P}(k-1)\,\underline{w}(k)\left[1 + \underline{m}^{T}(k)\,\underline{P}(k-1)\,\underline{w}(k)\right]^{-1}\underline{m}^{T}(k)\,\underline{P}(k-1) =$$
$$= \rho(k)\,\underline{P}(k)\,\underline{w}(k)\,\underline{m}^{T}(k)\,\underline{P}(k-1)$$

and with Eq. (B.2) follows

$$\underline{P}(k-1)\,\underline{w}(k)\left[1 + \underline{m}^{T}(k)\,\underline{P}(k-1)\,\underline{w}(k)\right]^{-1}\underline{m}^{T}(k)\,\underline{P}(k-1) = \underline{P}(k-1) - \rho(k)\underline{P}(k)$$

or

$$\underline{P}(k) = \frac{1}{\rho(k)}\{\underline{P}(k-1) - \underline{P}(k-1)\,\underline{w}(k)\left[1 + \underline{m}^T(k)\,\underline{P}(k-1)\,\underline{w}(k)\right]^{-1} \cdot \cdot\,\underline{m}^T(k)\,\underline{P}(k-1)\}\,.$$

From here we obtain with the correction vector

$$\underline{q}(k+1) = \underline{P}(k)\,\underline{w}(k+1)\left[1 + \underline{m}^T(k+1)\,\underline{P}(k)\,\underline{w}(k+1)\right]^{-1} \qquad (B.4)$$

the recursive expression for

$$\underline{P}(k+1) = \frac{1}{\rho(k+1)}\left[\underline{P}(k) - \underline{q}(k+1)\,\underline{m}^T(k+1)\,\underline{P}(k)\right]. \qquad (B.5)$$

Finally, the desired estimated parameter vector $\hat{\underline{p}}$ can be calculated from Eq.(2.60) using Eqs. (B.4) and (B.5) by

$$\hat{\underline{p}}(k+1) = \underline{P}(k+1)\rho(k+1)\left[\sum_{i=n+1}^{k}\prod_{j=1}^{k}\rho(j)\,\underline{w}(i)y(i) + \underline{w}(k+1)y(k+1)\right]$$

$$= \left[\underline{P}(k) - \underline{q}(k+1)\underline{m}^T(k+1)\underline{P}(k)\right]\left[\sum_{i=n+1}^{k}\prod_{j=i}^{k}\rho(j)\underline{w}(i)y(i) + \right.$$

$$\left. + \underline{w}(k+1)\,y(k+1)\right]$$

$$= \hat{\underline{p}}(k) + \underline{P}(k)\,\underline{w}(k+1)\,y(k+1) - \underline{q}(k+1)\,\underline{m}^T(k+1)\,\hat{\underline{p}}(k) -$$

$$- \underline{q}(k+1)\,\underline{m}^T(k+1)\,\underline{P}(k)\,\underline{w}(k+1)\,y(k+1)$$

$$= \hat{\underline{p}}(k) - \underline{q}(k+1)\,\underline{m}^T(k+1)\,\hat{\underline{p}}(k) +$$

$$+ \left[\underline{P}(k)\,\underline{w}(k+1) - \underline{P}(k)\,\underline{w}(k+1) + \underline{q}(k+1)\right]y(k+1)$$

$$= \hat{\underline{p}}(k) + \underline{q}(k+1)\,\left[y(k+1) - \underline{m}^T(k+1)\,\hat{\underline{p}}(k)\right]$$

and thus

$$\hat{\underline{p}}(k+1) = \hat{\underline{p}}(k) + \underline{q}(k+1)\,\hat{e}(k+1) \qquad (B.6)$$

where

$$\hat{e}(k+1) = y(k+1) - \underline{m}^T(k+1)\,\hat{\underline{p}}(k) \quad . \tag{B.7}$$

Thus Eqs. (B.4) and (B.7) represent the recursive version of the weighted IV-estimation algorithm.

REFERENCES

Akaike, H. (1969), Fitting autoregressive models for prediction. Ann. Inst. Statist. Math., 21, pp. 243-247.

Akaike, H. (1970), Statistical predictor identification. Ann. Inst. Statist. Math., 22, pp. 203-217.

Akaike, H. (1978), On newer statistical approaches to param. est. and structure determination. IFAC WK Helsinki, pp. 1877-1884.

Aström, K.J. (1968), Lectures on the ident. problem - the L.S. method. Lund Inst. of Techn., Rep. 6806.

Aström, K.J. (1970), Introduction to stochastic control theory. Academic Press, N.Y.

Aström, K.J., Källström. C. (1972), Ident. and modelling of ship dynamics. Lund Inst. of Techn., Rep. 7202.

Bauer, B. (1977), Parameterschätzverfahren zur on-line Ident. dynamischer Systeme im offenen und geschlossenen Regelkreis. Diss. Ruhr-Universität Bochum.

Baur, U. (1976), On-line Parameterschätzverfahren zur Identifikation linearer dynamischer Prozesse mit Prozeßrechnern - Entwicklung, Vergleich, Erprobung. Diss. Universität Stuttgart und PDV-Bericht KFK-PDV 65, Gesellschaft für Kernforschung, Karlsruhe.

Bingulac, S.P., Bassani (1976), CAD of linear systems using L-A-S language, Colloquio Franco-Brasileiro, Rio de Janeiro.

Bohlin, T. (1968), The ML method of ident., IBM Schweden, paper 18.191.

Boom, v.d., Emden, v.d. (1974), The determination of the orders of process and noise dynamics, Automatica, V.10, pp. 245-256.

Boom, v.d., A.J.W., Lemmens, W.J.M. (1977), Sater, an interactive program package, IFAC Symp. Trends in Aut. Control Education, Barcelona.

Bosch, P.P.J., Bruyn, P.M. (1977), The dedicated digital computer, IFAC Symp. Trends in Aut. Control Education, Barcelona.

Chow, J.C. (1972), On estimating the order of an autoregressive moving average process with uncertain observations. IEEE Tr. on AC-17, pp. 707-709.

Clarke, D.W. (1967), Generalized LS est. of the param. of a dynamical model. IFAC Symp. Prag, pap. 3.17.

Edmunds, J.M. (1979), Cambridge linear analysis and design programs, IFAC Symp. 'CAD of Control Systems', Zürich.

Eykhoff, P. (1967), Process param. and state est., IFAC Symp. Prag,pap.O.2.

Eykhoff, P. (1973), Identification and system parameter estimation. 3rd IFAC-Symposium, Hague.

Göhring, B. (1973), Erprobung statistischer Parameterschätzmethoden und Strukturprüfverfahren zur exp. Ident. von Regelsystemen. Diss. Universität Stuttgart.

Goodwin, G.C., Payne, R.L., Kabaila, P. (1976), On canonical forms and algorithms for multivariable time series analysis. IV. IFAC Symposium Tbilisi, USSR.

Gustavsson, I., (1972), A comparison of different methods for ident. of industrial processes. Automatica, 8, pp. 127-142.

Hastings-James, R., Sage, M.W. (1969), Recursive GLS procedure for on-line ident. of process param., Proc. IEE, 116, pp. 2057-2062.

Isermann, R. and others (1973), Comparison and evaluation of six on-line ident. and param. est. methods with three simulates processes. IFAC Symposium, The Hague, pap. E-1.

Isermann, R., Dymchiz, E. (1976), A software package for process computer aided control system design, IFAC/IFIP Symp. Computer Control, Talinn, USSR.

Isermann, R. (1979), Identification and system parameter estimation. 5th IFAC Symposium, Darmstadt.

Källström, C.G., Essebo, T., Aström, K.J. (1976), A computer program for maximum likelihood identification of linear, multivariable stochastic systems, IFAC Symp. Tbilisi, USSR.

Keviczky, L.K., Bânyâsz, Cs.M. (1976), Some new results on multiple input-multiple output identification methods, IFAC Symp. Tbilisi, USSR.

Mc Corkell, L. (1977), CAD system analysis and control for undergraduates, IFAC Symp. 'Trends in Aut. Control Education', Barcelona.

Niemann, R. (1971), A review of proc. ident. and param. est. techn. Int. J. Cont., 12, pp. 209-264.

Rajbman, V. (1976), Identification and system parameter estimation. 4th IFAC-Symposium, Tbilisi, USSR.

Richalet, J., Foigel, J.K. (1979), Self-adapting IDCOM. 5th IFAC-Symp. Darmstadt.

Saridis, G.N. (1974), Comparison of six on-line ident. algorithms, Automatica, 10, pp. 69-79.

Schmid, Chr., Unbehauen, H. (1979), KEDDC, a general purpose CAD software system for application in control engineering, IFAC Symp. SOCOCO, Prague.

Söderström, T. (1977), On model structure testing in system ident., Int. J. Control, 26, pp. 1-18.

Talmon, J.I. (1971), Approximated Gauss-Markov estimators and related schemes. Eindhoven Univ. of Techn., Rep. 71-E-17.

Unbehauen, H. (1973a), Übersicht über Methoden zur Ident. dynamischer Systeme. Regelungstechnik, 21, pp. 2-8.

Unbehauen, H., Göhring, B. (1973b), Application of different statistical tests for the determination of most accurate order of the model in param. est., IFAC Symp. Hague, p. TS-1.

Wellstead, P.E. (1978), An instrumental product moment test for model order estimation, Automatica, V.14, pp. 89-91.

Wieslander, J. (1976), IDPAC-user's guide, Lund Institute of Technology, Rep. 7605.

Wong, K.Y., Polak, E. (1967), Ident. of linear discrete time systems using IV method, IEEE Tr. AC, 12, pp.707-718.

Woodside, C.M. (1971), Estimation of the order of linear systems. Automatica, V.7, pp. 727-733.

Young, P.C. (1970), An instrumental variable method for realtime identification of a noisy process. Automatica, 6, pp. 271-286.

Young, P., Jakemann, A., Mc Murtrie, R. (1980), An instrumental variable method for model order identification, Automatica, V. 16, pp.281-294.

Some Identification Methods Using Measured Harmonic Responses

R. FILLOD
Laboratoire de Mécanique Appliquée associé
au CNRS, Faculté des Sciences, La Bouloie,
25030 BESANCON (France)

The experimental determination of eigenvalues and eigenmodes of the linear mechanical structures is a fundamental step in the study of the problems brought about by the vibrations of a structure. Indeed the knowledge of eigensolutions allows the obtaining of the dynamic response of the structure to any excitation. These eigensolutions are also necessary to determine the modifications which optimize the dynamical behavior of the structure.

We present in this paper some methods which allow to obtain experimentally these eigensolutions.

I - THEORETICAL RELATIONS

The dynamical behavior of a structure may be represented by a discretized linear model of the form :

$$\tilde{M}y^{\circ\circ} + \tilde{B}y^{\circ} + \tilde{K}y = \tilde{f} \qquad (1)$$

M, B and K are respectively the matrices of mass, damping and stiffness of order NxN. These matrices are unknown, but we admit that they do not depend on time.

$\tilde{y}$ and $\tilde{f}$ are the vectors of forced response and applied force (order N).

Later on, we will study only the stationary or pseudo-stationary solutions of (1). So, we note :

$$\tilde{f} = f\, e^{st} \quad : \quad \tilde{y} = y\, e^{st} \quad ; \quad s = j\omega \quad \text{or} \quad s = j\omega - a\omega \,.$$

The equation (1) takes the form which is independent of time :

$$(M s^2 + B s + K)\, y = f \,. \tag{2}$$

The structure is supposed to be stable and slightly damped. Let us denote :

$$H = M s^2 + B s + K \quad .$$

A forced response y takes formally the form :

$$y = H^{-1}\, f \quad . \tag{3}$$

To invert H, one utilizes a regular transformation which avoids the matricial inversion. H is a function of s. In the general, it is not possible to diagonalize simultaneously the three matrices M, B and K with a real transformation, but a transformation can be found which diagonalizes simultaneously M and K which leads to the two following expressions (7) and (12) for the forced response y.

I.1. Utilization of the eigensolutions of an associated problem [1,4]

We define the associated problem :

$$(K - \lambda_\nu\, M)\, y_\nu = 0 \quad , \tag{4}$$

and the adjoint problem : with $K^* = \bar{K}^T$; $M^* = \bar{M}^T$

$$(K^* - \bar{\lambda}_\nu\, M^*)\, y^*_\nu = 0 \,, \tag{5}$$

If $M^* = M$ and $K^* = K$, the associated problem (4) is a adjoint one.

Let : $Y = \begin{bmatrix} & \vdots & \\ & y_\nu & \\ & \vdots & \end{bmatrix}$, $Y^* = \begin{bmatrix} & \vdots & \\ & y^*_\nu & \\ & \vdots & \end{bmatrix}$, $\Lambda = \begin{bmatrix} \ddots & & \\ & \lambda_\nu & \\ & & \ddots \end{bmatrix}$.

These matrices verify the relations :

$\bar{Y}^{*T} M Y = E$; $\bar{Y}^{*T} K Y = \Lambda$; E : identity matrix,

the generalized damping matrix β is defined by : $\beta = \bar{Y}^{*T} B Y$.

The forced harmonic response can be developed in the set of eigen-modes :

$$y = Y (\Lambda + s^2 E + s\beta)^{-1} \bar{Y}^{*T} f \quad . \tag{6}$$

This expression is of great interest when β is a diagonal matrix, because we have to invert a diagonal matrix. We obtain for y the simple expression :

$$y = \sum_{\nu=1}^{N} \frac{y_\nu \bar{y}_\nu^{*T} f}{\lambda_\nu + s^2 + s\beta_{\nu\nu}} \quad . \tag{7}$$

The transfer function T_{ij} is defined as the response at point i to a unity force applied at point j :

$$T_{ij} = \sum_{\nu=1}^{N} \frac{y_{\nu i} \bar{y}_{\nu j}^{*T}}{\lambda_\nu + s^2 + s\beta_{\nu\nu}} \quad ;$$

when M, B and K are symmetric matrices (real or complex) it can be shown that [3] : $\bar{Y}^{*} = Y$. Then, $T_{ij} = T_{ji}$.

I.2. Utilization of the eigensolutions of the dissipative problem :

The expression (1) can be rewritten under the form :

$$L \tilde{z} + N \tilde{z}^{\circ} = \tilde{g} \, , \tag{8}$$

with : $\tilde{z} = \begin{bmatrix} \tilde{y} \\ \tilde{y}^{\circ} \end{bmatrix}$; $\tilde{g} = \begin{bmatrix} \tilde{f} \\ 0 \end{bmatrix}$; $L = \begin{bmatrix} K & 0 \\ 0 & -M \end{bmatrix}$; $N = \begin{bmatrix} B & M \\ M & 0 \end{bmatrix}$.

L and N can be defined in several forms, but the chosen expression has the advantage of being symmetric when M, B and K are symmetric. For a harmonic excitation with the circular frequency $\omega (s=j\omega)$, the expression (8) takes the following form which is independent of time :

$$(L+sN)\, z = g \quad , \qquad \text{with : } z^T = \begin{bmatrix} y^T & sy^T \end{bmatrix}$$

We are interested by the solutions of the homogeneous problem :

$$(L + s_\nu N)\, z_\nu = 0 \tag{9}$$

and by those of the adjoint problem :

$$(L^* + s_\nu^* N^*)\, z_\nu^* = 0 \tag{10}$$

with : $L^* = \bar{L}^T$; $N^* = \bar{N}^T$; $s_\nu^* = \bar{s}_\nu$ and :

$s_\nu = j\omega_\nu - a_\nu\,\omega_\nu$. We suppose that : $0 < a_\nu \ll 1$.

Let : $Z = \begin{bmatrix} \vdots & z_\nu & \vdots \end{bmatrix}$; $Z^* = \begin{bmatrix} \vdots & z^*_\nu & \vdots \end{bmatrix}$; $S = \begin{bmatrix} \ddots & & \\ & s_\nu & \\ & & \ddots \end{bmatrix}$.

Z, Z^* and S are matrices of order 2Nx2N.

Eigenmodes can be normalized :

$$\overline{Z}^{*T}\, N\, Z = D \quad ; \quad \overline{Z}^{*T}\, L\, Z = -\, S\, D \ . \tag{11}$$

D is an arbitrary diagonal matrix. It is generally chosen as the identity matrix which leads to a simplified expression of the forced response z:

$$z = \sum_{\nu=1}^{2N} \frac{z_\nu\, \overline{z}_\nu^{*T} g}{s - s_\nu} \qquad \text{if } D_{\nu\nu} = 1.$$

Let : $z = \begin{bmatrix} y \\ sy \end{bmatrix}$; $z_\nu = \begin{bmatrix} {}_1y_\nu \\ s_\nu\, {}_1y_\nu \end{bmatrix}$; $z^*_\nu = \begin{bmatrix} {}_1y^*_\nu \\ s_\nu\, {}_1y^*_\nu \end{bmatrix}$.

The response y can be rewritten as :

$$y = \sum_{\nu=1}^{2N} \frac{{}_1y_\nu\ {}_1\overline{y}_\nu^{*T}\, f}{s - s_\nu} \ . \tag{12}$$

<u>If L and N are symmetric</u> : <u>$\overline{z}^*_\nu = z_\nu$</u>. The transfer function takes the form :

$$T_{ij} = \sum_{\nu=1}^{2N} \frac{{}_1y_\nu\ {}_1y_{\nu j}}{s - s_\nu} \ .$$

Moreover, if L and N are real matrices, then eigenvalues and eigenvectors are conjugate in pairs if the damping is small. The transfer function can be rewritten as :

$$T_{ij} = \sum_{1}^{N} \frac{A_{\nu ij}}{s-s_\nu} + \frac{\overline{A}_{\nu ij}}{s-\overline{s}_\nu} \ , \quad \text{where : } \ {}_1y_{\nu i}\ {}_1y_{\nu j} = A_{\nu ij}.$$

I.3. <u>Relationship between eigenvectors y_ν and ${}_1y_\nu$</u> [2,3] :

Hypothesis :

- M, B and K are real, symmetric and positive definite matrices
- the structure is slightly damped
- eigenvectors y_ν are normed such that : $y_\nu^T M\, y_\nu = 1$
- eigenvectors z_ν are normed by choosing : $D_{\nu\nu} = 2j\omega_\nu$,

$z_\nu^T\ N\ z_\nu = 2\, j\omega_\nu$, $\qquad \overline{z}^*_\nu = z_\nu$

The modal matrix Z has the following form :

$$Z = \begin{array}{|c|c|}\hline {}_1Y & {}_1\overline{Y} \\ \hline {}_1Y\ {}_1S & \overline{{}_1Y\ {}_1S} \\ \hline\end{array} \quad ; \quad {}_1Y = \begin{array}{|c|}\hline \quad \vdots\ {}_1y_\nu\ \vdots \quad \\ \hline\end{array} \quad ;$$

$$D = \begin{array}{|c|c|}\hline \diagdown d & 0 \\ \hline 0 & \diagdown \overline{d} \\ \hline\end{array} \quad ; \quad S = \begin{array}{|c|c|}\hline \diagdown S_1 & 0 \\ \hline 0 & \diagdown \overline{S}_1 \\ \hline\end{array} \quad .$$

From (11.1) we deduce : $${}_1Y^TM\ {}_1Y\ S_1 + S_1\ {}_1Y^TM\ {}_1Y + {}_1Y^TB\ {}_1Y = d$$
$${}_1\overline{Y}^TM\ {}_1Y\ S_1 + \overline{S}_1\ {}_1\overline{Y}^TM\ {}_1Y + {}_1\overline{Y}^TB\ {}_1Y = 0 \qquad (13)$$

with : $d = 2j\ \Omega$, $\Omega_{\nu\nu} = \omega_\nu$, $\Omega_{\sigma\nu} = 0$ if $\sigma \neq \nu$.

We try to obtain the relationships between eigenvectors y_ν of the conservative system (4) and eigenvectors ${}_1y_\nu$ of the dissipative system (2), by introducing the damping B into (4) as a perturbation.

Let : $S_1 = j\Omega + V$; $| V_{\nu\nu} | << \omega_\nu$,
${}_1Y = Y\ (E + jQ)$; $| Q_{\nu\sigma} | << 1$.

By introducing these expressions into (13), one obtains the following relations from which Q and V are deduced to within the second order :

$$\beta = -\ 2V + Q\Omega + \Omega\ Q^T \quad \text{and} \quad : \quad Q^T\Omega + \Omega Q = 0. \qquad (14)$$

From (14) it can be deduced that :

1°) $Q_{\nu\nu} = 0$: the imaginary part of an eigenvector ${}_1y_\nu$ is orthogonal to its real part with respect to the mass matrix. Indeed, let : ${}_1y_\nu = y_r + jy_i$, with :

${}_1y_\nu = y_\nu + j\ Y\ Q_\nu$ and $Q_{\nu\nu} = 0$, consequently : $y_r^T\ M\ y_i = 0$.

2°) V is a real matrix : the perturbations of eigenvalues are only real to within the second order.

3°) β and V being real matrices, it leads that Q is a real matrix. It follows that the real part of an eigenvector ${}_1y_\nu$ is equal to the corresponding real eigenvector of the associated conservative system.

4°) The chosen norm : $d_\nu = 2j\omega_\nu$ minimizes the imaginary part of the eigenvectors ${}_1y_\nu$ (in the sense of the associated kinetic energy). It is easy to show that the scalar :

$$\alpha = \frac{y_i^T\ My_i}{y_r^T\ My_r} \quad \text{is minimum when : } d_\nu = 2j\ \omega_\nu\ .$$

II - IDENTIFICATION BY APPROPRIATION

II.1. Definition of appropriation

Appropriation consists of finding which system of forces must be applied to a structure in order to obtain a response proportional to an eigenmode of the associated conservative system.

Most experimental methods are based on the fact that all points of the structure have the same phase angle when the excitation is appropriate.

In the simplest methods, the experimenter proceeds tentatively by modifying the level of excitating forces and the frequency as best he can in order to minimise the force and velocity phase difference for all points. A method based on this principle was tested by LEWIS and WRISLEY[5] at the M.I.T. in 1950.

II.1.1. Preliminary considerations

In forced harmonic conditions, the movement of a discrete linear system is defined by the equation :

$$(K - \omega^2 M + j\omega B)\ y = f\ .$$

M, B, K are real, symmetric, positive definite.

To the damped initial system, we can associate the conservative system defined by the eigenproblem :

$$(K - \lambda M)\ \dot{y} = 0 \quad , \quad \lambda = \omega^2 \ ,$$

from which we deduce the modal matrix Y of eigenvectors y_ν and the spectral matrix Λ of eigenvalues λ_ν.

According as the eigenvectors are arbitrarily normed (x_ν) or relatively to the mass matrix (y_ν), we obtain the classical relations :

$$X^T M X = \mu \quad , \quad X^T K X = \mu\,\Lambda = \gamma \quad , \quad Y^T M Y = E \quad , \quad Y^T K Y = \Lambda \ ,$$

where μ is the diagonal matrix of the generalized masses, γ that of the generalized stiffness.

The forced displacements of the damped system can be decomposed on the basis of the eigenvectors y_ν or x_ν of the associated conservative system. Then, we obtain :

$$y(\lambda) = Y\Omega\ Y^T f \quad , \quad \text{or} : \ y(\lambda) = X\,\tilde{\Omega}\ X^T f \ ,$$

$$\text{with } \Omega = (\Lambda-\lambda E+j\sqrt{\lambda}\beta)^{-1} \quad \text{or} : \ \tilde{\Omega} = (\gamma-\lambda\mu+j\sqrt{\lambda}b)^{-1} \ ,$$

$$\beta = Y^T B Y \quad , \quad b = X^T B X = \mu^{1/2}\ \beta\ \mu^{1/2} \ ,$$

where β and b are the normed and non-normed generalized damping matrices.

II.1.2. Determination of appropriate forces

It is known [6] that a linear self-adjoint system can be appropriate to an eigen circular frequency ω_ν with a system of forces all in phase (or anti phase). For example, at the circular frequency ω_ν, the movement is described by :

$$(K-\lambda_\nu M + j\sqrt{\lambda_\nu}B)\ x_\nu = f_\nu \ , \qquad \lambda_\nu = \omega_\nu^2 \ , \qquad e_\nu = \begin{bmatrix} 0 \\ \vdots \\ 0 \\ 1 \\ 0 \\ \vdots \end{bmatrix} \leftarrow \nu$$

$$f_\nu = j\sqrt{\lambda_\nu}X^{-1T}\ b \quad , \quad e_\nu = j\sqrt{\lambda_\nu}\ Y^{-1T}B\ e_\nu \ . \tag{16}$$

If b is diagonal (Basile's hypothesis verified [7]), this force f_ν will excite an eigenmode which is out of phase in relation to the excitation whatever the frequency is.

Let : $\quad y = X\ c$.

At the circular frequency $\omega = \sqrt{\lambda}$, the response y to the appropriate excitation will be defined by :

$$c = j\sqrt{\lambda_\nu}\ \frac{b_{\nu\nu}}{\gamma_{\nu\nu}-\lambda\mu_{\nu\nu}+j\sqrt{\lambda}b_{\nu\nu}}\ e_\nu \ .$$

All the points of the structure vibrate in phase (phase résonance). The phase difference ϕ between the movement and excitation force is defined by :

$$tg\,\phi = \frac{\sqrt{\lambda}\; b_{\nu\nu}}{\lambda\mu_{\nu\nu}-\gamma_{\nu\nu}} = \frac{\sqrt{\lambda}\;\beta_{\nu\nu}}{\lambda-\lambda_\nu} .$$

Reciprocally, assuming that a real force vector f is found, whose response to any circular frequency ω close to ω_ν is :

$$y = Y\,\Omega\,Y^T . f = Gf + jHf = y_1 + j\,y_2 ,$$

with : $y_1 = k\,y_2$, and k = real constant,

then f is an appropriate force to an eigenvector, and we have :

$$k = \frac{\lambda - \lambda_\nu}{\sqrt{\lambda}\;\beta_{\nu\nu}} , \qquad \lambda = \omega^2 \quad ; \; \lambda_\nu = \omega_\nu^2 .$$

II.1.3. Form of the appropriate forces such that : $y_1 = ky_2$ when Basile's hypothesis is not verified

In this case, in order to obtain $y_1 = k\,y_2$ at $\lambda \neq \lambda_\nu$, complex forces $F_R + jF_i$ must be applied. We can show[3] that the generalized forces appropriated to the νth eigenmode defined by :

$$X^T(F_R + jF_i) = f_R + j\,f_i$$

are of the form :

$$f_R^T = -\sqrt{\lambda}\;\boxed{b_{1\nu}\,,\, b_{2\nu}\,,\,\ldots\ldots,\; b_{\nu\nu}\;(1+k^2)\,,\,\ldots,\, b_{n\nu}} ,$$

$$f_i^T = k\sqrt{\lambda}\;\boxed{b_{1\nu}\,,\, b_{2\nu}\,,\,\ldots\ldots\;,\,0\;,\,\ldots,\, b_{n\nu}} .$$

In practice $b_{i\nu}$ $(i\neq\nu) \ll b_{\nu\nu}$; furthermore in the proximity of $\lambda \simeq \lambda_\nu$, $k \simeq 0$. Consequently, the imaginary force is negligible compared to the real force.

The above results show that an appropriate force must be applied at all points of the structure, which is practically not realizable. As a matter of fact, forces will be best appropriate to a mode if they do not excite the neighbouring modes. It follows that appropriate forces are completely changed when the number of excitation points is modified.

II.2. DECK's method[6] :

The method developped by Deck at ONERA is an iterative method which permits to determine by successive approximations, the forces appropriate to one mode and the corresponding eigenfrequency.

Let : $y = y_1 + jy_2$ the displacement vector (of order N) of the structure due to p forces f_j with the same phasis and circular frequency. These forces are represented by the vector f of order p.

At each point k, the component y_{1k} of the displacement, in phase with the excitation force is picked up. y_{1k} must be zero if the appropriate forces are applied to the eigenfrequency. Then, we try to minimize

$$R = \sum_{k=1}^{n} | y_{1k} | \qquad \text{where : } n >> p.$$

If a force f_j is modified without modifying the frequency, each component y_{1k} of y_1 is a linear function of f_j. The evolution of its modulus r_{1k} versus f_j will be a curve like a V. Then the following form is obtained for R :

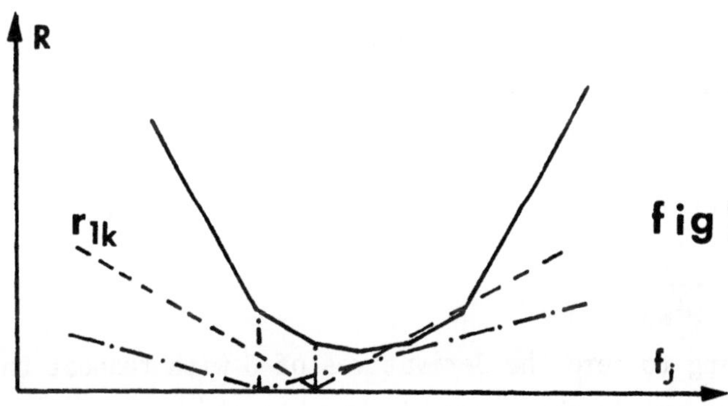

fig 1

The force f_j is chosen to minimize R and then the excitation frequency is varied such that R will be a minimum. This procedure is repeated with each component f_j until the mean difference of phase ϕ_m will be sufficiently small :

$$\operatorname{tg} \phi_m = \frac{\sum_{k=1}^{N} |y_{1k}|}{\sum_{k=1}^{N} |y_{2k}|} \qquad (y_k = y_{1k} + j\, y_{2k}) \quad .$$

The eigenmode shape is then directly measured on the structure. The method is efficient. It can be easily automatized. It gives generally good results. On the other hand, the user is never certain that all the modes have been isolated.

II.3. CLERC's method[7] :

CLERC suggests a systematic method which permits to obtain the appropriate forces to one mode, by minimizing y_1 at the phase resonance.

p forces in p different points are applied successively to the structure with the same circular frequency ω close to an eigen circular frequency ω_ν . To simplify, we will suppose that all the forces are normed to 1. The p forced responses y^i on n points (n>>p) are picked up and arranged under matrix form resulting in matrices V (nxp) for the in phase response and W (nxp) for the quadrature response.

$$V = \begin{bmatrix} & \vdots & \\ & y_1^i & \\ & \vdots & \end{bmatrix} \quad ; \quad W = \begin{bmatrix} & \vdots & \\ & y_2^i & \\ & \vdots & \end{bmatrix} \quad ; \quad y^i = y_1^i + j\, y_2^i \ .$$

If we apply to the structure, the force : f = a (a being any vector of order p), the structure's response will be : $y = (V+jW)a$.

So, we try to obtain the real vector a and the excitation circular frequency ω which minimizes Va.

The parameter δ which is minimized by a is chosen as :

$$\delta = \frac{a^T\, V^T V a}{a^T\, W^T W a} \ .$$

By equaling to zero the derivatives of δ with respect to the components of a, one obtain the eigenvalue problem :

$$(V^T V - \delta\ W^T W)\ a = 0 \ ,$$

from which the smaller eigenvalue δ_1 and the corresponding vector a_1 are calculated. By varying the excitation frequency step by step, we obtain for δ_1 versus ω, the following curve :

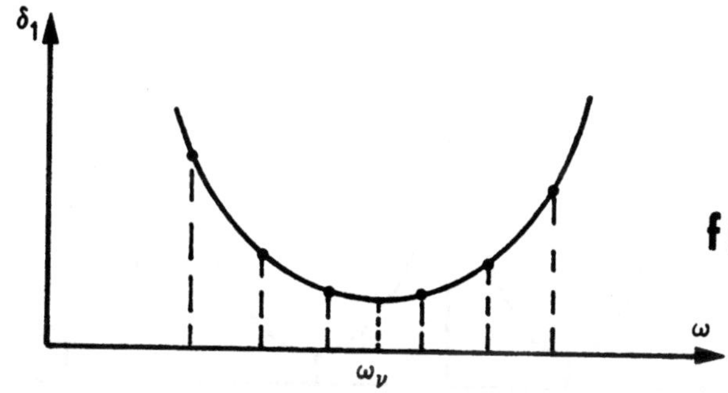

fig 2

δ_1 is minimum for $\omega = \omega_\nu$. The corresponding vector $a_{1\nu}$ is the appropriate force $f_\nu = a_{1\nu}$. The eigenmode y_ν can be calculated by the product Wa_ν . This method is systematic, its permits to detect and appropriate eigenmodes with very close eigenfrequencies. On the other hand it requires too many measurements.

II.4. LMA method [8,3,10] :

The method utilizes the same matrices V and W obtained at the circular frequency ω close to one or several eigen circular frequencies ω_ν. We take advantage of the property : $y_1 = k\, y_2$ when applied forces are appropriate to one eigenmode. Therefore we determine the real vectors a such that :

$$V\, a = k\, W\, a \, .$$

Since it is not possible to realize an exact appropriation with a restricted number of forces, we try to minimize $Va - kWa = \varepsilon$ with respect to the amplitude of the movement. Then we minimize. $\Delta = \dfrac{\varepsilon^T \varepsilon}{a^T (V^T V + W^T W) a}$.

By cancelling the derivatives of Δ with respect to the components of a, one obtains the eigenvalue problem :

$$\boxed{\begin{aligned} &(G - \Delta H)\, a = 0 \\ &\text{and : } k = \frac{a^T W^T V a}{a^T\, W^T W a} \end{aligned}} \quad . \qquad \begin{aligned} \text{with : } G &= V^T V - k\, (W^T V + V^T W) + k^2\, W^T W \, , \quad (16) \\ H &= W^T W + V^T V \, . \end{aligned}$$

It is sufficient to determine the values of k which minimize the smallest eigenvalue Δ in the eigenproblem (16). These values can easily be obtained by an iterative process[3]

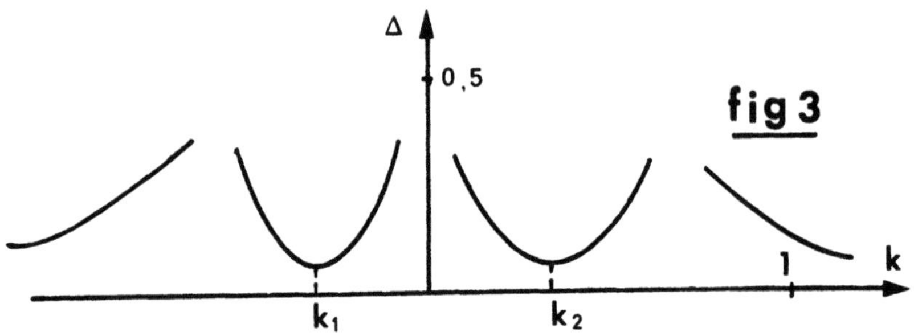

By plotting $\Delta = f(k)$, we get the diagramm (fig. 3).

For each value k_ν of k for which Δ is minimum, there is a corresponding force vector a_ν which is the appropriate force f_ν to the νth eigenmode.

II.5. Determination of generalized parameters [8,10,3] :

If we apply to the structure the appropriate force, we can measure directly the eigenvector y_ν. Then, we leave to determine ω_ν, $\mu_{\nu\nu}$, $\beta_{\nu\nu}$.

The extradiagonal elements $\beta_{\sigma\nu}$ have a negligible effect upon the dynamical behaviour if the eigenfrequencies are not too close. Therefore there is often no interest to determine them. The following references on this subject can be consulted.[3,22,2]

Various methods were suggested to obtain generalized parameters,[11,12,6] but it seems that it is the "complex power" method which is the most accurate.

Appropriate forces f_ν are applied to the structure at an excitation frequency which is varied step by step around the eigenfrequency. Forced responses y proportionnal to y_ν are picked up and "the complex power" P is calculated :

$$P = \bar{y}^T f_\nu = P' + jP'' \quad ; \quad P' = \text{real}(P) \quad ; \quad P'' = \text{Img}(P).$$

Replace f_ν by $f_\nu = (K-\omega^2 M + j\omega B)\, y$ in the above expression

$P = \bar{y}^T (K - \omega^2 M + j\omega B)\, y.$

Let : $y = y_\nu\, (c + jd)$.

For $\omega = \omega_\nu$, we will obtain : $c = 0$ and $d = -1$;
it follows immediatly :

$$P' = (c^2 + d^2)\,(\omega_\nu^2 - \omega^2)\,\mu_{\nu\nu} \;;\quad \mu_{\nu\nu} = -\left[\frac{dP'}{d\omega^2}\right]_{\omega=\omega_\nu}$$

$$P'' = \omega\,(c^2+d^2)\,b_{\nu\nu} \qquad ;\quad b_{\nu\nu} = \left[\frac{P''}{\omega}\right]_{\omega=\omega_\nu} = \mu_{\nu\nu}\,\beta_{\nu\nu}$$

In practice the tangent to P' in the neighbouring of $\omega = \omega_\nu$ varies quickly. Therefore $\mu_{\nu\nu}$ is obtained with very poor accuracy (Fig. 4). $\mu_{\nu\nu}$ can be better determined without any complementary measures : for this purpose the scalar k is plotted versus ω^2 :

$$k = \frac{y_1^T\, y_2}{y_2^T\, y_2} = \frac{\omega^2-\omega_\nu^2}{\omega\,\beta_{\nu\nu}} \quad ; \quad \left(\frac{dk}{d\omega^2}\right)_{\omega_\nu} = \frac{1}{\omega_\nu\beta_{\nu\nu}}$$

k is obtained by utilizing the data of all the pick-ups of the structure, then its calculated value is very accurate. For $\omega = \omega_\nu$, k is zero and the function k (ω^2) is quasi-linear in the neighbouring of $\omega = \omega_\nu$

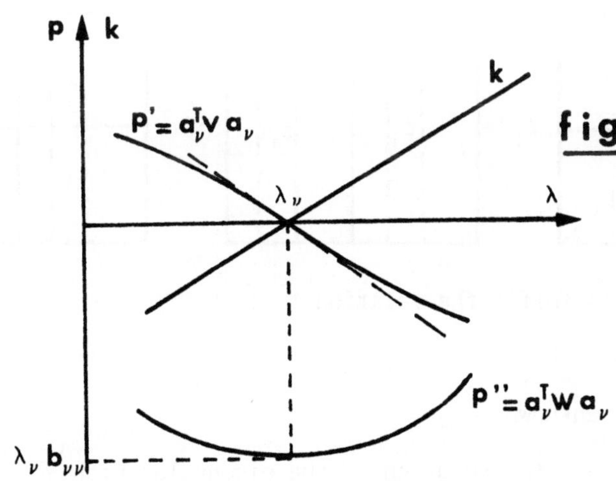

fig 4

III. IDENTIFICATION OF EIGENSOLUTIONS BY GALERKIN TECHNIQUE [13] :

III.1. Problem formulation :

This method only needs the measurement of two forced responses near each eigenfrequency of the structure. In order to simplify, we suppose that K, M and B are real and symmetric matrices, but the same method can be applied for more general matrices. We suppose that eigenfrequencies have been previously located. We successively apply to the structure a set of p forces f_i with circular frequencies ω_i close to eigen circular frequencies ω_ν.

The associated responses y_i verify the relations :

$$(L + s_i\, N)\, z_i = g_i \quad , \quad i = 1 \text{ to } p \quad , \; s_i = j\,\omega_i \; ,$$

$$(L + \bar{s}_i\, N)\, \bar{z}_i = \bar{g}_i \; ,$$

$$s_i = j\,\omega_i \quad , \quad z_i = \begin{bmatrix} y_i \\ s_i y_i \end{bmatrix} \quad , \quad g_i = \begin{bmatrix} f_i \\ 0 \end{bmatrix} \quad .$$

The responses y_i are obtained from n points (n > p) and z_i is a (2nx1) matrix.

Results can be arranged under matrix form :

$$S_1 = \begin{bmatrix} \ddots\, s_i \,\ddots & 0 \\ 0 & \ddots\, \bar{s}_i \,\ddots \end{bmatrix} \qquad Z_1 = \begin{bmatrix} \cdots\; z_i \;\cdots & \cdots\; \bar{z}_i \;\cdots \end{bmatrix} \qquad F = \begin{bmatrix} \cdots\; f_i \;\cdots & \cdots\; \bar{f}_i \;\cdots \\ \cdots\; 0 \;\cdots & \cdots\; 0 \;\cdots \end{bmatrix}$$

These matrices fulfil the relation :

$$L\,Z_1 + N\,Z_1\,S_1 = F \; , \tag{20}$$

where L and N are unknown.

We try to obtain the solution of the eigenvalue problem

$$(L + s_\nu\, N)\, z_\nu = 0 \; , \tag{21}$$

in the form : $z_\nu = Z_1 \quad c_\nu$, (22)

where c_ν is a column matrix.

By introducing (22) into (21) and multiplying by Y_1^T one obtains :

$$(Z_1^T \, L \, Z_1 + s_\nu \, Z_1^T \, N \, Z_1) \, c_\nu = 0 \tag{23}$$

Using : $\gamma = Z_1^T \, L \, Z_1$ and : $\mu = Z_1^T \, N \, Z_1$, we have :

$$(\gamma + s_\nu \, \mu) \, c_\nu = 0 \ . \tag{24}$$

(24) is an eigenvalue problem. From the eigensolutions s_ν , c_ν , one obtains the eigenvectors of (21) :

$$z_\nu = Z_1 \, c_\nu \ .$$

This method is similar to the Galerkin technique.

In our case, the operators L and N are unknown, but it is possible[3] to obtain the matrices γ and μ by measuring the harmonic responses and the applied forces.

IV. METHOD OF DIFFERENCES OF FORCED RESPONSES[3] :

This elementary method gives complex eigensolutions of structure by utilizing three harmonic responses for each identified mode. The method can be applied when the distance between two eigenfrequencies is larger than two or three times the 3db bandwith and when measurements are slightly disturbed.

We suppose that the harmonic forced response y_i of the structure can be developed on the set of the complex eigenvectors ${}_1y_\nu$:

$$y_i = \sum_{\nu=1}^{N} \frac{{}_1y_\nu \, {}_1y_\nu^T f}{s_i - s_\nu} \quad , \quad s_\nu = j\omega_\nu - a_\nu \, \omega_\nu .$$

In practice, the force will be applied on only one point k, then :

$${}_1y_\nu^T \; f = {}_1y_{\nu k} \; f_k = c_\nu .$$

The same force f is applied to the structure three times at three circular frequencies close to the isolated eigen circular frequency ω_ν (fig. 5). One may write :

$$y_i = \frac{{}_1y_\nu c_\nu}{s_i - s_\nu} + R_i \quad ; \quad R_i = \sum_{\sigma \neq \nu} \frac{{}_1y_{\sigma 1} y_\sigma^T f}{s_i - s_\sigma} \quad .$$

Where R_i is a function of ω_i which slowly varies in the neighbouring of ω_ν.

$$\text{Let : } \Delta y_{ij} = y_i - y_j = \frac{s_j - s_i}{(s_i - s_\nu)(s_j - s_\nu)} \; {}_1y_\nu \, c_\nu + R_i - R_j \qquad i \neq j = 1,2,3.$$

If ω_ν is an isolated eigencircular frequency, $R_i - R_j$ can be neglected. Δy_{ij} will be obtained with a good accuracy if circular frequencies ω_i are judiciously chosen as shown in figure 5.

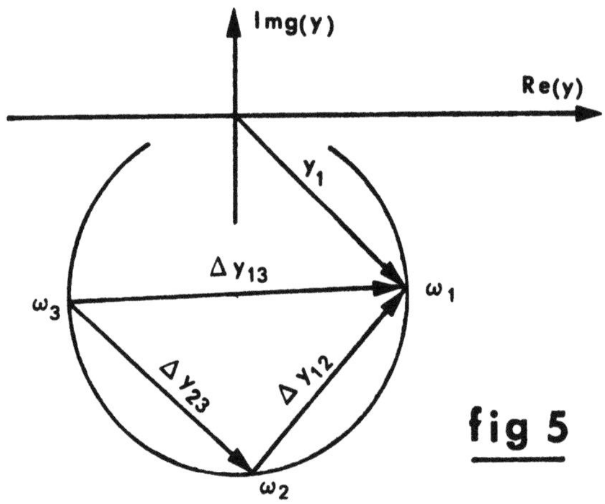

fig 5

Determination of eigensolutions : We suppose that $R_i - R_j \simeq 0$.

It is possible to calculate :

$$\Delta y_{ij}^T f = d_{ij} = c_\nu^2 \; \frac{s_j - s_i}{(s_i - s_\nu)(s_j - s_\nu)} \quad .$$

For i = 1, j = 2,3 ; one obtain 2 equations to determine c_ν and s_ν.

$$s_\nu = \frac{s_3 - \alpha s_2}{1 - \alpha} \quad ; \qquad c_\nu^2 = d_{ij} \; \frac{(s_i - s_\nu)(s_j - s_\nu)}{s_j - s_i} \quad ,$$

$$\alpha = \frac{d_{12}}{d_{13}} \, \frac{s_3 - s_1}{s_2 - s_1} \quad \text{. We deduce : } \quad {}_1y_\nu = \frac{c_\nu}{d_{ij}} \Delta y_{ij}, \qquad i \neq j = 1,2,3 \; .$$

${}_1y_\nu$ verifies the relation : $\quad 2s_\nu \, {}_1y_\nu^T M \, {}_1y_\nu + {}_1y_\nu^T B \, {}_1y_\nu = 1.$

Each eigenmode is calculated three times. Thus it is easy to see that is the quality of the given results. Let : $s_\nu = j\omega_\nu - a_\nu \omega_\nu$, ω_σ an eigen circular frequency close to ω_ν.

The method can be applied if $\frac{\omega_\nu - \omega_\sigma}{\omega_\nu} > 6\, a_\nu$.

If eigenfrequencies are closer together, the proposed method may be applied in two steps. The first one that has just been described, gives a first approximation for eigenmodes. It allows for the neglected terms R_i and R_j to be evaluated and the method is iterated by replacing Δy_{ij} by : $\Delta y_{ij} - R_i + R_j$.

With this method, we can calculate the eigensolutions one after the other, during dynamic tests and it gives possibility for an immediate survey upon the quality of measurements. However, another method must be employed when eigenmodes are coupled too much together.

V. TRANSFER FUNCTION METHOD :

We suppose that K, B and M are real symmetric positive definite. A forced harmonic response y takes the form :

$$y = \left(\sum_{\nu=1}^{N} \frac{{}_1y_\nu \, {}_1y_\nu^T}{s - s_\nu} + \frac{{}_1\bar{y}_\nu \, {}_1\bar{y}_\nu^T}{s - \bar{s}_\nu} \right) f = \Gamma . f \qquad ; \quad \Gamma = H^{-1} \tag{25}$$

Because of the particular structure of Γ, the components of the eigenvector can be determined from only one row or one column. It will be sufficient to apply the force at only one point j and to pick up the displacements y_i on the N points i of the structure.

$$\text{Let : } \Gamma_{ij} = \frac{P(s)}{Q(s)} = \sum_{\nu=1}^{N} \left(\frac{A_{\nu ij}}{s-s_\nu} + \frac{\bar{A}_{\nu ij}}{s-\bar{s}_\nu} \right) \quad ; \tag{26}$$

$A_{\nu ij} = {}_1y_{\nu i} \, {}_1y_{\nu j}$ (force applied at point j).

It can be shown that : $\sum_{\nu=1}^{N} {}_1y_\nu \cdot {}_1y_\nu^T + {}_1\bar{y}_\nu \, {}_1\bar{y}_\nu^T = 0.$

Therefore, each term Γ_{ij} of the Γ matrix is the ratio of two polynomials respectively of order 2N-2 with respect to s for the numerator and 2N for the denominator. In practice, only $n < N$ modes are identified: Q(s) will be a polynomial or order 2n will respect to s, but the order of P(s) is unknown. Its order p is judiciously chosen between 2n-2 and 2n+1.

The determination of eigensolutions may be done in the following manner : let us denote by $\Gamma^{*}(s_k)$ the experimental values of Γ_{ij} obtained by applying an unitary force at point j for a set of r excitation circular frequencies ω_k ($r \gg 2n$).

The coefficients for polynomials P (s) and Q(s) are determined so that the difference between experimental values $\Gamma^{*}(s_k)$ and calculated values $\frac{P(s_k)}{Q(s_k)}$ is minimized.

For this purpose the following references may be consulted[15,16,17]. After having determined the polynomials coefficients, eigenvalues s_ν are obtained by solving : $Q(s_\nu) = 0$. Then the values of $A_{\nu ij}$ are extracted from (26) by multiplying each member by $(s-s_\nu)$ and putting $s = s_\nu$:

$$A_{\nu ij} = \frac{P(s_\nu)}{R(s_\nu)} \quad \text{and :} \quad R(s_\nu) = \frac{Q(s_\nu)}{s - s_\nu} = (s-\bar{s}_\nu)\Pi_{\sigma \neq \nu}(s-s_\sigma) .$$

The eigenvectors components are immediately obtained by the expression :

$$y_{\nu i} = \frac{A_{\nu ij}}{\sqrt{A_{\nu ij}}} \qquad \text{(force applied at point j).}$$

This method leads to some numerical problems particularly for the determination of P and Q coefficients. More, when a limited number of modes are identified, it is not easy to choose the degree of P. Furthermore, the coefficients for P are imposed to be real, which is not mathematical justified.

The method utilized by the L.M.A. also takes advantage of the expression (25) but with a very different point of view. In practice, only a few modes of neighbouring frequencies can be simultaneously isolated, and one operates in limited band frequencies.

We have seen that the response can be written as :

$$y = \left(\sum_\nu \frac{{}_1y_\nu \, {}_1y_\nu^T}{s - s_\nu} + \sum_\nu \frac{{}_1\bar{y}_\nu \, {}_1\bar{y}_\nu^T}{s - \bar{s}_\nu}\right) f$$

given that : $s - \bar{s}_\nu = j\omega + a_\nu\omega_\nu + j\omega_\nu$,

we will choose therefore the approached developement of the response at point i :

$$y_i = u_i + sv_i + \sum_{\nu=1}^{n} \frac{t_{\nu i}}{s-s_\nu} \qquad (27)$$

where : u_i, v_i are complex constants

$t_{\nu i}$ will also be a complex constant if the excitation forces are independent of the excitation frequency.

- Smoothing technique

The problem is to obtain the unknowns : u_i, v_i, $t_{\nu i}$, s_ν ; ν=1 to p.

For each excitation circular frequency ω_k the relation (27) can be written as :

$$f(q_j,\omega_k) = y_i-u_i-j\omega_k v_i - \sum_{\nu=1}^{n} \frac{t_{\nu i}}{j\omega_k - s_\nu} \simeq 0 \quad , \qquad k = 1 \text{ to } p,$$

where the q_j's represent the (2n+2) unknowns. The redundant system of p equations with (2n+2) complex unknowns can be solved by a gradient technique associated to a least squares method.

The quality of the smoothing can immediately be verified by comparing the measured values y_i to the calculated values obtained from the relation (12).

If the experimental values are disturbed by noise, the smoothing can give incoherent results on various points. By comparing the measured and calculated y_i, it is possible to eliminate the points for which the analysis is defective. The other points for which the estimation is correct are used to calculate an average of the eigenvalues . Then the iterated process is repeated again and in the second calculation the eigenvalues are fixed to the average one.

- Determination of the eigenvectors

If the excitation force is applied only at point i, ${}_1y_\nu$ can be easily obtained from the $t_{\nu j}$. If the amplitude of the force is normed to 1, we obtain :

$${}_1y_\nu^T = \frac{1}{\sqrt{t_{\nu i}}} \left[t_{\nu 1}, \; \dots \; t_{\nu j}, \; \dots \right] \quad .$$

In case of a multipoint excitation, the smoothing of the response at point i gives the scalars :

$$t_{\nu i} = {}_1y_{\nu i} \cdot {}_1y_\nu^T f ,$$

noting that : $c_\nu = {}_1y_\nu^T f$; $u = \{u_i\}$; $v = \{v_i\}$, then :

$$y - u - sv = \sum_{\nu=1}^{n} \frac{{}_1y_\nu \cdot c_\nu}{s - s_\nu} ,$$

$$f^T (y - u - sv) = \sum_{\nu=1}^{n} \frac{c_\nu^2}{s-s_\nu} ,$$

this relation being available for any s. It is possible to deduce relations giving the c_ν, so :

$${}_1y_\nu^T = \frac{1}{c_\nu} [t_{\nu 1}, \dots t_{\nu j}, \dots] .$$

- Non-linear structure identification

1 - Measurements at constant amplitude

It can be considered that stiffness and damping non-linearities are dependent on the amplitude and the mode shape of the structure. Close to an isolated eigenfrequency, the mode shape will be nearly invariable if the amplitude is forced to a constant value. The non-linearities will not appear and the measurement will be coherent with the linear model used for the analysis. By modifying the amplitude, one obtains the evolution of the frequency and damping as a function of the amplitude. So a non-linear structure may be correctly identified with a linear analysis method.

In the case of close modes this method can always be applied if each mode is successively isolated. To do this, the excitation must be appropriated to the isolated mode and this requires a multipoint excitation.

Figure (6) shows the measured and calculated results obtained for a non-linear structure whose measurements were performed up at constant amplitude. It can be seen that Nyquist diagrams are quasi perfect circles, as they would be on a linear structure.

Figure (7) shows exactly the same experiment on the same structure but this time, at constant force. It can be seen that the circles have got out of shape and a jump phenomenon appears. Besides, the smoothing of these transfer functions with a linear programme leads to a wrong coïncidence between, calculated and measured values.

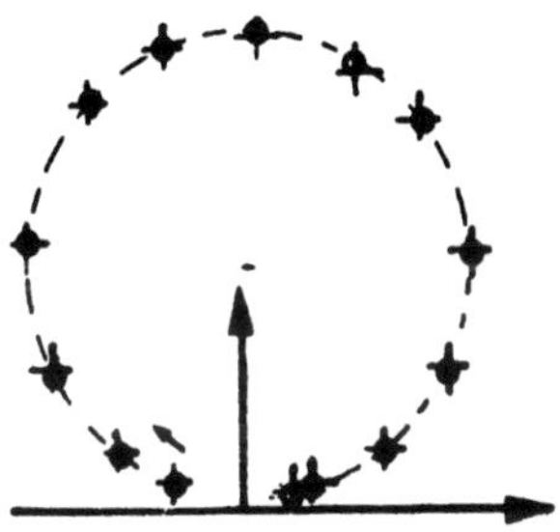

TRANSFERT 2 6

s_1= -2,766+j 1777,41

t_1=0,391 E-06+j 0,169 E-03

Fig. 6 : Constant amplitude
Linear model

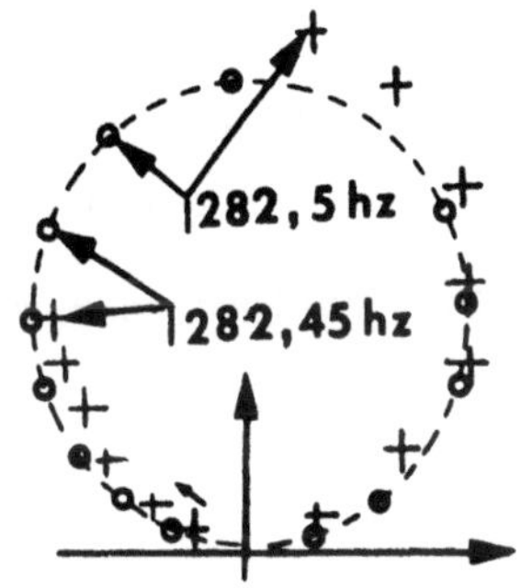

TRANSFERT 2 6

s_1= -1,908+j 1775,05

t_1=0,206 E-05 +j 0,140 E-03

Fig. 7 : Constant force
Linear model

0 Calculated values + Measured values

2 - Measurement at constant force

If the non-linearity is introduced as a function of the amplitude the mobility at point i for an isolated mode can be represented by :

$$y_i = u_i + sv_i + \frac{y_{\nu i} \cdot y_\nu^T f}{s - s_\nu (1 + \alpha |A|^\gamma)} \qquad (28)$$

where : A depends on the amplitude : $A = (\sum_i y_i^2)^{1/2}$

α is a complex constant

γ is a real constant

This relation is obtained from the approximation called "of the first harmonic" for a one degree of freedom non-linear system. The relation (28) will allow us to obtain a good approximation of the dynamic behaviour of the structure for isolated modes and weak non-linearities. To apply this method the motion of the structure must be quasi-sinusoïdal if the applied force is sinusoïdal.

Taking into account these assumptions, the smoothing of the transfer functions will be obtained by employing the method already described. The parameter α will be assimilated to a parameter q_j which has to be calculated. γ is an imposed parameter determinated by successive approximations in order to obtain the better correspondance between calculated and measured values. It can be noticed that the value of the exponent γ can be determined only in the case of weak damping such that the amplitude A varies at least of 2 decades during the experiment at constant forces.

The results obtained on a non-linear structure are shown in fig.8.

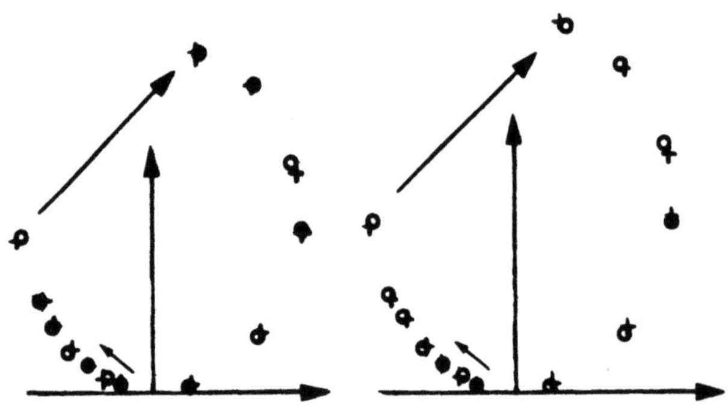

TRANSFERT 2 6	TRANSFERT 2 7
s_1=-2,645+j,1778,51	s_1=-2,668+j,1778,46
t_1=0,730 E-05 +j0,146E-03	t_1=0,784 E-05 + j,0,157 E-03
α =-0,714 E-02+j0,367E-01	α =-0,733 E-02 +j,0,363 E-01

FIG. 8 : Non linear model γ = 1,5

Results presented Fig. 6,7,8 were been obtained with two beams connected by a joint with a non-linear stiffness characteristic. Fig.9.

fig 9

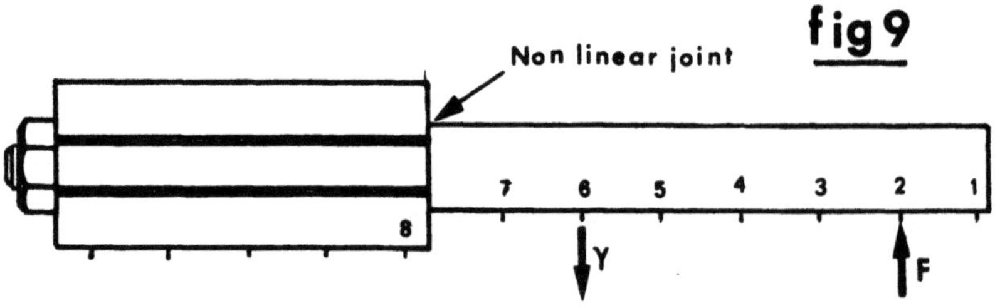

CONCLUSION

During identification procedures some difficulties arise due to non-linearity of mechanical structures and measurement noise. When measurements are greatly affected by the noise, smoothing of transfer function must be prefered. On the other hand for low level noise experiments, simpler methods can be used which require only a very limited number of measurements for each eigenmode.

Il all cases the non-linearities effects must be eliminated by operating at constant amplitude near each eigenfrequency and, if it is necessary, by the application of appropriated forces for the concerned eigenmode.

REFERENCES

1 - Danek, O., Dynamika Stroju, Strojni Fakulta CVUT, Praha (1977), Ustav Termomechaniky ČSAV Puskinovo 9, 16000 PRAHA 6 Bubenec.

2 - Hasselman,I.K., "Damping synthesis from substructure tests" AIAA Paper n° 74387, 15 p.

3 - Fillod, R., "Contribution à l'identification des structures mécaniques linéaires", Thèse de doctorat ès Sciences, Labo. de Méca. Appliquée, Fac.Sciences de Besançon, juin 1980.

4 - Fraeys de Veubeke, Gerardin, "Structural dynamic" CISM, Udine (1972) (Springer Verlag).

5 - Lewis, R. and Wrisley, D.L., "A system for the excitation of pure natural modes of complex structure" Journal of Aeronautical Sciences, nov. 1950 pp. 705-722.

6 - Deck, A., "Méthode automatique d'appropriation des forces d'excitation dans l'essai au sol d'une structure d'avion" EUROMECH 22, Dynamics of Mechanics, New Caste upontyne (1970) 31 p. ; ref. ONERA 870.

"Contribution à l'étude d'une méthode semi-automatique d'appropriation" Note Technique, ONERA n° 129 (1968)

7 - Clerc, D., "Sur l'appropriation des forces d'excitation lors des essais de vibrations en régime harmonique", Recherche Aérospatiale n°87 (1962) p. 55-58.

'Méthode de recherche des modes propres par calcul de l'excitation harmonique optimum d'après les résultats bruts d'essais de vibrations" Note technique ONERA n° 119 (1967), 57 p.

8 - Fillod, R., Piranda, J., "Etude d'une méthode d'appropriation des systèmes linéaires" Proceedings of the 8ème Conférence of Machines, Liblice, Prague (1973).

Research method of the Eigenmodes and generalized elements of a linear mechanical structure" 48th shock and vibration Bull. (1978) part. 3, pp. 5-12.

9 - Nguyen, X.T., "Restitution par calcul des modes propres à partir d'excitations non appropriées" Note technique ONERA n° 1975-9.

10- Bonneau, F., "Determination des caractéristiques vibratoires d'une structure à l'aide de l'expression de la puissance complexe fournie" Recherche Aérospatiale n° 130, juin (1969) pp. 45-51.

11- Kennedy, C., Pancu, C.D.P., "Use of vectors in vibration Measurement and Analysis" Journal of the Aeronautical Sciences, vol. 14, Novembre (1947) pp. 603-6025.

12- Sillard, H., "Sur l'appropriation des forces d'excitation dans l'essai de vibration harmonique" ONERA, 1er trimestre (1974) n° 459, Note technique ONERA n° 208, 52 p.

13- Fillod, R., Piranda J., "Identification of eigensolutions by Galerkin technique" Publication ASME (1979) DET 35

14- De Vries, G., Beatrix, C., "Les procédés généraux de mesure des caractéristiques vibratoires des structures linéaires faiblement amorties" Progress in Aeronautical Sciences, vol. 9 (1968) Pergamon Press

15- Dat, R., Meurzec, J.L., "Exploitation par lissage mathématique des mesures d'admittance d'un système linéaire" La Recherche Aerospatiale n° 4 (1972) pp. 209-215.

16 - Akamatsu, Y., Darras, B., "Lissage par les moindres carrés. Application aux formes propres d'une structure" La Recherche Aerospatiale n° 13 (1969).

17 - Miramand, N. et Billaud, "Identification des paramètres des structures mécaniques" Thèse de 3ème cycle, Compiègne (1975).

Many references concerning Identification Problems in structural dynamics can be found in the following articles :

18 - Morosow, G., "Exciter force apportioning for modal vibration testing using incomplete excitation" University of Colorado at Boulder, Ph. D. 1977, Engineering civil (128 références).

19 - Rades, M., "Analysis techniques of experimental frequency response data" The Shock and Vibrations Digest vol. 11 n° 2 (1979) pp. 15-22 (84 références).

20 - System identification of vibrating structure, Mathematical models from test data, presented at 1972 winter Annual Meeting of the ASME ; collection of papers edited by W.D. PILKEY, ASME 1972, 200 p.

21 - Klosterman, A., "On the experimental determination and use of modal representation of dynamic characteristics" University of Cincinnati (Doctor of Philosophy) 1971 (55 références).

22 - Coupry, G., "Mesure des amortissements généralisés non diagonaux d'une structure lors d'un essai au sol de vibration" La Recherche Aerospatiale (1977) n° 4, pp. 239-244.

IDENTIFICATION METHODS II
CONCEPTS AND TECHNIQUES

Dr. E. Breitbach
Deutsche Forschungs- und Versuchsanstalt
für Luft- und Raumfahrt e.V.
Institut für Aeroelastik
Bunsenstr.10, D-3400 Göttingen
W. Germany

1. INTRODUCTION

Sophisticated technology, particularly as developed in aerospace engineering, is penetrating more and more into other areas such as ground transportation (by rail, road and water), off-shore oil source exploration, and energy engineering. This ever-growing tendency has prompted a dramatically increasing demand for structural dynamics techniques.

One class of important tools in the general field of structural dynamics is known by the general term

Parametric system identification (PSI).

More specifically, the terms

Modal survey testing (MST) and
Ground vibration testing (GVT)

designate experimental concepts whereas

Finite element analysis (FEA)

labels the most renowned theoretical method.

The principle purpose of PSI is to determine the dynamic characteristics of elastodynamic systems as a basis for establishing the dynamic equations of motion. Furthermore, the experimental methods have turned out to be a very useful means to

- *Verify and update mathematical models such as those resulting from FEA*
- *Investigate damping effects, structural nonlinearities and the like.*

As far as experimental concepts are concerned, the improvement of existing MST techniques and creation of new methods have been emphasized in the last decade. This work has been supported and accelerated by the tremendous progress in computer technology, electronics and measuring techniques.

Modern MST methods can generally be categorized into two groups:

- *Phase resonance - direct experimental determination*
- *Phase separation - indirect experimental-numerical approaches.*

This lecture can deal with only a limited number of MST methods representing each of these categories. Such limitation does not imply any prejudice against other methods, but aims at a more profound insight into and understanding of the physical principles of MST rather than giving a broad but superficial survey. Furthermore, there is no mention of a special group of largely computerized MST techniques, usually termed *modal analysis* which entails merely aiming at the same modal parameters as determined by other GVT or MST approaches.

These vigorously commercialized *modal analysis* techniques use the capability of FFT (Fast Fourier Transform) programs and other statistical methods to gain reliable transfer functions as basic data for the determination of modal characteristics. In this context it should be mentioned that having the transfer functions in hand is only the first step toward a complete parametric mathematical model in terms of modal data. The virtual art of PSI starts at the point where these modal data have to be extracted from the transfer functions no matter what kind of PSI approach may be applied.

The methods with which this lecture will be concerned are published in Refs.[1],[2],[3],[4], and [5] and are termed for handy reference by the author's name or affiliation.

To obtain a synoptical view of the present state of the art of MST techniques, reference can be made to [6], which contains the most comprehensive bibliography with summarizing comments on the features of each method.

2. FUNDAMENTAL RELATIONS

The fundamental equations of any MST method start without exception at the simplified condition of structural linearity. Then, the equations of motion, formulated in terms of physical deflections, can be written in matrix notation as follows:

$$A\,\ddot{u}(t) + B\,\dot{u}(t) + C\,u(t) = P(t) \qquad (1)$$

where

A	mass matrix
B	viscous damping matrix
C	stiffness matrix
P(t)	column matrix of external forces
u(t)	column matrix of physical deflections; $\dot{u}(t)$ and $\ddot{u}(t)$ are first- and second-order differentials with respect to time t .

Developing the arbitrary deflection vector $u(t)$ in a series expansion of the normal modes Φ_r , $(r = 1,2,\ldots,n)$, of the conservative undamped system results in the modal transformation

$$u(t) = \Phi\, q(t) \qquad (2)$$

where

Φ modal matrix containing normal modes Φ_r as columns

$q(t)$ column vector of generalized coordinates $q_r(t)$.

Substituting Eq.(2) into Eq.(1) and premultiplying by the transposed Φ^T leads to the equations of motion formulated in terms of the generalized coordinates q_r

$$M\,\ddot{q}(t) + D\,\dot{q}(t) + K\,q(t) = \Phi^T\,P(t) \quad . \tag{3}$$

Because of the orthogonality conditions

$$\Phi_r^T\,A\,\Phi_s = \begin{cases} M_r , & \text{if } s = r \\ 0 , & \text{if } s \neq r \end{cases} \tag{4a}$$

and

$$\Phi_r^T\,C\,\Phi_s = \begin{cases} K_r , & \text{if } s = r \\ 0 , & \text{if } s \neq r \end{cases} \tag{4b}$$

the matrices M and K are diagonal with the generalized masses M_r and the generalized stiffnesses K_r as diagonal elements, whereas $D = \Phi^T\,B\,\Phi$ is generally nondiagonal due to damping coupling.

By assuming simple harmonic motions

$$q(t) = q\,e^{j\omega t} \quad , \quad j = \sqrt{-1} \tag{5a}$$

in response to an harmonic excitation

$$P(t) = P\,e^{j\omega t} \quad , \tag{5b}$$

Eq.(3) reduces to

$$(-\,\omega^2\,M + j\omega D + K)\,q = \Phi^T\,P \tag{6}$$

where ω is the circular frequency of the external excitation. Solving Eq.(6) for $D = 0$ leads to the circular normal frequencies of the conservative system

$$\omega_r = \sqrt{\frac{K_r}{M_r}} \quad . \tag{7}$$

By using Eqs.(2) and (7) and, for the sake of a simplifying approximation, by assuming D to be diagonal, Eq.(6) can be transformed to

$$u = \Phi F \Phi^T P \tag{8}$$

where F is a diagonal admittance matrix, the diagonal of which is occupied by the generalized dynamic flexibility coefficients

$$F_r = \frac{1}{M_r(\omega_r^2-\omega^2) + j\omega D_r} \quad . \tag{9}$$

By using component notation, Eq.(8) can be written altenatively

$$u = \sum_{r=1}^{n} \Phi_r \, (F_r \, \Phi_r^T \, P) \quad . \tag{10}$$

It has to be mentioned that a dynamic system can be described by the Eqs. (8) and (10) if, and only if, damping coupling effects are negligible.

If damping coupling cannot be neglected it may be sensible to replace in Eq.(2) the real normal modes Φ_r of the conservative system by the complex normal modes Ψ_r of the system with damping coupling, leading to the modal transformation

$$u(t) = \Psi \, q(t) \quad , \tag{11a}$$

with the complex modal matrix Ψ containing the normal modes Ψ_r ,

$(r = 1,2,\ldots,n)$ of the system with damping coupling. Then the equations of motion can be written as follows:

$$\overline{M}\,\ddot{q}(t) + \overline{D}\,\dot{q}(t) + \overline{K}\,q(t) = \Psi^T P(t) \tag{11b}$$

where

$\overline{M}$ diagonal matrix of generalized masses $\overline{M}_r = \Psi_r^T A \Psi_r$

$\overline{K}$ diagonal matrix of generalized stiffnesses $\overline{K}_r = \Psi_r^T C \Psi_r$.

Unlike matrix D defined in Eq.(3) matrix $\overline{D} = \Psi^T B \Psi$ is also diagonal even if damping coupling occurs. Furthermore, matrices $\overline{M}$, $\overline{D}$ and $\overline{K}$ are generally complex whereas M , D and K are real.

Let the dynamic response vector u be expressed similar to Eqs.(8) through (10) in terms of the complex normal modes Ψ_r , then it follows that

$$u = \sum_{r=1}^{n} \Psi_r(\overline{F}_r \Psi_r^T P) \tag{12}$$

where

$$\overline{F}_r = \frac{1}{\overline{K}_r - \omega^2\overline{M}_r + j\omega\overline{D}_r} \;. \tag{13}$$

The above-derived equations are the basic relations to understand the implications and features of the MST methods dealt with in the following sections.

3. DFVLR PROCEDURE - AN IMPROVED PHASE RESONANCE TECHNIQUE

The phase resonance technique usually termed *classical* is a direct experimental approach which has been used for many decades to measure the modal characteristics pertaining to the undamped conservative system, the dynamic behavior of which can be described by the Eqs.(2) through (10).

Decomposing force vector P and response vector q into their real and imaginary parts

$$P = P' + jP'' \tag{14a}$$

$$q = q' + jq'' \tag{14b}$$

and using the input forces as a phase reference with the imaginary part being zero

$$P'' = 0 \tag{14c}$$

converts the equation of motion (6) into

$$(-\omega^2 M + K)(q' + jq'') + j\omega D(q' + jq'') = \Phi^T P' \tag{15}$$

The separation of Eq.(15) into its real and imaginary terms yields

$$(-\omega^2 M + K)\, q' - \omega D q'' = \Phi^T P' \tag{16a}$$

$$(-\omega^2 M + K)\, q'' + \omega D q' = 0 \quad . \tag{16b}$$

Eq.(16b) is satisfied for any normal frequency ω_r resulting from

$$\det(K - \omega^2 M) = 0 \tag{17}$$

and for

$$q' = 0 \tag{18a}$$

which because of Eq.(2) is equivalent to

$$u' = 0 \quad . \tag{18b}$$

Application of Eq.(18) to Eq.(16a) leads to

$$- \omega D q'' = \Phi^T P' \tag{19}$$

which is the general condition to be satisfied by the external forces in order to excite the normal modes Φ_r of the undamped conservative system.

For the special case of separately exciting a single mode Φ_r at $\omega = \omega_r$, condition (19) reduces to

$$\omega_r \begin{Bmatrix} D_{1r} \\ D_{2r} \\ \vdots \\ D_{nr} \end{Bmatrix} = \Phi^T P_r \quad . \tag{20}$$

The so-called *phase resonance criterion* defined by the above derived equations can be expressed in words as follows:

> A group of external phase-coherent forces $P = P'$ and the exciting frequency ω have to be changed in the vicinity of a normal frequency yet to be determined until condition (18b) is satisfied.

This iterative procedure converges to the modal data

$$\begin{aligned} P' &= P_r \\ \omega &= \omega_r \\ u'' &= \Phi_r \end{aligned} \tag{21}$$

and correspondingly to a phase shift $\varphi \equiv \pm \pi/2$ between force and response vector.

Application of the phase resonance criterion during an MST requires a steady and simultaneous visual monitoring of as many measuring points as possible (usually several hundred). Concretely speaking, decisions must be made continuously as to whether the successive changes of the independent test variables ω and P result in an overall improvement of the phase criterion. It is obvious that this time-consuming decision process implies a higher inclination to error the larger the number of measuring points. Coping with these difficulties is only possible by replacing the human factor of visual monitoring by means of a computerized process which comprehensively combines the individual observations at all measuring points in one single significant value termed *indicator function*. This indicator function[2,3] has been implemented in the DFVLR test concept and successfully applied to a large number of tests since about 1973. To get an idea of the physical background of this indicator function, symbolized by Δ , we start by considering one of the measuring points on an aircraft structure, as sketched in Figure 1.

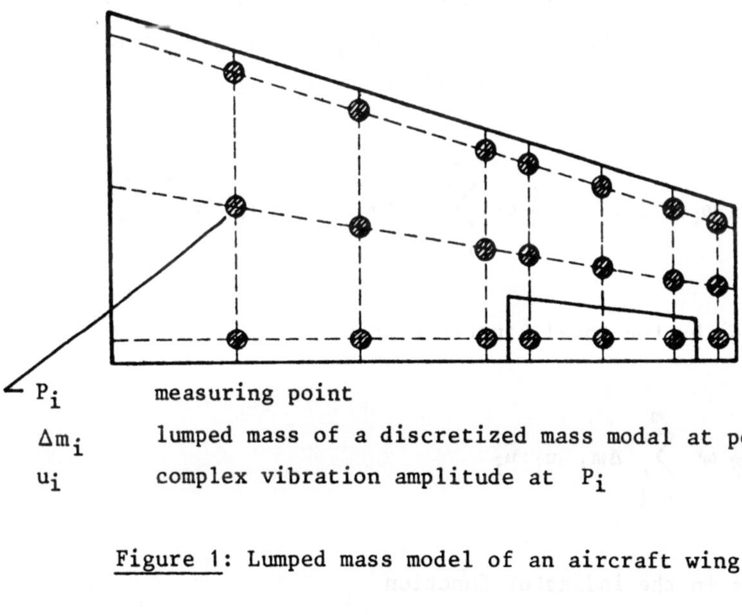

P_i measuring point

Δm_i lumped mass of a discretized mass modal at point P_i

u_i complex vibration amplitude at P_i

Figure 1: Lumped mass model of an aircraft wing.

Let Δm_i be a lumped mass representing the continuously distributed mass in the vicinity of measuring point P_i $(i = 1,2,\ldots,m)$ and let $u_i = u_i' + ju_i''$ be the corresponding complex vibration amplitude, then the kinematic energy of this structural element can be written as

$$\Delta E_{ki} = \frac{1}{2}\,\omega^2\,\Delta m_i\, u_i\, \bar{u}_i \tag{22}$$

where $\bar{u}_i$ denotes the conjugate complex of u_i . The deviation from the phase resonance criterion (18) can be formulated in a slightly different way by the equation

$$|\sin\varphi_i| = \frac{|u_i{}'|}{|u_i|} = \frac{1}{2}\,\frac{|\bar{u}_i - u_i|}{\sqrt{u_i\,\bar{u}_i}} \tag{23}$$

where φ_i denotes the phase shift between the exciting forces and the dynamic response at point P_i . It is obvious that in satisfying the phase resonance criterion a phase shift has more significance the higher the appertaining local energy value. This fact can be taken into account by multiplying Eq.(23) by the local energy ΔE_{ki} and by subsequently obtaining a global value by summation of all measuring points:

$$\sum_{i=1}^{m} |\sin\varphi_i|\,\Delta E_{ki} = \frac{1}{4}\,\omega^2 \sum_{i=1}^{m} \Delta m_i\,|\bar{u}_i - u_i|\,\sqrt{u_i\,\bar{u}_i} \quad . \tag{24}$$

Dividing Eq.(24) by the total kinetic energy

$$E_k = \frac{1}{2}\,\omega^2 \sum_{i=1}^{m} \Delta m_i\, u_i\,\bar{u}_i \tag{25}$$

finally results in the indicator function

$$\Delta = \frac{1}{2}\,\frac{\sum_{i=1}^{m} \Delta m_i\,|\bar{u}_i - u_i|\,\sqrt{u_i\,\bar{u}_i}}{\sum_{i=1}^{m} \Delta m_i\, u_i\,\bar{u}_i} \quad , \tag{26}$$

an especially advantageous feature of which is given by the inequality condition

$$1 \geqq \Delta \geqq 0 \tag{27}$$

with Δ converging to zero if condition (18) is close to fulfillment. Experience has shown that the phase resonance criterion is sufficiently fulfilled for $\Delta \leqq 0.1$.

Furthermore, the indicator function can be used to cope with the problem of identifying the different normal modes in the frequency range of interest. A practical example of such an identification process is illustrated in Figure 2 showing an arbitrary selection of several indicator functions calculated in a process computer and plotted versus frequency, each function being assigned to a different exciter configuration. The subsequent isolation of the normal modes can be facilitated considerably by selecting those indicator functions with Δ closest to zero as start configurations to find appropriate force distributions.

Simultaneous to the computerized on-line processing and display of the indicator function the dynamic deflections of the structure are displayable on a screen showing alternatively either the in-phase or out-of-phase response components u_i'' or u_i' , respectively. Due to the high speed data reduction there is no noticeable time delay between test engineer-induced changes of the test variables ω and P' and the corresponding changes of the indicator function and the dynamic response displayed on the screen. Moreover it can simply be shown that the shape of the indicator function, if applied to a linear system, is independent of any force level alternations as far as the force distribution is kept invariant. Hence it follows that this feature can in turn be used as a very sensitive detector of strength and type of nonlinear effects. Once a normal mode has been identified and isolated the normal mode shape may be measured via acceleration pick-ups. Then the output signals are decomposed by vector component resolvers, converted into digital values, normalized in a process computer and the imaginary parts are plotted as shown in Figure 3 for the example of an aircraft with external stores.

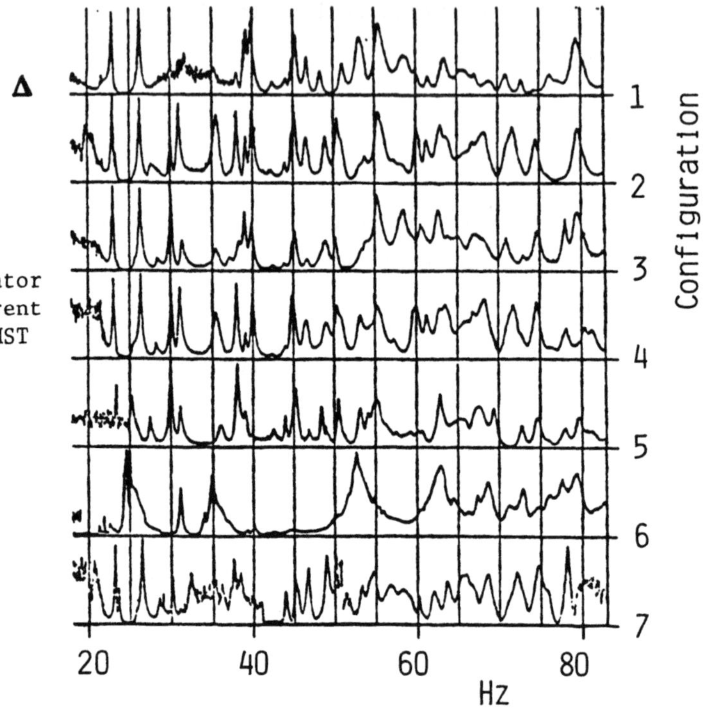

Figure 2: A set of indicator functions in seven different exciter configurations; MST on MAROTS satellite.

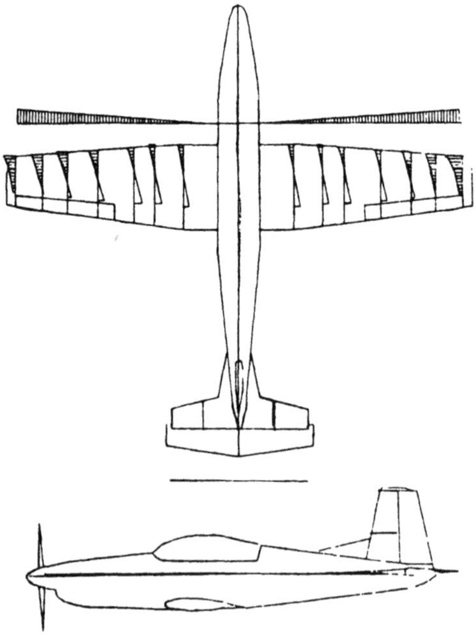

Figure 3: A typical mode shape of an airplane with external stores, f = 39.7 Hz .

As for the experimental determination of the generalized masses and the damping coefficients, various publications may be cited.[1,2] Thus, the generalized masses can be determined by

- *Adding small masses and/or stiffnesses and measuring the resulting normal frequency shifts*
- *Adding small forces in quadrature* ($\pm \frac{\pi}{2}$ *with respect to the regular exciting forces) and measuring the resulting normal frequency shifts*
- *Determining damping (directly for instance through decay rate measurements) and input energy (by measuring forces and amplitudes at the excitation points).*

Damping may be determined by

- *Measuring decay rates averaged over as many measuring points as possible*
- *Determining input energy by measuring forces* P' *and amplitudes* u'' *at the excitation points*
- *Measuring the gradient* $\left. \frac{\partial u_i'}{\partial \omega} \right|_{\omega = \omega_r}$.

It is worth mentioning that there is a growing tendency in the last few years to replace the experimental methods for the determination of generalized masses by a numerical procedure. This procedure uses the measured normal modes Φ_r and a reliable mass model which is usually a by-product of a finite element model. This method, basically defined by Eq.(4a), delivers not only the generalized masses M_r as diagonal elements of matrix $\Phi^T A \Phi$, but also an orthogonality measure in terms of the off-diagonal elements and therewith an accuracy indication of the measured normal modes. Table 1 shows the result of such an orthogonality check on a real airplane.

To give an idea of the considerable progress made since 1966 by steady improvement of the phase resonance technique, Table 2 compares 3 typical test campaigns on an aircraft, especially with respect to test time. Besides tremendous time savings, measuring accuracy has also been largely enhanced.[7,8]

Mode No.	1	2	3	4	5	6	7	8	9
1	1.00								
2	0.01	1.00							
3	-0.05	-0.07	1.00						
4	0.00	-0.04	-0.02	1.00					
5	-0.05	0.07	-0.05	-0.07	1.00				
6	0.02	0.04	0.07	-0.03	0.09	1.00			
7	0.01	0.01	0.02	-0.08	0.01	0.13	1.00		
8	-0.05	-0.10	-0.12	-0.02	-0.11	-0.05	-0.01	1.00	
9	-0.05	0.02	-0.00	0.04	0.06	0.05	0.11	0.11	1.00

Table 1: Orthogonality check for the first 9 normal modes of an airplane.

Year	Configuration	Number of Measured Eigenmodes	Test Duration
1966	clean wing	21	6 weeks
1974	5 configurations with 2-6 external stores	110	6 weeks
1978	clean wing and 5 configurations with 2-6 external stores	125	10 days

Table 2: Development of test time effort since 1966, taken from tests on three different aircraft.

resonance technique based on the Eqs.(11) through (13), formulated in terms of the complex normal modes Ψ_r of the system with damping coupling. A special advantage of this approach is that the modal transformation

$$u(t) = \Psi\, q(t) \tag{28}$$

transforms not only the mass and stiffness matrices but also the damping matrix, even though damping coupling is involved, into diagonal matrices.

The viscous-type damping law used in the foregoing sections is very suitable as long as the equations of motion are formulated in the time domain. Once the equations of motion are transformed into the frequency domain it is much more convenient to define damping as the imaginary part of a complex stiffness. Thus, the damping term $j\omega\overline{D}_r$ in Eq.(13) has to be replaced by jD_r or in relation to $\overline{K}_r$

$$\frac{\overline{D}_r}{\overline{K}_r} = \overline{\gamma}_r \quad . \tag{29}$$

By normalizing the normal mode shapes Ψ_r in such a manner that the generalized mass matrix converts to the unity matrix

$$I = \Psi^T A\, \Psi \tag{30}$$

and by using the abbreviations

$$\Psi_r^N = \Psi_r^T P\, \Psi_r \tag{31a}$$

$$\mu = \omega^2 \tag{31b}$$

$$\lambda_r = \overline{\omega}_r^2(1+j\overline{\gamma}_r) \quad , \tag{31c}$$

Eq.(12) reduces to

Figure 4 shows a block diagram illustrating the MST setup as currently used in the DFVLR-Institute of Aeroelasticity. The DFVLR MST facility is installed in a 20-ft. mobile container. The electronic part provides among other things about 400 measuring channels and up to 10 controllable electrodynamic exciters of various force levels.

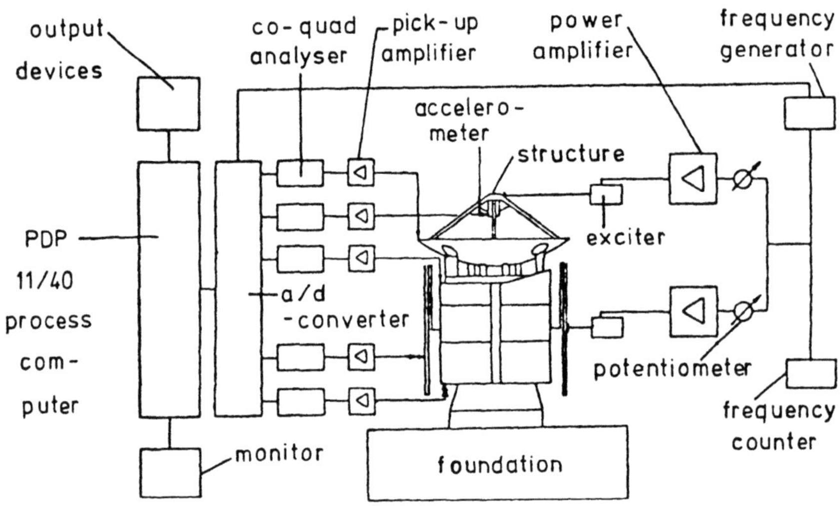

Figure 4: Block diagram of the MST test setup used at DFVLR.

4. NATKE'S METHOD

Natke's method, a typical representative of the so-called *phase separation technique*, consists of an experimental part followed by an analytical part. The experimental part provides the complex dynamic response (transfer function) from which the normal mode data are extracted in the analytical part by means of a sophisticated numerical approach. Natke's method,[3] which is especially dealt with in this section, uses only one exciter configuration kept invariant during one test run in the frequency range of interest. Furthermore, this method is unlike the phase

$$u = \sum_{r=1}^{n} \frac{\psi_r^N}{\lambda_r - \mu} \quad . \tag{32}$$

Applying matrix notation to Eq.(32) yields

$$u = \psi^N t \tag{33}$$

where ψ^N is an $m \times n$ matrix and t is a column matrix of order $n \times 1$:

$$t = \left\{ \frac{1}{\lambda_1 - \mu}, \frac{1}{\lambda_2 - \mu}, \ldots, \frac{1}{\lambda_n - \mu} \right\}^T \quad . \tag{34}$$

As illustrated in Figure 5 for the example of $|u|$, the dynamic response u due to the invariant force vector P is measured twice in the vicinity of every normal frequency $\bar{\omega}_r$, once at the frequency ω_r^α and once at ω_{n+r}^β , $(r = 1,2,\ldots,n)$.

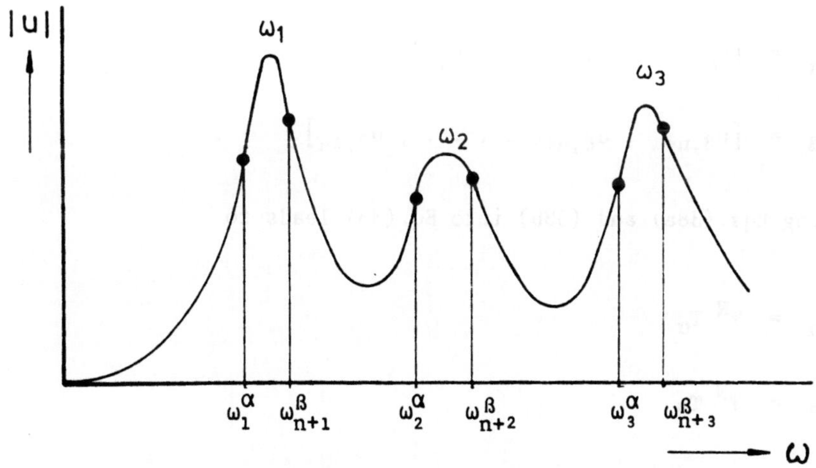

Figure 5: Measuring scheme - Natke's method.

The values of these frequencies, separated into the following two sets of n frequencies each

$$\omega_1^\alpha \; , \; \omega_2^\alpha \; , \; \ldots \; , \; \omega_n^\alpha \tag{35a}$$

$$\omega_{n+1}^\beta \; , \; \omega_{n+2}^\beta \; , \; \ldots \; , \; \omega_{2n}^\beta \tag{35b}$$

are defined by

$$\left.\begin{array}{ll} \omega_r^\alpha \sim \overline{\omega}_r & \\ \omega_{n+r}^\beta \sim \overline{\omega}_r & (r = 1,2,\ldots,n) \end{array}\right\} \tag{36}$$

but

$$\omega_r^\alpha \neq \omega_{n+r}^\beta \qquad (r = 1,2,\ldots,n) \quad . \tag{37}$$

We obtain in accordance with the two sets of measuring frequencies two groups of transfer function values $u_{\alpha r}$ and $u_{\beta,n+r}$ which can be arranged in the matrices

$$U_\alpha = [u_{\alpha 1} \; , \; u_{\alpha 2} \; , \; \ldots \; , \; u_{\alpha n}] \tag{38a}$$

$$U_\beta = [u_{\beta,n+1} \; , \; u_{\beta,n+2} \; , \; \ldots \; , \; u_{\beta,2n}] \quad . \tag{38b}$$

Inserting Eqs.(38a) and (38b) into Eq.(33) leads to

$$U_\alpha = \Psi^N \, T_\alpha \tag{39a}$$

$$U_\beta = \Psi^N \, T_\beta \tag{39b}$$

where

$$T_\alpha = [t_{\alpha 1} \; , \; t_{\alpha 2} \; , \; \ldots \; , \; t_{\alpha n}] \tag{40a}$$

$$T_\beta = [t_{\beta,n+1}\ ,\ t_{\beta,n+2}\ ,\ \ldots\ ,\ t_{\beta,2n}] \quad . \tag{40b}$$

On the condition that the exciting force vector P remains invariant, the matrix Ψ^N can be eliminated from Eq.(39) resulting in

$$U_\alpha\, T_\alpha^{-1} = U_\beta\, T_\beta^{-1} \quad . \tag{41}$$

Premultiplication of Eq.(41) by matrix U_α^T and subsequently by the inverse of $U_\alpha^T\ U_\alpha$ leads to

$$T_\alpha^{-1} = (U_\alpha^T\ U_\alpha)^{-1}\, U_\alpha^T\ U_\beta\, T_\beta^{-1} \tag{42}$$

from which T_β can be expressed explicitly as follows:

$$T_\beta = T_\alpha\, W \quad . \tag{43}$$

The procedure leading to Eq.(43) is well known in control theory under the name of the method of auxiliary variables. In Eq.(43)

$$W = (U_\alpha^T\ U_\alpha)^{-1}\, U_\alpha^T\ U_\beta \tag{44}$$

is a complex matrix of order $n \times n$ containing the measured dynamic response vectors u at the frequencies $\overset{\alpha}{\omega}_r$ and $\overset{\beta}{\omega}_{n+r}$. The matrix element at the intersection of the r-th row and the s-th column of matrix T_β has the form

$$\frac{1}{\lambda_r - \mu_{n+s}} = \frac{w_{1s}}{\lambda_r - \mu_1} + \frac{w_{2s}}{\lambda_r - \mu_2} + \ldots \quad . \tag{45}$$

Multiplying Eq.(45) by its common denominator results in the polynomial of order n :

$$\prod_{k=1}^{n} (\lambda_r - \mu_k) - (\lambda_r - \mu_{n+s}) \sum_{k=1}^{n} \prod_{\substack{\sigma \neq 1 \\ \sigma \neq k}}^{n} (\lambda_r - \mu_\sigma) w_{ks} = 0 \qquad (46)$$

$$(r,s = 1,2,\ldots,n) \quad .$$

Solving this polynomial iteratively in the vicinity of ω_r^α and ω_{n+r}^β leads to the eigenvalue λ_r . In accordance with Eq.(30) the normal frequency $\overline{\omega}_r$ and the damping loss angle $\overline{\gamma}_r = \overline{D}_r/\overline{K}_r$ can be determined from

$$\left.\begin{aligned} \overline{\omega}_r &= \sqrt{\lambda_r'} \\ \overline{\gamma}_r &= \frac{\lambda_r''}{\lambda_r'} \end{aligned}\right\} \quad . \qquad (47)$$

It is obvious that for the determination of any λ_r a number of $s = 1,2,\ldots,n$ polynomials is available, meaning that λ_r can be determined more accurately by taking the average of n different values.

The normal modes can be calculated alternatively from Eq.(39a) or (39b):

$$\psi^N = U_\alpha T_\alpha^{-1} \qquad (48a)$$

$$\psi^N = U_\beta T_\beta^{-1} \quad . \qquad (48b)$$

Let the arbitrarily normalized normal modes be expressed by

$$\Psi_r = \eta_r \Psi_r^N \qquad (49)$$

where η_r denotes a normalizing factor, then the dynamic response u can be expressed according to Eq.(32) by

$$u = \sum_{r=1}^{n} \frac{\Psi_r^N}{\lambda_r - \mu} \tag{50a}$$

or according to Eqs.(12) and (13) by

$$u = \sum_{r=1}^{n} \frac{(\Psi_r^N)^T \; P \; \Psi_r^T}{M_r^N \; (\lambda_r - \mu)} \quad . \tag{50b}$$

Setting Eq.(50a) equal to (50b) and comparing their coefficients results in the generalized mass

$$M_r^N = (\Psi_r^N)^T \quad P \quad , \tag{51}$$

the determination of which requires measurement of the exciting forces P. To obtain the generalized mass $\overline{M}_r$ relating to $\Psi_r = \eta \, \Psi_r^N$, Eq.(51) must be multiplied by η. Hence it follows that

$$\overline{M}_r^N = \eta(\Psi_r^N)^T \quad P \quad . \tag{52}$$

5. ANGÉLINI'S METHOD

Angélini's method[4] is evidently aiming at the modal data of the conservative undamped system which are identical to the modal data of the damped system without damping coupling. The basic relations of this method are described by Eqs.(2), (5a) and (6). Premultiplying Eq.(6) by q^T and taking into account that according to Eq.(2)

$$q^T \, \Phi^T = u^T \tag{53}$$

leads to

$$q^T(-\omega^2 M + K + jD)q = u^T P \quad , \tag{54}$$

where the damping coefficients in contrast to Eq.(6) are defined again in terms of the imaginary parts of complex stiffnesses. Suppose a set of n response vectors contained in the $m \times n$ matrix U is excited by a set of linearly independent force vectors arranged in the $m \times n$ matrix R, then Eq.(54) can be extended to

$$Q^T(-\omega^2 M + K + jD)Q = U^T R \quad , \tag{55}$$

where the $n \times n$ matrix Q contains the generalized coordinates q as columns. The n different exciter configurations have to be chosen prior to the test in such a way that all the normal modes in a frequency range of interest can be distinctly excited. The response matrix can be measured at any desired frequency ω.

Decomposing the matrices of Eq.(55) into Hermitian and anti-Hermitian terms results in

$$Q^T(-\omega^2 M + K)Q = H \tag{56}$$

$$jQ^T D Q = A \tag{57}$$

where the matrices H and A are the Hermitian and the anti-Hermitian of the measured matrix $U^T R$. The eigenvalues and eigenvectors of matrix H can be calculated as functions of ω. Condition $\omega = \omega_r$ is fulfilled if the corresponding eigenvector of matrix H becomes zero and the eigenvector is

$$Q(\omega_r)\, v_r = e_r = \begin{Bmatrix} 0 \\ \vdots \\ 1 \\ \vdots \\ 0 \end{Bmatrix} \quad . \tag{58}$$

Multiplication of the equation

$$U = \Phi Q \tag{59}$$

by v_r leads to

$$U(\omega_r)\, v_r = \Phi\, Q(\omega_r)\, v_r \tag{60}$$

which results as a consequence of Eq.(58) in the r-th normal mode vector

$$\Phi_r = U(\omega_r)\, v_r \quad . \tag{61}$$

Let a quadratic expression $S_1(\omega)$ be formulated as follows:

$$S_1(\omega) = v_r^T\, H(\omega)\, v_r \tag{62}$$

then the generalized mass M_r can be derived by differentiating $S_1(\omega)$ with respect to ω :

$$\left. \frac{dS_1(\omega)}{d\omega} \right|_{\omega = \omega_r} = -\, 2\omega_r\, M_r \quad . \tag{63}$$

The damping factor D_r can be determined from the quadratic expression

$$S_2(\omega) = v_r^T\, A\, v_r \tag{64}$$

of the anti-Hermitian A as follows:

$$\left. S_2(\omega) \right|_{\omega = \omega_r} = jD_r \quad . \tag{65}$$

Because of Eq.(64) only the diagonal elements of matrix D can be determined. Off-diagonal elements resulting from damping coupling are neglected.

6. WITTMEYER'S METHOD

The newest form of this method is published in Ref.[5]. Therefore, no reference must be made to two older papers by the same author. The derivation of this method starts at the fundamental relations (8) through (10) formulated in terms of the normal modes Φ_r of the undamped system. Consequently, damping coupling effects are neglected.

It is presumed that approximate values $\hat{\omega}_r$ for the normal frequencies ω_r as well as an estimated value $\hat{P}_r$ for the appropriate exciter configuration P_r according to Eqs.(19) through (21) are available. By applying force vector $\hat{P}_r$ at the frequencies $\omega_{r,\nu} = \sqrt{\mu_{r,\nu}}$, ($\nu = 1,2$; $\hat{\nu} = 3,4$; $\ell = 5$) in the vicinity of $\hat{\omega}_r = \sqrt{\hat{\mu}_r}$, an equivalent number of complex response vectors $u_{r,\nu}$ can be measured. The measuring scheme is shown in Figure 6 for the special case of 5 frequencies.

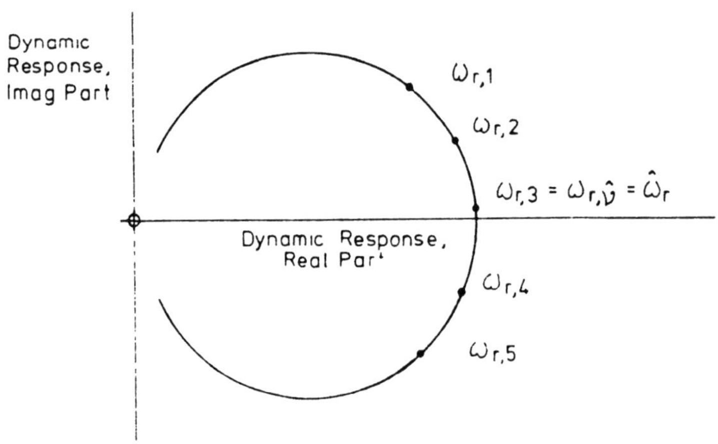

Figure 6: Measuring scheme - Wittmeyer's method; $\ell = 5$ measuring frequencies.

Premultiplying $u_{r,\nu}$ and $\hat{u}_r$ by an arbitrary vector Y^T yet to be defined and forming the quotients

$$R_{r,\nu} = \frac{Y^T u_{r,\nu}}{Y^T \hat{u}_r} \quad , \qquad (\nu = 1,2,\dots,\ell) \tag{66}$$

leads with Eqs.(8) through (10) to

$$R_{r,\nu} = \frac{Y^T \sum\limits_{s=1}^{n} \dfrac{\Phi_s^T \hat{P}_r}{M_s(\lambda_s - \mu_{r,\nu})} \Phi_s}{Y^T \sum\limits_{s=1}^{n} \dfrac{\Phi_s^T \hat{P}_r}{M_s(\lambda_s - \hat{\mu}_r)} \Phi_s} \quad , \qquad (\nu = 1,2,\dots,\ell) \; . \tag{67}$$

This equation can be simplified to

$$R_{r,\nu} = \frac{\dfrac{Y^T \Phi_r^T \hat{P}_r \Phi_r}{M_r(\lambda_r - \mu_{r,\nu})} + \varepsilon}{\dfrac{Y^T \Phi_r^T \hat{P}_r \Phi_r}{M_r(\lambda_r - \hat{\mu}_r)} + \varepsilon} \tag{68}$$

$$= \frac{\lambda_r - \hat{\mu}_r}{\lambda_r - \mu_{r,\nu}} \; \frac{1 + \varepsilon(\lambda_r - \mu_{r,\nu})}{1 + \varepsilon(\lambda_r - \hat{\mu}_r)} \quad , \qquad (\nu = 1,2,\dots,\ell)$$

because term ε is negligibly small if the exciting vector $\hat{P}_r$ excites mainly the normal mode Φ_r with only small portions of adjacent normal modes. Eq.(68) may be further transformed to

$$R_{r,\nu}\left(1 + \frac{\hat{\mu}_r - \mu_{r,\nu}}{\lambda_r - \hat{\mu}_r}\right) = 1 + \varepsilon(\hat{\mu}_r - \mu_{r,\nu}) \; , \qquad (\nu = 1,2,\dots,\ell) \tag{69}$$

containing the two unknowns ε and $\kappa = \lambda_r - \hat{\mu}_r$ which can be determined if a set of two values $R_{r,\nu}$ and $\hat{\mu}_r - \mu_{r,\nu}$ is measured. For $\ell > 2$, Eq.(69) leads to an overdetermination with respect to the terms κ and ε. In this case the method of least squares can be used advantageously. Then, the terms ω_r and γ_r can be determined as follows:

$$\omega_r = (\hat{\omega}_r^2 + \kappa')^{\frac{1}{2}} \tag{70a}$$

$$\frac{\gamma_r}{\omega_r^2} = \kappa'' \quad . \tag{70b}$$

Exciting the dynamic system at ω_r with the excitation vector $\hat{P}_r$ results in an approximate normal mode $\tilde{\Phi}_r \approx u_r$. After changing the damping law by replacing ωD_r in Eq.(9) by D_r, the dynamic response u_r can be expressed in accordance with Eq.(10) as follows:

$$u_r = \sum_{s=1}^{n} \frac{\Phi_s^T \hat{P}_r}{M_s(\lambda_s - \mu_r)} \Phi_s \tag{71}$$

where λ_r und μ_r are defined in the Eqs.(31b) and (31c). By separating Eq.(71) into two parts such that

$$u_r = \frac{\Phi_r^T \hat{P}_r}{M_r(\lambda_r - \mu_r)} \Phi_r + \sum_{\substack{s=1 \\ s \neq r}}^{n} \frac{\Phi_s^T \hat{P}_r}{M_s(\lambda_s - \mu_r)} \Phi_s \quad , \tag{72}$$

it can be shown that at $\omega = \omega_r$ the right-hand term of Eq.(72) converges to zero, thus

$$u_r \approx \Phi_r \frac{\Phi_r^T \hat{P}_r}{M_r \omega_r^2 \gamma_r} \tag{73}$$

if $\hat{P}_r$, ω_r and γ_r are close to the correct values. It is obvious that M_r is yet to be determined. Normalizing the normal modes on the condition that

$$\frac{\Phi_r^T \; P_r}{M_r \; \omega_r^2 \; \gamma_r} = 1 \tag{74}$$

results in the approximate generalized mass

$$\tilde{M}_r = \frac{u_r^T \; \hat{P}_r}{\omega_r^2 \; \gamma_r} \tag{75}$$

where

$$\begin{aligned} \tilde{\Phi}_r &\approx u_r \\ P_r &\approx \hat{P}_r \end{aligned} \quad . \tag{76}$$

The terms ω_r and γ_r can be taken from Eq.(70). The approach described above can also be used to find approximations for the modal data ω_s , γ_s , $\tilde{M}_s$ and $\tilde{\Phi}_s$ of the other normal modes $(s = 1,2,\ldots,n \; ; \; s \neq r)$, by means of which an improved normal mode $\overline{\Phi}_r$ can be found by changing (72) into

$$\overline{\Phi}_r = u_r - \sum_{\substack{s=1 \\ s\neq r}}^{n} \frac{u_s^T \; P_r}{\tilde{M}_s(\lambda_s - \mu_r)} u_s \quad . \tag{77}$$

Correspondingly, an improved value of the generalized mass can also be determined from

$$\overline{M}_r = \frac{\overline{\Phi}_r \; P_r}{\omega_r^2 \; D_r} \quad . \tag{78}$$

This process can be applied iteratively until the outcome no longer changes.

There remains the problem of finding a well-appropriated force vector P_r. This is rather simple in a case without so-called frequency clusters. By considering Eq.(71) it is almost self-explanatory that the best approximation for $\hat{P}_r$ is determined by

$$\hat{P}_r = A\,\Phi_r^T \quad . \tag{79}$$

However, if strong interaction occurs between two or more normal modes in a relatively small frequency band, more sophisticated approaches to find optimal force distributions are required. Exciting the system at the frequencies $\hat{\omega}_r = \sqrt{\hat{\mu}_r}$ and the neighboring frequency $\hat{\omega}_{r+1} = \sqrt{\hat{\mu}_{r+1}}$ by the excitation vector P_r yet to be determined results in the dynamic responses $\hat{u}_r$ and $\hat{u}_{r+1}$, see Eq.(71). Forming the power terms

$$P_r^T\ u_r(\lambda_r - \hat{\mu}_r) \tag{80}$$

$$P_r^T\ u_{r+1}(\lambda_r - \hat{\mu}_{r+1}) \tag{81}$$

and subtracting the second term from the first leads after some minor transformations to the expression

$$\left.\begin{aligned} G &= P_r^T \left(\hat{u}_r - \frac{\lambda_r - \hat{\mu}_{r+1}}{\lambda_r - \hat{\mu}_r}\ \hat{u}_{r+1}\right) \\ &= \frac{1}{\lambda_r - \hat{\mu}_r} \sum_{\substack{s=1 \\ s\neq r}}^{n} \frac{(\Phi_s^T\ P_r)^2}{M_s}\ \frac{(\hat{\mu}_r - \hat{\mu}_{r+1})(\lambda_r - \lambda_s)}{(\lambda_s - \hat{\mu}_r)(\lambda_s - \hat{\mu}_{r+1})} \end{aligned}\right\} \quad . \tag{82}$$

If P_r guarantees an optimum excitation of the r-th normal mode with only negligibly small contributions of the normal modes located nearby, the right-hand term of Eq.(82) becomes zero resulting in the eigenvalue problem

$$G = P_r^T (\hat{u}_r - \kappa \hat{u}_{r+1}) = 0 \tag{83}$$

where the eigenvalue is defined by

$$\kappa = \frac{\lambda_r - \hat{\mu}_{r+1}}{\lambda_r - \hat{\mu}_r} \quad . \tag{84}$$

Development of the excitation vector into the series expansion

$$P_{r+\ell-1} = z_{\ell 1} P_a + z_{\ell 2} P_b + z_{\ell 3} P_c + \ldots \quad ; \quad \ell = 1,2,3 \tag{85}$$

corresponds to the dynamic response

$$\hat{u}_{r+\ell-1} = z_{\ell 1} \hat{u}_{r,a} + z_{\ell 2} \hat{u}_{r,b} + z_{\ell 3} \hat{u}_{r,c} \quad ; \quad \ell = 1,2,3 \tag{86}$$

at the frequencies $\hat{\omega}_r$, $\hat{\omega}_{r+1}$ and $\hat{\omega}_{r+2}$. By inserting Eq.(86) into Eq.(83) one obtains an eigenvalue problem the solution of which results in the eigenvalues κ_ℓ and the eigenvectors z_ℓ . Then an improved value of the normal frequency can be obtained from

$$\lambda_r = \frac{\kappa_\ell \hat{\mu}_r - \hat{\mu}_{r+1}}{\kappa_\ell - 1} \tag{87}$$

and

$$\hat{\hat{\omega}}_r = \sqrt{\lambda_r} \quad . \tag{88}$$

Solution of the eigenvalue problem Eq.(83) delivers not only improved values $\hat{\hat{\omega}}_r$ and $\hat{\hat{P}}_r$ but also improved normal frequencies and excitation vectors of the adjacent normal modes.

7. CONCLUDING REMARKS

The MST methods described represent just a small though important section of the present state of the art within this special area of structural dynamics. The most experience is still made with the classical phase resonance method or improved derivatives of this approach, such as the DFVLR method. The considerable shortcomings this approach might have had in the past in comparison with newer concepts have largely been compensated for meanwhile. Especially regarding the problem of test time, the break-even point has been reached at least. Another essential advantage of the phase resonance concept is that the normal mode characteristics to be determined are measured immediately without any subsequent calculations. Thus, error accumulation due for instance to nonlinear effects are negligibly small. However one virtually inherent problem of this concept is caused by the requirement that the isolation of a normal mode generally requires accessibility to all structural points not only for the sensors but also for finding out an appropriate exciting force configuration.

As is well known, there may be structures such as aircraft engines, control devices, payload parts and fuel tanks to parts of which exciters cannot be attached. Thus, in some cases, inaccessibility may prevent the isolation of a pure normal mode. Natke's and Angélini's methods clearly overcome this problem by measuring only the complex dynamic response and subsequently evaluating it by analytical means. Angélini's method is formulated in terms of the normal modes of the dampingless conservative system. This rough approximation may entail serious inaccuracies in testing real structures with inherent damping coupling. In contrast, Natke's method is based on the complex normal modes of the damped system with damping coupling delivering a consistent set of normal mode data without any neglecting simplifications or approximations.

Wittmeyer's method, more conservative, i.e. closer to the classical phase resonance technique, also searches for the normal modes of the undamped system and requires structural accessibility to almost the same extent as the phase resonance method to find appropriate exciter configurations. Like the phase resonance technique but unlike Natke's and Angélini's methods, Wittmeyer's concept is able to cope with more immediately nonlinear effects.

Unlike the DFVLR approach as well as Angélini's and Wittmeyer's methods, Natke's method does not deliver the modal data of the conservative system, but the complex modal data of the real system usually characterized by damping coupling. The complex modal data cannot be used immediately for the verification and correction of finite element models. For this reason, results of Natke's method have to be transformed into the modal data of the corresponding conservative system.

Each of the concepts dealt with in this lecture has advantages and shortcomings. Considerable emphasis has been and has still to be placed on the elaboration of an optimum MST concept by combining not only the above-mentioned methods but also taking into account the tremendously large amount of worthwhile concepts and ideas in this ever-challending field of structural engineering.

8. REFERENCES

1. Breitbach, E., A semi-automatic modal survey test technique for complex aircraft and spacecraft structures. *Proceedings of the IIIrd ESRO-Testing-Symp.*, ESRO-SP-99, 518.
2. Breitbach, E., Neuere Entwicklungen auf dem Gebiet des Standschwingungsversuchs an Luft- und Raumfahrtkonstruktionen, *VDI-Bericht No.221*, 33, 1974.
3. Natke, H.G., Die Berechnung der Eigenschwingungskenngrößen eines gedämpften Systems aus den Ergebnissen eines Schwingungsversuchs in einer Erregerkonfiguration, *Jahrbuch 1971 der DGLR*, 98.
4. Angélini, J.J., Darras, B., Détermination des modes propres de l'avion RF 8 à partir d'un essai de vibration an sol avec excitation non appropriée, *Note Technique NT 1/1984 RY*, ONERA, August 1973.
5. Wittmeyer, H., Standschwingungsversuch einer Struktur mit Dämpfungskoppelung und Frequenznachbarschaft. *ZfW No.24*, 139, 1976.
6. Natke, H.G., Survey of European ground and flight vibration test methods, *Proc. of Aerospace Engineering and Manufacturing Meeting*, San Diego, Calif., Nov.29 - Dec.2, 1976.
7. Niedbal, N., A simplified ground vibration test procedure for sailplanes and light aircraft. *AGARD-CP-278*, 1979.

8. Niedbal, N., Obtaining normal mode parameters from modal survey tests, *Proc. of the XXX. Congress of the Inter'l Astron. Fed.*, Munich, Sept.17-22, 1979.

INDIRECT IDENTIFICATION METHODS I : ADJUSTMENT OF MATHEMATICAL MODELS BY THE RESULTS OF VIBRATIONS TESTS : USING EIGENSOLUTIONS

Dr. G. LALLEMENT, Prof.
University of Besançon, LABORATOIRE DE MECANIQUE APPLIQUEE, associé au CNRS
25030 - BESANCON CEDEX (France)

I - Introduction

This lecture is devoted to the study of some adjustment procedures for the models of mechanical structures. This adjustment is obtained by the results of the tests done on a prototype of the structure and is based on the comparison of the calculated and identified eigensolutions. Its aim is to improve the precision with which the state equation permit to predit the dynamical behaviour. This lecture is limited to the case of linear mechanical structures, asymptotically stable, which can be represented by discrete models and constant, symmetrical, simple structure state matrices. The extension to the case of non symmetrical matrices is possible but first of all imply, for some of the proposed methods, the identification of the two sets of right and left eigenvectors.[1,2]

2 - Notes on the characteristics of the eigensolutions

We consider a discretised linear mechanical structure admitting as a model :

$$My^{\bullet\bullet}(t) + By^{\bullet}(t) + [K+jK'] \, y(t) = f(t) \,, \tag{1}$$

where : $y(N\mathrm{x}1)$ is the generalized displacement vector ; $f(t) = fe^{j\alpha t}$ the generalized force vector of the external excitation ; M,B,K,K' : constant, real, symmetric, NxN matrices with : M and K positive definite, B and K' non negative. For the autonomous system, the solutions are of the form : $y(t) = y\,e^{st}$ and we designate by s_ν the eigenvalue and by y_ν the associated eigenvector, $1 \leqslant \nu \leqslant 2N$. These eigensolutions are regrouped in the diagonal complex spectral matrix $S(2N\mathrm{x}2N) = \mathrm{diag}\,\{s_\nu\}$ and in the complex modal matrix $Y\,(N\mathrm{x}2N)$ such that : $Ye_\nu \triangleq y_\nu$ where e_ν represent the νth column of the unit matrix E_{2N}. It can be proved that the state, spectral and modal matrices satisfy :

$$M\,Y\,S^2 + B\,Y\,S + \{K+jK'\}\,Y = 0 \,, \tag{2}$$

$$S\,Y^T\,MY + Y^T\,MYS + Y^T\,BY = N \,, \tag{3}$$

$$S\,Y^T\,MYS - Y^T\,[K+jK']\,Y = NS \,, \tag{4}$$

where : Y^T is the transpose of Y ; $N(2N\mathrm{x}2N)$ a diagonal matrix determined by the choice of the eigenvectors norms. Therefore we suppose that the considered mechanical system is, in the mathematical sense, of a simple structure and has 2N different eigenvalues.

3 - Direct problem and perturbations equations.

We suppose that the associated conservative structure has been represented by a conservative discret model obtained by a finite element discretisation and is characterized by the square (NxN), constant, symmetric, positive definite matrices of inertia $M^{(c)}$ and stiffness $K^{(c)}$ and satisfies in free vibrations :

$$M^{(c)}\,y^{\bullet\bullet}(t) + K^{(c)}\,y(t) = 0 \,. \tag{5}$$

We regroupe the set of the N eigensolutions of : $[K^{(c)} - \lambda M^{(c)}]y = 0$ in the real spectral diagonal matrix $\Lambda^{(c)}\,(N\mathrm{x}N) = \mathrm{diag}\,\{\lambda_\nu^c\}$ and in the real modal matrix $Y^{(c)}\,(N\mathrm{x}N)$. These matrices are related together by :

$$K^{(c)}Y^{(c)} - M^{(c)}Y^{(c)}\Lambda^{(c)} = 0 \,, \tag{6}$$

$$Y^{T(c)} M^{(c)} Y^{(c)} = E, \tag{7}$$

$$Y^{T(c)} K^{(c)} Y^{(c)} = \Lambda^{(c)}. \tag{8}$$

The matrices $M^{(c)}$ and $K^{(c)}$ represent the initial estimation of the unknown matrices M and K introduced in (1). By definition we have : $M \triangleq M^{(c)} + \Delta M$; $K \triangleq K^{(c)} + \Delta K$, and we admit that : $\|\Delta M\| \ll \|M^{(c)}\|$; $\|\Delta K\| \ll \|K^{(c)}\|$. The non conservative forces $B\dot{y}$ and $K'y$ are supposed "small" compared to the conservative forces. The autonomous system of (1) is consider as the system (5) perturbated by the four matrices of perturbations ΔM ; ΔK ; B ; K' which are about the size of the scalar ε, $\varepsilon \in \langle 0;1 \rangle$; $\varepsilon \ll 1$ is considered as a small parameter. In order to establish the relations between the eigensolutions of (1) and (5) let :

$$S \triangleq V + S^{(0)} + jW, \tag{9}$$

$$Y \triangleq Y^{(0)} + \underline{R} + jU \triangleq Y^{(0)} \left[E + \underline{A} + jQ\right], \tag{10}$$

where : $\underline{R}$ and U are real Nx2N matrices ; $\underline{A}$ and Q are real, diagonal in sub-matrices, 2Nx2N matrices ; V and W real, diagonal of order 2N. These matrices are partitioned in :

$$Y = \begin{bmatrix} {}_1Y & {}_2Y \end{bmatrix}; \quad Y^{(0)} = \begin{bmatrix} Y^{(c)} & Y^{(c)} \end{bmatrix}; \quad \underline{R} = \begin{bmatrix} {}_1R & {}_2R \end{bmatrix}; \quad U = \begin{bmatrix} {}_1U & {}_2U \end{bmatrix};$$

$$S = \begin{bmatrix} {}_1S & 0 \\ 0 & {}_2S \end{bmatrix}; \quad S^{(0)} = \begin{bmatrix} j\Lambda^{(c)1/2} & 0 \\ 0 & -j\Lambda^{(c)1/2} \end{bmatrix}; \quad V = \begin{bmatrix} {}_1V & 0 \\ 0 & {}_2V \end{bmatrix}; \quad W = \begin{bmatrix} {}_1W & 0 \\ 0 & {}_2W \end{bmatrix} \tag{10'}$$

$$\underline{A} = \begin{bmatrix} {}_1A & 0 \\ 0 & {}_2A \end{bmatrix}; \quad Q = \begin{bmatrix} {}_1Q & 0 \\ 0 & {}_2Q \end{bmatrix},$$

where all the introduced sub-matrices are square of order N.

Using the equations (2), (6), (7) and (8), neglecting the second order small quantities in ε^2 and after premultiplication by $Y^{T(c)}$ we obtain :

$$\Delta\tilde{K} - \Delta\tilde{M}\,\Lambda^{(c)} + \Lambda^{(c)}\,{}_1A - {}_1A\,\Lambda^{(c)} - 2\Lambda^{(c)1/2}\,{}_1W = 0, \tag{11}$$

$$\Delta\tilde{K} - \Delta\tilde{M}\,\Lambda^{(c)} + \Lambda^{(c)}\,{}_2A - {}_2A\,\Lambda^{(c)} + 2\Lambda^{(c)1/2}\,{}_2W = 0, \tag{12}$$

$$\tilde{K}' + \tilde{B}\,\Lambda^{(c)1/2} + \Lambda^{(c)}\,{}_1Q - {}_1Q\,\Lambda^{(c)} + 2\Lambda^{(c)1/2}\,{}_1V = 0. \tag{13}$$

$$\tilde{K}' - \tilde{B}\Lambda^{(c)1/2} + \Lambda^{(c)}{}_2Q - {}_2Q\Lambda^{(c)} - 2\Lambda^{(c)1/2}{}_2V = 0\ , \tag{14}$$

where : $\tilde{Z} \triangleq Y^{T(c)}\ Z\ Y^{(c)}$. These relations lead to the following comments: a) the transition of the relations in $index_1$ to those in $index_2$ is done by replacing $\Lambda^{(c)1/2}$ to $-\Lambda^{(c)1/2}$; b) the matrices ${}_iW$; ${}_iR$; ${}_iA$, i = 1,2 only depend on ΔK and ΔM whereas the matrices ${}_iV$; ${}_iU$ and ${}_iQ$, i = 1,2 only depend on B and K' ; c) the equations (11) to (14) determine ${}_iV$; ${}_iW$ and the extra diagonal elements of the ${}_iA$ and ${}_iQ$ matrices, i = 1,2. Moroever : ${}_2W = -{}_1W$; Extra diag. $[A_1]$ = Extra diag. $[A_2]$. In ${}_iA$ and ${}_iQ$ matrices, i = 1,2, the diagonal elements ${}_ia_{\nu\nu}$ and ${}_iq_{\nu\nu}$ depend on the choice of the eigenvector ${}_iy_\nu$ norm. We choose a norm which coincide with the norm introducedin (7) when B = 0 = K' :

${}_iy_\nu^T\ M\ {}_iy_\nu = 1$, from which : ${}_ia_{\nu\nu} = -\frac{1}{2}\Delta\tilde{M}_{\nu\nu}$; ${}_iq_{\nu\nu} = 0$; i = 1,2 ; $1 \leqslant \nu \leqslant N$.

Therefore we obtain : ${}_1A = {}_2A \triangleq A$; ${}_1R = {}_2R \triangleq R$ and for the norm matrix N(2Nx2N) : ${}_1N = -{}_2N = [{}_1V - {}_2V] + 2j\ [\Lambda^{(c)1/2} + {}_1W]$. (15)

By introducing the relations (9)(10) and (15) into the orthonormality relations (3),(4) we obtain, taking into account (7) and (8), 12 equations completing the 4 equations (11) to (14). These 16 equations are dependent and we will only give here the main equations divided in two groups of three relations :

- Group corresponding to the conservative design variables :

$$\Delta KY^{(c)} - \Delta MY^{(c)}\Lambda^{(c)} = L, \text{ where : } L \triangleq M^{(c)}R\Lambda^{(c)} - K^{(c)}R + 2M^{(c)}Y^{(c)}\Lambda^{(c)1/2}{}_1W, \tag{16}$$

$$A + A^T + \Delta\tilde{M} = 0\ , \tag{17}$$

$$\Lambda^{(c)}A + A^T\Lambda^{(c)} + \Delta\tilde{K} = 2\Lambda^{(c)1/2}{}_1W\ . \tag{18}$$

- Group corresponding to the non-conservative design variables

$$2B\ Y^{(c)}\Lambda^{(c)1/2} = {}_1T - {}_2T\ , \tag{19}$$

$$2K'\ Y^{(c)} = {}_1T + {}_2T\ , \tag{20}$$

where : ${}_iT \triangleq M^{(c)}\ {}_iU\Lambda^{(c)} - K^{(c)}\ {}_iU + (-1)^i\ 2M^{(c)}Y^{(c)}\Lambda^{(c)1/2}\ {}_iV$,

$$2\tilde{B}\ \Lambda^{(c)1/2} = -2\Lambda^{(c)1/2}\left[{}_1V + {}_2V\right] + \left[{}_1Q\Lambda^{(c)} - \Lambda^{(c)}{}_1Q\right] - \left[{}_2Q\Lambda^{(c)} - \Lambda^{(c)}{}_2Q\right] \tag{21}$$

$$2\tilde{K}' = -2\Lambda^{(c)1/2}\left[{}_1V - {}_2V\right] + \left[{}_1Q\Lambda^{(c)} - \Lambda^{(c)}{}_1Q\right] + \left[{}_2Q\Lambda^{(c)} - \Lambda^{(c)}{}_2Q\right] \tag{22}$$

These relations give the solution of the direct perturbation problem. They have been developped in the general case. In practice only one type of non conservative terms ($B \neq 0$; $K' = 0$ or $B=0$; $K' \neq 0$) will be introduced to prevent redondancies between those unknown quantities. According to that choice, simplifications are then introduced in the general relations.

4 - Definition of the model adjustment problem (inverse problem of the above paragraph).

The resulting data of the calculations and of the measurements are :
- the initial mathematical model $M^{(c)}$; $K^{(c)}$ and its $n(n \ll N)$ first eigensolutions : $\Lambda_1^{(c)}$ (nxn) ; $Y_1^{(c)}$ (Nxn),
- the 2n first eigensolutions identified on the structure : ${}_iS_1$ (nxn) ; ${}_iY_1$ (Nxn), i = 1,2. We will consider later the case where ${}_iY_1$ is partially identified.

From these data, we want to calculate the matrices ΔM ; ΔK ; B ; K' in a way that the non conservative model characterized by the state matrices : $M^{(m)} = M^{(c)} + \Delta M$; $K^{(m)} = K^{(c)} + \Delta K$; B ; K' admits as eigensolutions those which have been identified. In most cases, this inverse problem is strongly under determined. The two groups of equations (16) to (18) and (19) to (22) show that, neglecting second order terms, we discern a separation into two independent inverse sub-problems which will be treated by similar technique. Independently of this fact, we will treat only the adjustment problem related to the conservative design variables.

5 - Adjustment of the conservative design variables based on the matrices $\Delta\Lambda$ and A and on the spectral and modal sensitivity.

Starting from equations (9), (10), notations (10'), taking into account ${}_1W = - {}_2W$; ${}_1R = {}_2R \triangleq R$ and neglecting the second order small quantities in ε^2, the eigensolutions of the conservative structure associated to the real one (abr : CSA) are represented by :

$$R({}_1S^2)=R({}_2S^2)=-[\Lambda^{(c)}+2\Lambda^{(c)\,1/2}\,{}_1W] \triangleq [\Lambda^{(c)}+\Delta\Lambda\ ] \triangleq -\Lambda^{(m)}, \tag{23}$$

$$R({}_1Y\ ) = R({}_2Y\ ) \triangleq Y^{(c)} + R \triangleq Y^{(c)}+\Delta Y \triangleq Y^{(m)}. \tag{24}$$

With these simplified notations (23) and (24), the complete set of the N eigensolutions $\Lambda^{(m)}$ (NxN) ; $Y^{(m)}$ (NxN) of the CSA satisfies :

$$[M^{(c)}+\Delta M]\ Y^{(m)}\Lambda^{(m)} = [K^{(c)}+\Delta K]\ Y^{(m)}, \tag{25}$$

$$Y^{T(m)}[M^{(c)}+\Delta M]\ Y^{(m)} = E\ , \tag{26}$$

$$Y^{T(m)}[K^{(c)}+\Delta K]\ Y^{(m)} = \Lambda^{(m)}. \tag{27}$$

Then, we obtain for the three relations (16) to (18) :

$$\Delta K\ Y^{(c)}-\Delta M\ Y^{(c)}\Lambda^{(c)} = L;\quad \text{where : } L \triangleq M^{(c)}\Delta Y\Lambda^{(c)}-K^{(c)}\Delta Y+M^{(c)}Y^{(c)}\Delta\Lambda \tag{28}$$

$$A + A^T + \Delta\tilde{M} = 0\ ;\ \Lambda^{(c)}A + A^T\,\Lambda^{(c)} + \Delta\tilde{K} = \Delta\Lambda, \qquad (29),(30)$$

where : $\Delta Y \triangleq Y^{(m)} - Y^{(c)} \triangleq Y^{(c)}\ A$.

We introduce the n identified eigensolutions $\Lambda_1^{(m)}$ (nxn) ; $Y_1^{(m)}$ (Nxn) by the partitioning in submatrices :

$$Y^{(k)}(N\text{x}N) = [Y_1^{(k)} \,\vdots\, Y_2^{(k)}]\ ;\ \Delta Y(N\text{x}N) = [\Delta Y_1 \,\vdots\, \Delta Y_2]\ ;$$

$$\Lambda^{(k)}(N\text{x}N) = \begin{bmatrix} \Lambda_1^{(k)} & 0 \\ 0 & \Lambda_2^{(k)} \end{bmatrix}\ ;\quad A(N\text{x}N) = \begin{bmatrix} A_{11} & A_{12} \\ A_{21} & A_{22} \end{bmatrix}\quad \text{where : } k = e,m; A_{11}(n\text{x}n)$$

is based on the three last equations and the three following stages :

1st stage : Matching of eigensolutions[3,4] : association by pair of the "similar" calculated and identified eigensolutions. Then we arrange them in the same column order in the submatrices with index $[\ \ ]_1^{(c)}$ and $[\ \ ]_1^{(m)}$.

2nd stage : Reduction of the under determination of the adjustment problem by localization of the dominant submatrices within ΔM and ΔK.

3rd stage : Adjustment of the conservative design variables of the initial model intervening only in the previous dominant submatrices. That means the calculation of the matrices ΔM ; ΔK which satisfies in the least squares sense the relations (25) to (27).

5.1 - Localisation of the dominant modelling errors. We use in a qualitative way the following equations : (28) for the simultaneous localization of the dominant submatrices within ΔM and ΔK ; (29) for the localization within ΔM ; (30) for the localization within ΔK. In (29) and (30) we introduce the approximation : $A_{12} = 0 = A_{21}$, implying the representation of $Y_1^{(m)}$ by its projection on the modal subbasis $Y_1^{(c)}$:

$$Y_1^{(m)} \approx Y_1^{(c)} \; [E+A_{11}] \tag{31}$$

With the above approximation, the relations (29) and (30) can be written :

$$\Delta M \; Y_1^{(c)} = L_2, \text{ where : } L_2 \triangleq - M^{(c)} \; Y_1^{(c)} \; [A_{11}+A_{11}^T], \tag{32}$$

$$\Delta K \; Y_1^{(c)} = L_3, \text{ where : } L_3 \triangleq M^{(c)} \; Y_1^{(c)} \; [\Delta\Lambda_1-(\Lambda_1^{(c)}A_{11} + A_{11}^T\Lambda_1^{(c)})]. \tag{33}$$

To these two equations, we associate the equation obtained from (28):

$$\Delta KY_1^{(c)}-\Delta MY_1^{(c)}\Lambda_1^{(c)}=L_1 \text{ where: } L_1\triangleq M^{(c)}\Delta Y_1\Lambda_1^{(c)}-K^{(c)}\Delta Y_1+M^{(c)}Y_1^{(c)}\Delta\Lambda_1 \tag{34}$$

The three equations (32) to (34) are mutually dependent (equation (34) can be got from (32) and (34) using the relation $\Delta Y_1=Y_1^C A_{11}$). They have the same general pattern : $\Omega_i \emptyset_i = L_i$, i = 1,2,3, and will be used for the qualitative localization of the dominant submatrices in ΔM and ΔK. Generally, it correspond to a dominant submatrix in Ω_i a group of lines relating to the same dof also dominant in the matrix L_i (fig.a). Conversely, if we detect in the matrix L_i the dominant (or group of) lines we can localize by this manner the dominant submatrices in Ω_i, i = 1,2,3 (it means in ΔM, ΔK or ΔM and ΔK). As an example (fig.b) for two dominant groups of lines in L_2 we can deduce, taking into account of the symmetry of ΔM, that can exists four dominant submatrices in ΔM.

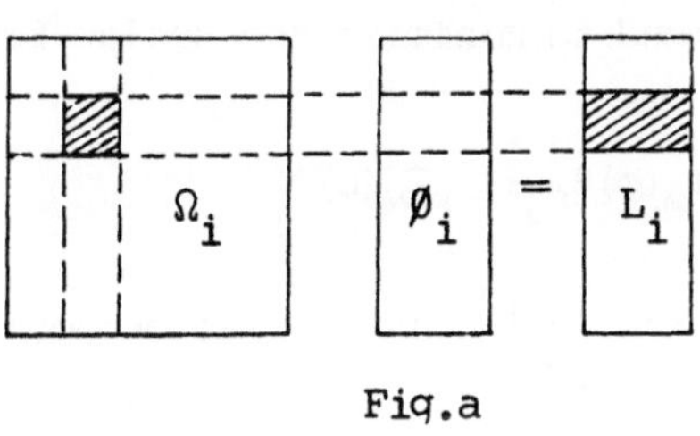

Fig.a

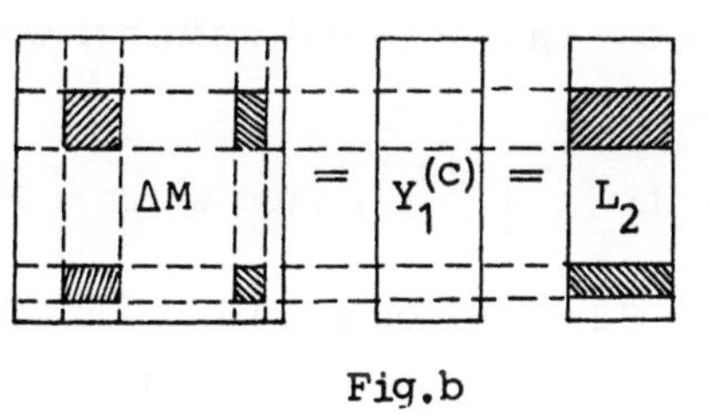

Fig.b

In practice we construct from a weighted sum of the n columns of L_i a vector l_i(Nx1). The components of this vector are representative of the relative importance of the N lines of the matrix L_i^4. The design variables intervening in a submatrix, that means in a group of dof, will be only taken into account in the adjustment procedure developped in 5.2 if the components l_{ji} of those dof are greater than an arbitrary chosen limit. We examine now the calculation of the matrix A_{11}(nxn).

When the modal submatrix $Y_1^{(m)}$ (Nxn) is totally identified, we deduce from orthonormality relations (7) and (8) : $A_{11}=Y_1^{T(c)}M^{(c)}Y_1^{(m)} - E$ (35) This A_{11} matrix simultaneously minimize the quadratic forms $\varepsilon_\nu^T M^{(c)} \varepsilon_\nu$ and $\varepsilon_\nu^T K^{(c)} \varepsilon_\nu$ where the vector of error : $\varepsilon_\nu = [\Delta Y_1 - Y_1^{(c)} A_{11}]\, e_\nu$, $1 \leqslant \nu \leqslant n$. When $Y_1^{(m)}$ is only partially identified and known by c(N>c>n) of its rows, we have to solve the over-determined problem : $\Delta Y_1^* = Y_1^{(c)*} A_{11}$, where the matrices $Y_1^{(c)*}$ (cxn) and ΔY_1^*(cxn) are only constructed from the c identified dof. We look for a pseudosolution of this equation minimizing the matrix of error $\varepsilon = \Delta Y_1^* - Y_1^{(c)*} A_{11}$. It satisfies: $A_{11} = [Y_1^{(c)*}]^{\dagger} \Delta Y_1^*$, where the notation † indicates the pseudo-inverse of the considered matrix. By definition, we have for the following : $F \triangleq Y_1^{(c)*}$.

- The classical least square method leads to $F^\dagger$ such that : $\|\varepsilon_\nu\|^2$, where $\varepsilon_\nu \triangleq \varepsilon e_\nu$, $1 \leqslant \nu \leqslant n$, is minimum and this problem is written :

$$\mathrm{Min}_{A_{11}}\ g = \mathrm{tr}\,[\varepsilon^T \varepsilon], \text{ and leads to : } F^\dagger = [F^T F]^{-1} F^T. \qquad (36)$$

If the matrix F of rank n is numerically ill conditioned, we recommend to apply the Singular Value Decomposition method.

- In order to maintain a certain homogeneity with (35) and to maintain the same physical meaning, we recommend to minimize the scalar function :

$$g' = \mathrm{tr}\,[\varepsilon^T M^{(c)*} \varepsilon], \text{ from which : } F^\dagger = [F^T M^{(c)*} F]^{-1} F^T M^{(c)*}.$$

- Simultaneous minimization of $\|\varepsilon_\nu\|^2$ and $\|a_\nu\|^2$, $1 \leqslant \nu \leqslant n$. The objective function to minimized is : $g = \mathrm{tr}\,[\varepsilon^T \varepsilon] + \alpha\,\mathrm{tr}\,[A_{11}^T A_{11}]$, and we obtain :

$F^{\dagger} = [F^T F + \alpha E]^{-1} F^T$. By a progressive increase of the positive scalar α, we detect the unknown quantities which rapidly decrease to zero. This technique is equivalent to the regularisation process of a numerically ill conditioned matrix $F^T F$.

- A better representation of the identified matrix ΔY_1^* can be obtained by choosing m eigenvectors from $Y^{(c)}$ with : $n<m\leq c$ (larger sub-basis of projection). Then, we solve in the least squares sense the equation :

$$\Delta Y_1^* = Y_3^{(c)*} A' \quad \text{where} \quad : \quad Y_3^{(c)*} \ (c\times m) \text{ and } A'(m\times n) \ .$$

In that case, the approximation of ΔY_1 which is introduced in the right member of (34) is : $\Delta Y_1 (N\times n) = Y_3^C A'$.

Consequently, (34) is not strictly linearly dependant of (32) and (33).

5.2 - Adjustment of the conservative design variables by an inverse sensitivity method. Let p (qx1) represent the design variable vector. A variation Δp_i of the parameter p_i leads to the new eigensolutions matrices :

$$\Lambda^{(c)} + \Lambda_{,i} \Delta p_i + O\,(\Delta p_i)^2 \ ; \ Y^{(c)} + Y_{,i} \Delta p_i + O\,(\Delta p_i)^2 \ .$$

By partial differentiation of (7) and (8) with respect to p_i and the introduction of the change of basis : $Y_{,i} = Y^{(c)}\, {}_iA$, we obtain two sensitivity equations similar to (29) and (30) :

$${}_iA + {}_iA^T + Y^{T(c)} M_{,i} Y^{(c)} = 0 \ ; \ \Lambda^{(c)}\, {}_iA + {}_iA^T \Lambda^{(c)} + Y^{T(c)} K_{,i} Y^{(c)} = \Lambda_{,i} .$$

As previously, we are limited to n eigensolutions and the approximations introduced once again consist in only taking into account the submatrix ${}_iA_{11}$ (nxn). Neglecting the second order terms, the variation of the νth eigensolution is :

$$\Delta\lambda_\nu = \lambda_{\nu,i} \Delta p_i \ ,$$

$$\Delta y_\nu = Y^{(c)}\, {}_ia_\nu \Delta p_i = (Y_1^{(c)}\, {}_ia_{1\nu} + Y_2^{(c)}\, {}_ia_{2\nu}) \Delta p_i , \quad 1 \leq \nu \leq n. \qquad (37)$$

By comparison of the last equation with the eigenvector difference :

$\Delta y_\nu = Y^{(c)} A\, e_\nu \stackrel{\Delta}{=} Y^{(c)} a_\nu = Y_1^{(c)} a_{1\nu} + Y_2^{(c)} a_{2\nu}$, we deduce :

$a_\nu = {}_i a_\nu \, \Delta p_i$ or, from the previous partitioning and the hypothesis :

$$A_{21} \simeq 0 \quad : \qquad a_{1\nu} \simeq {}_i a_{1\nu}\, \Delta p_i, \quad 1 \leqslant \nu \leqslant n. \tag{38}$$

Then we regroup the n scalar equations (37) and the n vectorial equations (38) and write them in such a way that we only have no dimensional quantities :

$$z = S_i \frac{\Delta p_i}{p_i},$$

where : $z((n^2+n)\times 1) = \{\frac{\Delta\lambda_1}{\lambda_1}\ ;\ \ldots\ ;\ \frac{\Delta\lambda_\nu}{\lambda_\nu}\ ;\ \ldots\ ;\ \frac{\Delta\lambda_n}{\lambda_n}\ ; a_1^T ; \ldots ; a_\nu^T ; \ldots ; a_n^T\}^T$,

$S_i((n^2+n)\times 1) = p_i \{\frac{\lambda_{1,i}}{\lambda_1^{(c)}}\ ;\ldots;\ \frac{\lambda_{\nu,i}}{\lambda_\nu^{(c)}}\ ;\ldots;\ \frac{\lambda_{n,i}}{\lambda_n^{(c)}} ; {}_i a_1^T\ ;\ldots;\ {}_i a_\nu^T\ ;\ldots;\ {}_i a_n^T\}^T$.

For q design variables, changing the vector p to p + Δp, the corresponding variation of the νth eigensolution is defined by :

$$\Delta\lambda_\nu = \sum_{i=1}^{q} p_i\, \lambda_{\nu,i} \frac{\Delta p_i}{p_i}\ ;\ a_{1\nu} = \sum_{i=1}^{q} p_i\ {}_i a_{1\nu} \frac{\Delta p_i}{p_i},\ 1 \leqslant \nu \leqslant n.$$

These n scalar equations and n vectorial equations constitute the direct sensitivity equation : $z = S\, v$, (39)

where : $S((n^2+n)\times q) = [S_1 ; \ldots ; S_i ; \ldots ; S_q]$; $v(q\times 1) = \{\frac{\Delta p_1}{p_1} ; \ldots ; \frac{\Delta p_i}{p_i} ; \ldots ; \frac{\Delta p_q}{p_q}\}^T$

5.3 - Determination of the solution v. Three cases must be considered.

a) The under-determined case[5,6] ($q > n(n+1)$) corresponds to an inaccurate localization. Among the infinite number of solutions of (39), we generally admit to choose the smallest in the euclidean norm sense Therefore, the minimization problem is : Min.$g = \|v\|^2$ satisfying the $n(n+1)$ scalar constraint equations : $z - Sv = 0$. All the elements are real and S supposed to be of rank $n(n+1)$; the solution is :

$$v = S^T [S\, S^T]^{-1}\, z \tag{40}$$

b) The case : $q = n(n+1)$ leads to a unique solution if S is invertible.

When one (or few) too weakly sensitive design variable p_j has been selected, the realization of the solution v may present some difficulties. By investigating the norm of each column of the matrix S we can select this (or these) design variable and reject it from v. Then the problem becomes over-determined and is treated in (c).

c) For the over determined case ($q < n(n+1)$), one of the different methods proposed for the calculation of A_{11} can be used. We recommend the formulation : $\text{Min}_v\ g = \varepsilon^T P^2 \varepsilon + \alpha v^T v$ where : P is diagonal, positive definite, square of order $n(n+1)$; its elements p_ν are chosen according to the relative importance of each eigensolution : α ($\alpha > 0$) is an arbitrary chosen real scalar. The useful range for α can be determined from the conditionning number of $S^T P^2 S$.

The solution is : $$(v)_\alpha = [S^T P^2 S + \alpha E]^{-1} S^T P^2 z \tag{41}$$

After the selection of the interval of variation of α and of the step $\Delta\alpha$, we calculate the solutions $(v)_\alpha$, then the vectors $(z)_\alpha = S(v)_\alpha$. Among this set, we select the best solution $(v)_{\alpha^*}$. This choice consists in a compromise between the accuracy of the adjustment (given by the norm of the vector $(z)_{\alpha^*}-z$) and the amplitude of the design variable modifications (given by the norm of v_{α^*}). When the value of α^* is chosen, we have to calculate the $(\Delta M)_{\alpha^*}$ and $(\Delta K)_{\alpha^*}$ matrices :

$$(\Delta M)_{\alpha^*} = \sum_{i=1}^{q} p_i \frac{\partial M^{(c)}}{\partial p_i} \left(\frac{\Delta p_i}{p_i}\right)_{\alpha^*} \ ; \ (\Delta K)_{\alpha^*} = \sum_{i=1}^{q} p_i \frac{\partial K^{(c)}}{\partial p_i} \left(\frac{\Delta p_i}{p_i}\right)_{\alpha^*} \tag{42}$$

5.4 - Criticism of the method. a) Precise adjustment by taking as parameters the physical design variables p_i. b) Possibilities of iteration of the method. c) Selection of the areas imprecisely modeled permitting a convergence to the right solution. d) Evaluation of the respective contribution of ΔM and ΔK with the A and $\Delta\Lambda$ matrices by the application of the linear superposition principle :

In $\Delta\Lambda$: $\Delta\lambda_\nu$ due to $\Delta M = 2\lambda_\nu^{(c)} a_{\nu\nu}$; then, $\Delta\lambda_\nu$ due to ΔK by difference.

In A : $A = A$ (due to ΔM) $+ A$ (due to ΔK) $\triangleq A^{(\Delta M)} + A^{(\Delta K)}$, as elements of $A^{(\Delta M)}$: $a_{\nu\nu}^{(\Delta M)} = a_{\nu\nu}$; $a_{\nu\sigma}^{(\Delta M)} = \dfrac{\lambda_\sigma^{(c)}\,[a_{\nu\sigma}+a_{\sigma\nu}]}{\lambda_\sigma^{(c)} - \lambda_\nu^{(c)}}$, $\nu\neq\sigma$

as elements of $A^{(\Delta K)} = -A^{T(\Delta K)}$: $a_{\nu\nu}^{(\Delta K)}=0$; $a_{\nu\sigma}^{(\Delta K)} = \dfrac{\lambda_\nu^{(c)} a_{\nu\sigma}+\lambda_\sigma^{(c)} a_{\sigma\nu}}{\lambda_\nu^{(c)} - \lambda_\sigma^{(c)}}$, $\nu\neq\sigma$

e) difficulties for the evaluation of ${}_iA_{11}$ in the case of closely spaced eigenvalues. f) Important influence of identified eigenvector norm errors due to parametrical non-linearities in damping terms. If we characterize these errors by the unknown diagonal matrix ε(nxn), we introduce them by changing $Y_1^{(m)}$ to : $Y'^{(m)}_1 = Y_1^{(m)} D$, where : $D = E + \varepsilon$.

It can be shown that the matrices of correction due to these norm errors are : $\Delta M = J\,[D^{-2} - E\,]\,J^T$; $\Delta K = J\,[D^{-2}-E\,]\,\Lambda_1^{(m)} J^T$, where : $J \triangleq M^{(c)} Y_1^{(m)}$. These norm errors lead to an erroneous model taking into consideration that the diagonal elements ε_ν may be of the order $\varepsilon_\nu \simeq 0.5$. For $\varepsilon\neq 0$, we introduce the change of basis :

$Y'^{(m)}_1 = Y_1^{(c)}\,[E + A'_{11}] + Y_2^{(c)}\,A'_{21}$, and the relations between A'_{ij} and A_{ij} are : $A'_{11} = A_{11}\,[E+\varepsilon]+\varepsilon$; $A'_{21} = A_{21}\,[E+\varepsilon]$.

* For $\varepsilon_\nu \leqslant 0.2$, $1 \leqslant \nu \leqslant n$, the diagonal elements of A'_{11} are seriously erroneous and cannot be used more. The only useful elements are the extra diagonal elements of A_{11} and A_{21}.

* For $\varepsilon_\nu > 0.2$ the only useful equations are in (39) those concerning the differences of eigenvalues. The methods proposed in the next paragraph can then be applied.

6 - Adjustment of conservative design variables based on sub-structures affinity and utilisation of the orthogonality and equilibrium equations.

In order to reduce the number of design variables we propose to realize a global adjustment by imposing to the unknown ΔM and ΔK matrices:

$$\Delta M = \sum_{j=1}^{v} M_j^{(c)} \frac{\Delta p_j}{p_j}\ ; \ \Delta K = \sum_{i=1}^{u} K_i^{(c)} \frac{\Delta p_i}{p_i}\ , \qquad (43)$$

where the matrices $M_j^{(c)}$ and $K_j^{(c)}$ square of order N define the jth substructure.

6.1 - Equations related to the differences of eigenvalues and to the conditions of orthogonality. Introducing the identified eigenvectors affected by the norm errors :

$Y_1^{'(m)} = Y^{(c)}\ C'$ where : $Y_1^{'(m)}$ (Nxn) ; $Y^{(c)}$ (Nxm) ; C' (mxn) and $N \geqslant m \geqslant n$.

If $Y_1^{'(m)}$ is partially identified, C' is calculated by one of the methods proposed for the evaluation of A_{11}. From (26) and (27) we obtain :

$$C^{T'}\ [E + \Delta\tilde{M}]\ C' = D^2 \quad ; \quad C^{T'}\ [\Lambda^{(c)} + \Delta\tilde{K}]\ C' = \Lambda^{(m)} D^2,$$

where : $\Delta\tilde{M}(m \times m) \triangleq Y^{T(c)}\ \Delta M\ Y^{(c)}$; $\Delta\tilde{K}(m \times m) \triangleq Y^{T(c)}\ \Delta K\ Y^{(c)}$.

As the diagonal matrix D(nxn) is unknown, the extra-diagonal elements are the only useful ones, i.e. :

$$\text{Extra Diag. } [\Sigma_j\ C^{T'}\ \tilde{M}_j\ C'\ \frac{\Delta p_j}{p_j}] = - \text{ Extra Diag. } [C^{T'}\ C'] \ ,$$

which can be transformed in : $J_1\ v_1 = z_1$, (44)

and : $$\text{Extra Diag. } [\Sigma_i\ C^{T'}\ \tilde{K}_i\ C'\ \frac{\Delta p_i}{p_i}] = - \text{ Extra Diag. } [C^{T'} \Lambda^{(c)} C'],$$

also transformed in : $J_2 v_2 = z_2$ (45)

Taking into account the symmetry of the matrices, these two relations represent n(n-1) scalar equations. The differences of eigenvalues permit to obtain n additional independent equations from the diagonal terms of (29) and (30) :

$$\text{Diag. } [\Sigma_i\ \tilde{K}_i\ \frac{\Delta p_i}{p_i} - \Sigma_j\ \tilde{M}_j\ \Lambda^{(c)}\ \frac{\Delta p_j}{p_j}] = \text{Diag. } [\Delta\Lambda] \ ,$$

which is transformed in : $J_3 v_1 + J_4 v_2 = z_3$. (46)

The set of these n^2 linear equations is regrouped in : $Jz = v$,

where : $z = \{\ldots ; \frac{\Delta p_j}{p_j} ; \ldots ; \ldots ; \frac{\Delta p_i}{p_i} ; \ldots\}$;

which will be solved for z by one of the methods proposed in 5.3.

Criticism of the method : a) concise formulation ; unnecessary matching process ; errors of norm without any consequence ; b) the localization can still be applied from (28) ; c) smaller number of equations and specially quadratic dependence with respect to the elements of C'. The components of the C' columns corresponding to the projection on the remote eigenvectors are frequently unaccurately defined. As a consequence, in the matrices of the type : $C^{T'} \tilde{M}_j C'$; $C^{T'} C'$..., the remote elements from the principal diagonal will also be unaccurately calculated and the corresponding equations will have to be evicted.

6.2 - Equilibrium equations. They are represented by the homogeneous equations : $[K^{(c)}+\Delta K]Y'^{(m)}-[M^{(c)}+\Delta M]Y'^{(m)}\Lambda^{(m)} = 0$, unaffected by the errors of norm. By introducing the previous notations we obtain :

$$\Delta M\, Y^{(c)}\, C'\, \Lambda^{(m)} - \Delta K\, Y^{(c)}\, C' = K^{(c)}Y^{(c)}C' - M^{(c)}Y^{(c)}C'\Lambda^{(m)} \tag{47}$$

In order to reduce the influence of the errors of measurement, an aggregation is obtained by premultiplying by $Y^{(c)T}$:

$$\Sigma_j \tilde{M}_j\, C'\, \Lambda^{(m)} \frac{\Delta p_j}{p_j} - \Sigma_i \tilde{K}_i\, C' \frac{\Delta p_i}{p_i} = \Lambda^{(c)}\, C' - C'\, \Lambda^{(m)} ,$$

where : $\Lambda^{(c)}$ (mxn) ; $\Lambda^{(m)}$ (nxn) ; C'(mxn). This last relation is, as in 6.1: $Uz = w$, and represents mxn scalar equations.

Such a method is similar to the ONERA[7] and Fradellos - Evans' ones[8,9]. In these three methods we simultaneously minimize with respect to the unknown quantities z_j, a set of n $h_\nu \varepsilon_\nu$ vectors :

$$[F_\nu + \sum_{j=1}^{q} A_j z_j]\, h_\nu\, y_\nu^{(m)} = h_\nu \varepsilon_\nu \quad , \qquad 1 \leqslant \nu \leqslant n,$$

where : $F_\nu \triangleq [K^{(c)} - \lambda_\nu^{(m)} M^{(c)}]$; h_ν a weighting scalar concerning the eigenvector $y_\nu^{(m)}$; $h_\nu \varepsilon_\nu$(Nx1) represents the external generalized forces which, in order to insure the equilibrium of the structure, have to be applied after the adjustment process.

7 - Conclusion.

These methods of adjustment can be extended to mechanical structures modeled by non-symmetric matrices. Moroever, they can be adapted to the dynamical synthesis problems formulated in natural and modal coordinates[10] Up to our knowledge, the existing applications are limited and does not furnish a sufficient basis for the evaluation and the comparative criticism of the different adjustment methods.

REFERENCES

1 - DANEK, Problèmes d'identification. Applications aux machines, EUROMECH 131, *Identitication problems in structural dynamics,* 24-27 juin 1980, Besançon (France) (to be published in the proceedings of the Colloquium)

2 - FILLOD R. Contribution à l'identification des structures mécaniques linéaires, *Thèse Docteur ès Sciences,* Laboratoire de Mécanique Appliquée, Faculté des Sciences de Besançon, 1980

3 - BUGEAT L.P., LALLEMENT G., Methods of matching of calculated and identified eigensolutions, *Proceedings of the XIIth Conference on Dynamics of Machines,* Slovak Academy of Sciences, Institut of Machine Mechanics, 80931 Bratislava, Dubravska Cesta, CSSR, 81, 1979

4 - BUGEAT L.P., Ajustement des caractéristiques dynamiques d'un modèle discret conservatif non gyroscopique au comportement identifié sur structure, *Thèse Docteur-Ingénieur,* Faculté des Sciences de Besançon, Besançon 1978 - L.R. BUGEAT, FILLOD R., LALLEMENT G., PIRANDA J., Ajustement of a conservative non gyroscopic mathematical model from measurement, *The Shock and Vibration Bulletin,* Part.3, 71, sept.1978

5 - DONE G.T.S., Adjustment of a rotor model to achieve agreement between calculated and experimental natural frequencies, *Journal of Mechanical Engineering Science,* vol.21, n°6, 389, 1979

6 - CHEN J.C., GARBA J.A., Matrix perturbation for analytical model improvement, *Proceedings of the AIAA/ASME/AHS 20th Structures,* Structural Dynamics and Materials Conferences, St Louis, 428, April 1979

7 - BERGER H., CHAQUIN J.P., OHAYON R., Une méthode d'identification de modèles de structures complexes utilisant des résultats d'essais de vibration, EUROMECH 131, *Identification problems in Structural Dynamics,* 24-27 juin 1980, Besançon (France)

8 - FRADELLOS G., EVANS F.J. - Improvement of dynamic models by inverse eigenproperty assigment, *Appl. Math. Modelling,* vol.2,123, June 1978

9 - LALLEMENT G., Recalage de modèles mathématiques de structures mécaniques ; CEA-EDF, *Cycle de Conférences sur les vibrations dans le domaine industriel,* Jouy en Josas, octobre 1979

10 - BOUGHANEM M., Identification paramétrique de modèles conservatifs de structures mécaniques, *Thèse Doct. Ing.,* Université de Franche-Comté, Besançon (France) octobre 1979

BIBLIOGRAPHY

- NATKE H.G., Indirect identification methods III : Correction of the results of system analysis by results of identification - a Survey, *identification of vibrating structures,* CISM Udine, 1980

- WHITE C.W., MAYTUM B.D., Eigensolution sensitivity to parametric model perturbations, *The Shock and Vibration Bulletin,* n°46, part.5, 123, 1976

- WADA B.K., GARBA J.A., CHAN J.C., Development and correlation : viking orbiter analytical dynamic model with modal test, *The Shock and Vibration Bulletin,* n°44, part.2, 125, 1974

- GARBA J.A., WADA B.K., Application of perturbation methods to improve analytical model correlation with test data, *ASME-SAE,* Aerospace Meeting, Los Angeles, paper n°770959, Nov.1977

- CHROSTOWSKI H.D., EVENSEN D.A., HASSELMAN T.K., Model verification of mixed dynamic systems, *Journ. of Mech. Design,* Vol.100, 266, April 1978.

CORRECTION OF ANALYTICAL MODELS FOR DAMPED LINEAR SYSTEMS USING EXPERIMENTALLY OBTAINED FORCED VIBRATION RESPONSES

H.-P. Felgenhauer
Curt-Risch-Institut
Universität Hannover

The dynamic behaviour of mechanical systems can usually be described by discrete analytical models which are obtained through an analytical systems analysis. Errors are mainly caused by the simplifications which are necessary in order to create a physical model of the continuum. They can possibly result from discretization as well as from deviations of the system parameter values. Discretization errors, which will not be regarded here, influence primarily the order of the model given be the number of degrees of freedom as well as by the number of modes. The deviations of the mass-, stiffness- and damping parameters are due to insufficient knowledge of the behaviour of the structure, the boundary and the connection elements. In general there is no difficulty in evaluating the inertia properties to a sufficient degree of accuracy. The evaluation of stiffness parameters is more difficult, and the evaluation of the damping parameters by means of an analytical systems analysis only is in general not possible. Thus in the derivation of damping matrices - which will be discussed later - use should be made of experimental data. Due to the above - mentioned error possibilities, a verification of the analytical model may be necessary, and if certain analysis - test correlation requirements are exceeded the model has to be adjusted.

The subject of this lecture is a method for the correction of system parameters of analytical models for damped linear systems using experimentally derived system responses. This was prepared as a dissertation by Felgenhauer.[1] Without limiting the general use of the method, it is assumed that the inertia matrix of the uncorrected analytical model is sufficiently accurate, such that the correction of the analytical model can be carried out by adjusting the stiffness- and damping parameters. Stiffness- and damping matrix of the uncorrected model thereby serve as starting matrices in the correction process. The possible interpretation of the correction procedure as method for the identification of complete damping matrices will also be discussed.

The correction of an analytical model by the use of experimental data can be viewed as one possibility of a synthesis between the analytical and the experimental systems analysis.

A schematic view of both kinds of systems analysis is given in Fig. 1.

Let us first regard the analytical systems analysis. Starting with a physical model of the continuum and proceeding with a mathematical model a discrete analytical model can be obtained which is well suited for numerical computations. In the frequency domain the equations of motion form an algebraic system of equations with constant coefficients

$$(-\Omega_r^2 \underline{M} + i\Omega_r \underline{C} + \underline{K})\, \underset{\sim}{Y}_r = \underset{\sim}{F}_r \tag{1}$$

for excitation frequencies Ω_r.

$\underline{M}$, $\underline{C}$, $\underline{K}$ are, respectively, n x n inertia -, viscous damping - and stiffness matrices where n is the number of degrees of freedom. The required response vector

$$\underset{\sim}{Y}_r = \underline{Z}_r^{-1} \underset{\sim}{F}_r \tag{2}$$

can be obtained via the inverse of the dynamic stiffness matrix

$$\underline{Z}_r = -\Omega_r^2 \underline{M} + i\Omega_r \underline{C} + \underline{K} . \tag{3}$$

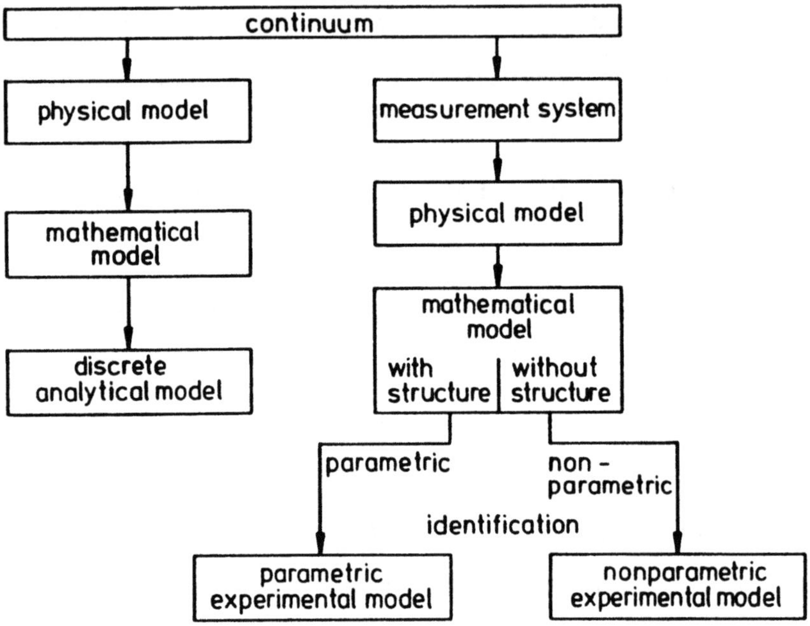

Fig. 1 Flow chart of the analytical and experimental system analysis

The experimental systems analysis proceeds from a physical model which also contains measurement and excitation devices via a mathematical model without structure to a discrete non-parametric experimental model. In case a mathematical model with structure is given, the system parameters can be identified by means of an appropriate identification algorithm.

The direct comparison of the identified parameters of a

parametric experimental model and equivalent values of the analytical model is the basis for some published correction methods.

The present correction method, however, is based on a comparison which is indirect with regard to the system parameters. It is carried out between the system response of the analytical model and experimentally obtained system responses of a nonparametric experimental model.

The reason for this approach is, that in practical applications it is quite cumbersome to allocate the identified and computed mode shapes. Furthermore, the parametric experimental model is, in general, incomplete in comparison with the analytical model. The incompleteness results from the fact that generally the number of degrees of freedom n of the analytical model is much larger than the number m of measurement points, which again is larger than the number of identified modes.

Further advantages of the indirect comparison are:

- a reduced experimental effort e.g. multipoint excitation is possible with forces for which equal phase angles are not required
- parametric identification is unnecessary.

The nonparametric experimental model used in the correction procedure consists of

$F(\Omega_r)$, m x k matrix of frequency dependent excitation vectors

and the m x k matrix of the complex frequency responses

$$\underline{Y}^M(\Omega_r) = \underset{\sim}{Y}^M_{R1} + i\,\underset{\sim}{Y}^M_{I1}, \;\ldots\ldots,\; \underset{\sim}{Y}^M_{Rk} + i\,\underset{\sim}{Y}^M_{Ik}$$

with $\underset{\sim}{Y}^M_{R1} = \underset{\sim}{Y}^M_R(\Omega_1)$ m x 1 - vector of the realparts

$\underset{\sim}{Y}^M_{I1} = \underset{\sim}{Y}^M_I(\Omega_1)$ m x 1 - vector of the imaginary parts.

The uppercase letters indicate values belonging to the experimental model (measurements), while the lower case letters R mark the real- and I the imaginary part. Indices 1 through k give the respective excitation frequencies. The excitation can be harmonic with k excitation frequencies. If periodic or transient excitation forces are used together with the Fast Fourier Transform of the input and output signals, the nonparametric experimental model consists of the real and imaginary parts at k selected frequencies.

The formulation of the correction problem is carried out for reasons of simplicity for an analytical model with viscous damping (excluding further nonconservative forces and gyroscopic effects), using a nonparametric experimental model consisting of complex frequency responses.

The assumptions are:

- linearity
- time invariant system behaviour
- the measured frequency responses contain only uncorrelated stochastic noise with zero mean value
- stiffness and damping matrices can be partitioned in substructure matrices
- the correction is possible by multiplication of the substructure matrices with the correction coefficients.

The solution contains:

- the selection of analysis-test correlation criteria
- the formulation of a correction strategy

and leads to the derivation of the correction-algorithm.

The residuals (analysis-test correlation criteria) for an indirect comparison is based on the complex difference of the frequency responses

$$\Delta \underset{\sim}{Y}(\Omega_r) = \underset{\sim}{Y}^K(\Omega_r) - \underset{\sim}{Y}^M(\Omega_r). \tag{4}$$

Upper case letters K indicate values of the corrected analytical model Fig. 2 shows the i-th components of the frequency response of the analytical model and the experimental model together with the difference $\Delta \underset{\sim}{Y}$ for an excitation frequency Ω_r.

Separating the differences into the real- and imaginary part leads, together with the introduction of frequency dependent diagonal weighting matrices ($\underline{G}_{R_r}$ for the real and $\underline{G}_{I_r}$ for the imaginary part), to the weighted difference of the frequency response

$$\underset{\sim}{\Delta Y}(\Omega_r) = \underline{G}_{Rr}\left\{\underset{\sim}{Y}^K_{Rr} - \underset{\sim}{Y}^M_{Rr}\right\} + i\,\underline{G}_{Ir}\left\{\underset{\sim}{Y}^K_{Ir} - \underset{\sim}{Y}^M_{Ir}\right\}. \tag{5}$$

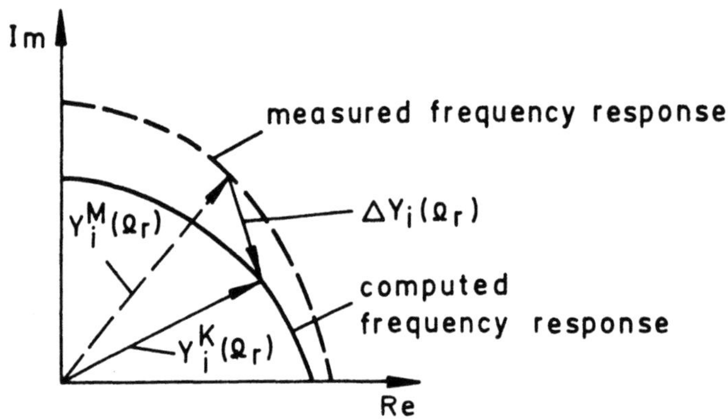

Fig. 2 Analysis-test correlation criteria

The chosen correction strategy follows a suggestion by Natke.[2]

The present correction is thus an extension of the above-mentioned method, including the adjustment of damping parameters. The assumptions made with regard to the correction strategy are:

- the inertia matrix is sufficiently exact in the mechanical sense

- it is possible to partition the damping and stiffness matrix in P substructure matrices

$$\underline{C} = \sum_{\sigma=1}^{P} \underline{C}_{\sigma} \quad \text{respectively} \quad \underline{K} = \sum_{\sigma=1}^{P} \underline{K}_{\sigma} \tag{6}$$

- the correction can be performed through multiplication of the substructure matrices with correction coefficients

$$\underline{C}^K = \sum_{\sigma=1}^{P} d_{\sigma} \underline{C}_{\sigma} \ , \quad \underline{K}^K = \sum_{\sigma=1}^{P} b_{\sigma} \underline{K}_{\sigma} \ . \tag{7}$$

The correction strategy concerning the substructures is a device which reduces the computational effort. In the limit it leads to a correction of each matrix element if the partitioning is carried out accordingly.

On the other hand, the partitioning is also justified in a physical sense when the substructures have different stiffness or damping properties.

The mathematical correction strategy can become incompatible with respect to the physical model of the experimental model in those cases where the correction is not carried out for each element.

Examples show, however, that even in these cases a correction of the analytical model may still be possible.

The declared goal of the method is to minimize the complex difference of the system responses. Based on the discussed analysis test correlation criterion and the correction strategy, the summation of the squares of a generalized Euclidian norm

$$\|\Delta \underset{\sim}{Y}_r\| = |\Delta \underset{\sim}{Y}_r| = \sqrt{\Delta \underset{\sim}{Y}_r^* \, \Delta \underset{\sim}{Y}_r}$$

over k excitation frequencies leads to the objective function

$$J(\underset{\sim}{b}, \underset{\sim}{d}) = \sum_{r=1}^{k} |\Delta \underset{\sim}{Y}_r|^2 \tag{8}$$

as a function of the correction coefficients.

Substitution of the weighted differences (5) in the objective function leads to the basic equation of the correction-method,

$$\begin{aligned} J(\underset{\sim}{b}, \underset{\sim}{d}) = \sum_{r=1}^{k} & \left\{ \underset{\sim}{Y}_{Rr}^{K} - \underset{\sim}{Y}_{Rr}^{M} \right\}^{T} \underline{G}_{Rr} \left\{ \underset{\sim}{Y}_{Rr}^{K} - \underset{\sim}{Y}_{Rr}^{M} \right\} \\ & + \left\{ \underset{\sim}{Y}_{Ir}^{K} - \underset{\sim}{Y}_{Ir}^{M} \right\}^{T} \underline{G}_{Ir} \left\{ \underset{\sim}{Y}_{Ir}^{K} - \underset{\sim}{Y}_{Ir}^{M} \right\} \overset{!}{=} \text{Min} \end{aligned} \tag{9}$$

which is given as a minimization problem.

The necessary condition states that the gradient of the objective function

$$\frac{\partial J}{\partial \underset{\sim}{b}} \Big/_{\substack{\underset{\sim}{b} = \hat{\underset{\sim}{b}} \\ \underset{\sim}{d} = \hat{\underset{\sim}{d}}}} = 0 \, , \qquad \frac{\partial J}{\partial \underset{\sim}{d}} \Big/_{\substack{\underset{\sim}{d} = \hat{\underset{\sim}{d}} \\ \underset{\sim}{b} = \hat{\underset{\sim}{b}}}} = 0$$

is equal to zero for the estimates $\hat{\underset{\sim}{b}}$ and $\hat{\underset{\sim}{d}}$ of the correction coefficients.

The analytically derived gradient consists of the partial derivatives of the objective function with respect to the stiffness correction coefficients b_σ

$$f_{1\sigma} = 2 \sum_{r=1}^{K} \left[(\underset{\sim}{Y}^K_{Rr} - \underset{\sim}{Y}^M_{Rr}) \; \underline{G}_{Rr} \frac{\partial \underset{\sim}{Y}^K_{Rr}}{\partial b_\sigma} + (\underset{\sim}{Y}^K_{Ir} - \underset{\sim}{Y}^M_{Ir}) \; \underline{G}_{Ir} \frac{\partial \underset{\sim}{Y}^K_{Ir}}{\partial b_\sigma} \right] \Bigg|_{\substack{\underset{\sim}{b} = \hat{\underset{\sim}{b}} \\ \underset{\sim}{d} = \hat{\underset{\sim}{d}}}} = 0 \tag{10}$$

and the partial derivatives with respect to the damping correction coefficients d_σ

$$f_{2\sigma} = 2 \sum_{r=1}^{K} \left[(\underset{\sim}{Y}^K_{Rr} - \underset{\sim}{Y}^M_{Rr}) \; \underline{G}_{Rr} \frac{\partial \underset{\sim}{Y}^K_{Rr}}{\partial d_\sigma} + (\underset{\sim}{Y}^K_{Ir} - \underset{\sim}{Y}^M_{Ir}) \; \underline{G}_{Ir} \frac{\partial Y^K_{Ir}}{\partial d_\sigma} \right] \Bigg|_{\substack{\underset{\sim}{b} = \hat{\underset{\sim}{b}} \\ \underset{\sim}{d} = \hat{\underset{\sim}{d}}}} = 0 \tag{11}$$

with the partial derivatives of the real and imaginary parts of the frequency responses

$$\frac{\partial \underset{\sim}{Y}^K_{Rr}}{\partial b_\sigma} = (\underline{B}^K \, \underline{K}_\sigma \, \underset{\sim}{Y}^K_{Ir} - \underline{A}^K \, \underline{K}_\sigma \, \underset{\sim}{Y}^K_{Rr}),$$

$$\frac{\partial \underset{\sim}{Y}^K_{Ir}}{\partial b_\sigma} = - (\underline{A}^K \, \underline{K}_\sigma \, \underset{\sim}{Y}^K_{Ir} + \underline{B}^K \, \underline{K}_\sigma \, \underset{\sim}{Y}^K_{Rr}$$

$$\frac{\partial \underset{\sim}{Y}^K_{Rr}}{\partial d_\sigma} = \Omega_r (\underline{A}^K \underline{C}_\sigma \underset{\sim}{Y}^K_{Ir} + \underline{B}^K \underline{C}_\sigma \underset{\sim}{Y}^K_{Rr}),$$

$$\frac{\partial \underset{\sim}{Y}^K_{Ir}}{\partial d_\sigma} = \Omega_r (\underline{B}^K \underline{C}_\sigma \underset{\sim}{Y}^K_{Ir} - \underline{A}^K \underline{C}_\sigma \underset{\sim}{Y}^K_{Rr}).$$

The matrices $\underline{A}^K$ and $\underline{B}^K$ represent the real and imaginary part of the inverse of the complex stiffness matrix (3) which are obtained by real-valued matrix operations.

If σ takes the values from 1 to P, a nonlinear homogenous system of equations is obtained for the gradient.

The equations are nonlinear, since the correction coefficients are contained in the matrices $\underline{A}^K$ and $\underline{B}^K$ as well as in the real and imaginary part of the computed frequency response.

The correction coefficients, which are included in the vector $\underset{\sim}{B}$ are computed iteratively

$$\underset{\sim}{B}_{i+1} = \begin{Bmatrix} \underset{\sim}{b} \\ \underset{\sim}{d} \end{Bmatrix}_{i+1} = \underset{\sim}{B}_i - \Delta \underset{\sim}{B}_i \qquad (12)$$

by means of a P-dimensional Newton-Algorithm.

The parameter increments ΔB_i are determined from the linear system of equations

$$\underline{S}\,(\underset{\sim}{B}_i)\,\Delta \underset{\sim}{B}_i = \underset{\sim}{f}(\underset{\sim}{B}_i) \tag{13}$$

with the gradient vector

$$\underset{\sim}{f}(\underset{\sim}{B}_i) = \left\{\begin{array}{c} f_{11}(\underset{\sim}{b},\underset{\sim}{d}) \\ \vdots \\ f_{1P}(\underset{\sim}{b},\underset{\sim}{d}) \\ \hline f_{21}(\underset{\sim}{b},\underset{\sim}{d}) \\ \vdots \\ f_{2P}(\underset{\sim}{b},\underset{\sim}{d}) \end{array}\right\}$$

and the numerically evaluated Hesse-Matrix

$$\underline{S}(\underset{\sim}{B}_i) = \left[\begin{array}{ccc|ccc} \frac{\partial f_{11}}{\partial b_1} & \cdots & \frac{\partial f_{11}}{\partial b_P} & \frac{\partial f_{11}}{\partial d_1} & \cdots & \frac{\partial f_{11}}{\partial d_P} \\ \vdots & & \vdots & \vdots & & \vdots \\ \frac{\partial f_{1P}}{\partial b_1} & \cdots & \frac{\partial f_{1P}}{\partial b_P} & \frac{\partial f_{1P}}{\partial d_1} & \cdots & \frac{\partial f_{1P}}{\partial d_P} \\ \hline \frac{\partial f_{21}}{\partial b_1} & \cdots & \frac{\partial f_{21}}{\partial b_P} & \frac{\partial f_{21}}{\partial d_1} & \cdots & \frac{\partial f_{21}}{\partial d_P} \\ \vdots & & \vdots & \vdots & & \vdots \\ \frac{\partial f_{2P}}{\partial b_1} & \cdots & \frac{\partial f_{2P}}{\partial b_P} & \frac{\partial f_{2P}}{\partial d_1} & \cdots & \frac{\partial f_{2P}}{\partial d_P} \end{array}\right]$$

The partitioning of the gradient vector and the functional matrix indicates that with the solution algorithm, in addition to the simultaneous correction of the stiffness and damping matrix, a separate correction of either one of the two matrices can be carried out. The results of correction examples show that a correction performed in 2 steps can be of advantage or even essential:

- during the 1st step a prior correction of the stiffness matrix takes place, while the damping matrix remains unchanged; the test analysis comparison is carried out for off-resonant excitation frequencies at which the influence of damping is small,
- the goal of the 2nd step is the correction of the damping matrix and a final correction of the stiffness matrix; the comparison is carried out for near-resonant frequencies at which the influence of damping is dominant.

The 2nd correction step carried out with the stiffness matrix held constant can be interpreted as the identification of a complete damping matrix for an analytical model for which the inertia and stiffness matrices are known in advance. A schematic view of the adaptive correction method is given in Fig. 3. The analytical model to be corrected is complemented by a nonparametric experimental model. The output of both models resulting from identical input is compared via the analysis test correlation criteria. New esti-

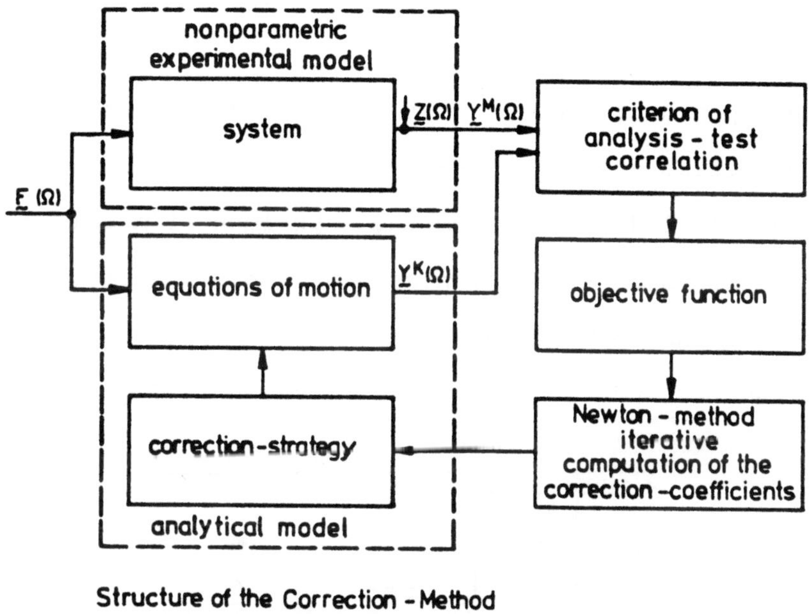

Fig. 3 Schematic View of Correction Method

mates for the correction coefficients, which are computed with the Newton algorithm, enter the analytical model by means of the correction strategy. The calculation is repeated until a convergence criteria is satisfied which checks the stationarity of the objective function. Finally, the results of two correction examples will be discussed. In the 1st example the analytical model of a damped discrete 6 degree of freedom system will be corrected. The frequency response of the experimental model will be simulated on the computer and can thus be obtained without experimental error.

The experimental model is shown in Fig. 4, where the exact parameters are indicated by the lower case E. The analytical model deviates from the experimental model in the given errors of the stiffness and damping parameters. According to the correction strategy, the structure will be partitioned in three substructures as shown in Fig. 4.

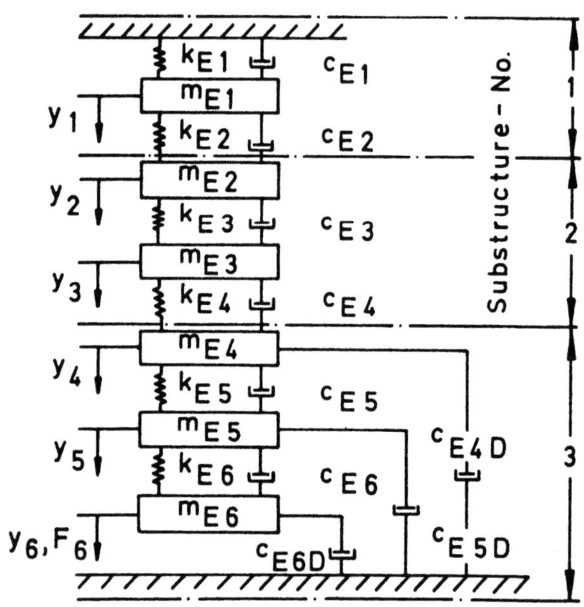

Fig. 4 Experimental Model

Corrections will be carried out for three analytical models which differ in the given errors of the parameters , as shown in Tab. 1. In order to be able to evaluate the influence of damping on the results of the correction, the modal damping of the experimental model is varied from 0,5% to 20%.

uncorrected analytical model	deviation of the start elements [%]					
	b_{10}	b_{20}	b_{30}	d_{10}	d_{20}	d_{30}
RM1	0	- 9	11	0	33	-23
RM2	-9	-16	25	0	33	-23
RM3	0	-30	30	-15	40	-35

Table 1 Uncorrected analytical models

Some of the results of the correction are given in Tab. 2. The used nonparametric experimental model consists of the excitation force F_3 or F_6, and the real and imaginary parts for the given frequency arrays. The array FREQ 6-1 contains 6 off-resonant frequencies - all other arrays contain near-resonance frequencies. It can be seen that for all three analytical models - the last figures mark the modal damping in percent - the correction method converges only for the models with higher damping. In these cases where the computation converges, the exact values for the

correction no.	analy-tical model	exci-tation force	excitation frequencies	no. of iterations
1	RM1005	F_3	FREQ6-1	no convergence
2	"	F_3	FREQ8-1	"
3	"	F_6	FREQ8-2	"
4	RM105	F_3	FREQ6-1	5
5	RM205	F_3	FREQ6-1	no convergence
6	RM210	F_3	FREQ6-1	"
7	RM220	F_3	FREQ6-1	8
8	RM305	F_6	FREQ4	6 (local minimum)
9	"	F_6	FREQ8-1	no convergence
10	"	F_6	FREQ8-2	"
11	RM310	F_6	FREQ4	5
12	RM320	F_6	FREQ4	7

FREQ6-1: .0637, .0955, .1432, .2546, .2706, .2944 Hz

FREQ8-1: .0300, .0500, .0637, .0955, .1040, .1250
.1660, .2000 Hz

FREQ8-2: .0380, .0430, .1090, .1170, .1740, .1880
.2250, .2546 Hz

FREQ4 : .0716, .0796, .1469, .2096 Hz

Table 2 Results of correction

correction coefficients are obtained. The only exception is the correction no.8,at which a local minimum is arrived with small deviations in the stiffness correction coefficients, but still with large deviations in the damping correction coefficients. The results of these examples lead to the conclusions that, after the introduction of additional damping, the models with small damping can also be corrected.

An increase of the damping, which greatly improves the success of the correction (as shown by the results in Tab. 2), can also be arrived at when real measurements are used.

To increase the modal damping, the measurements can be multiplied in the time domain by an exponential window, which results in an increase of modal damping in the frequency domain. After the correction has been completed, the artificially introduced damping can be removed from the adjusted damping matrix.

Results of correction no.11 are shown in Fig. 5, which contains plots of acceleration frequency responses $\ddot{Y}_6$ for the experimental model and for the analytical model before and after the correction.

The aim of the second example is to carry out a correction in a practice-oriented manner using measured frequency responses. In this example the analytical model of a labo-

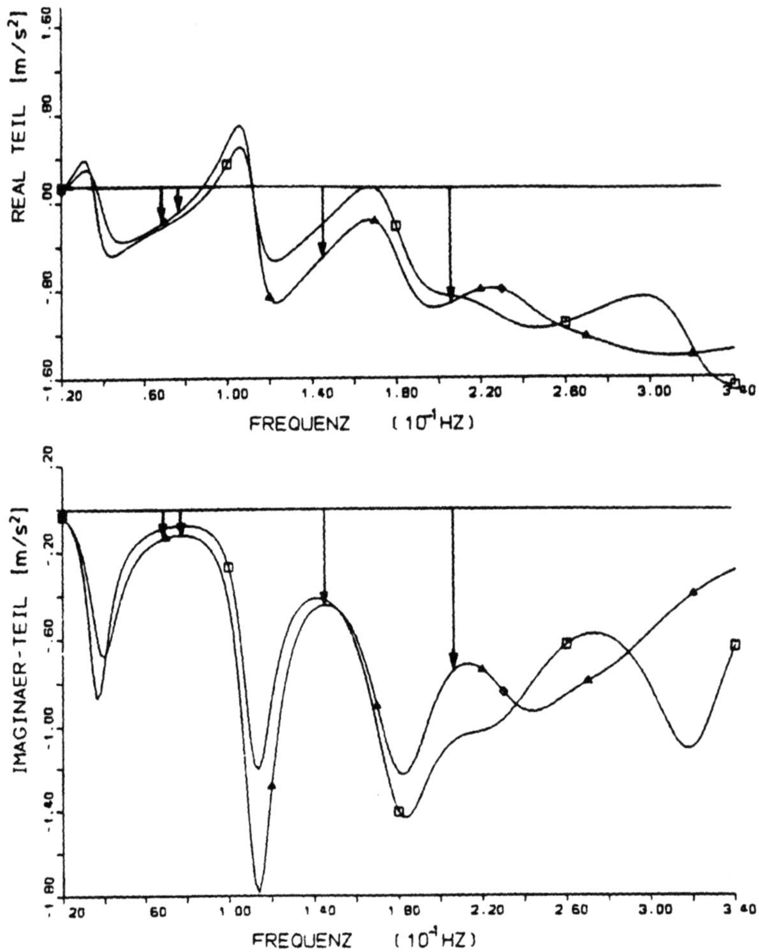

Fig. 5 Acceleration response $\ddot{Y}_6$ of the experimental model (VO), the uncorrected and corrected analytical model RM3 with 10% modal damping (RM310)

ratory model will be corrected. This consists of a cantilever beam and a support mass, as shown in Fig. 6.

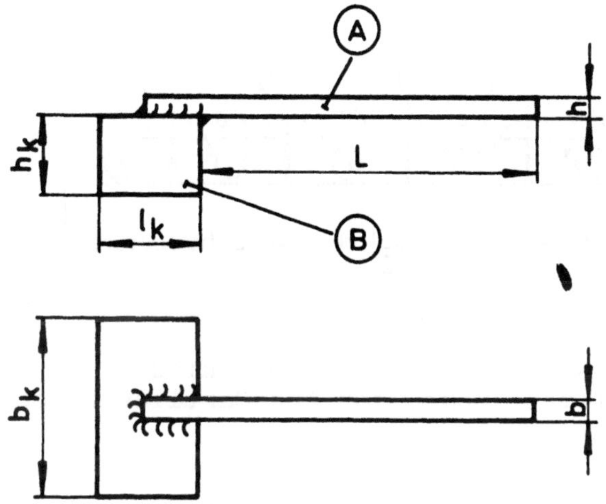

Material: St 370

Dimensions: $L=500mm$; $b=30mm$; $h=30mm$

$l_k=150mm$; $b_k=250mm$; $h_k=110mm$

Fig. 6 Cantilever beam model

Cantilever and mass are connected through weldings. The low structural damping of the steel model is increased by an analogous exponential window in order to permit the evaluation of the damping influence on the results of the correction. A finite element model with 6 translational and rotational degrees of freedom is partitioned in 3 substructures as shown in Fig. 7

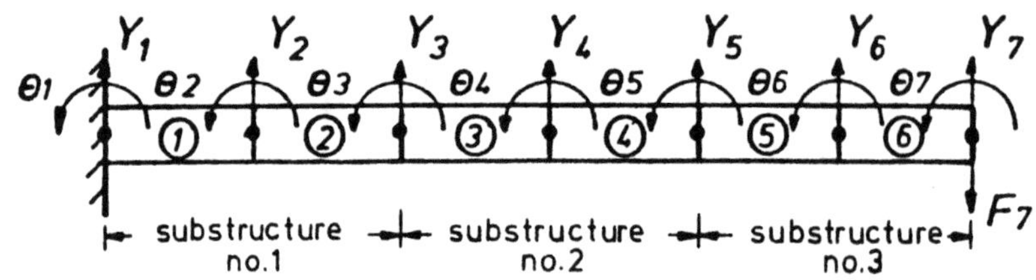

Fig. 7 Discretization of the Finite-Element Model for the Cantilever Beam

Inserting the boundary conditions with one end fixed results in a positive definite stiffness matrix. While the stiffness and the consistent inertia matrix can be constructed from the geometry and the material properties, this does not hold true for the damping matrix. Since no data based on experience are available, it has to be constructed with the use of experimentally obtained parameters. For the given structure, the assumption of the damping matrix being proportional,

$$\underline{C} = \alpha \underline{M} + \beta \underline{K} \tag{14}$$

seems to be justified, since the modes are well separated, as can be concluded from the measured frequency responses shown in Fig. 8.

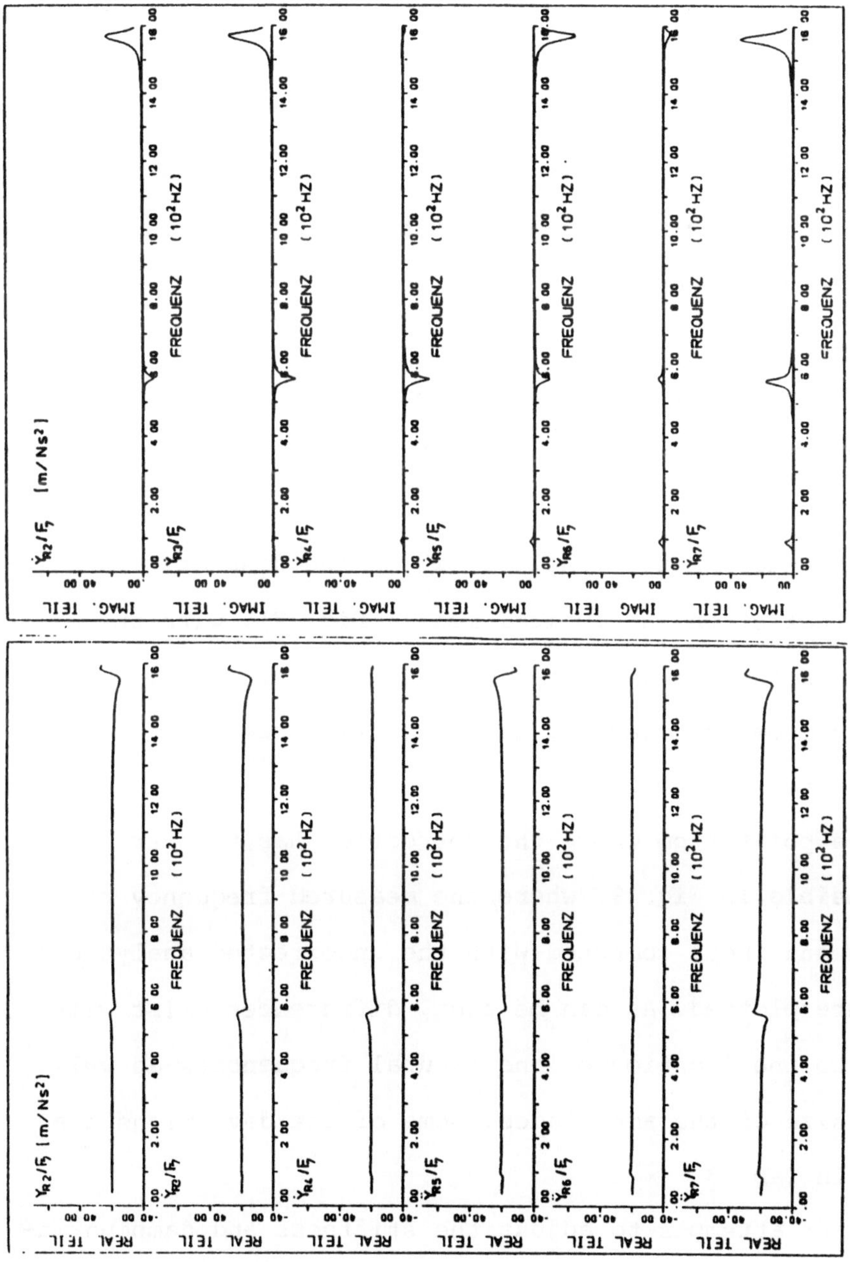

Fig. 8 Real- and imaginary part of measured frequency responses

The responses result from a force impact at coordinate no.7. Both constants α and β will be computed from the first three modal damping ratios which are obtained from the measurements using the method of half-power points. The results of the correction leads to the conclusion that the applied procedure can be used to construct a start damping matrix even in these cases where the modes are not so well separated. Since normally the damping correction coefficients vary for different substructures, the adjusted damping matrix is nonproportional and does not satisfy (14) any longer.

The experimental model for which the discretization equals that of the analytical model (Fig. 7), with the exception of the missing rotational degrees of freedom, is obtained from the measured frequency responses (Fig. 8) for the selected frequencies.

The point from where the correction has to start is made visible in Fig. 9, where the measured frequency responses and those computed with the uncorrected analytical model are plotted. As can be seen, differences exist with regard to the location of the natural frequencies as well as the size of the amplitudes. Some of the deviations are listed in Tab. 3.

First attempts to adjust the stiffness and damping matrices simultaneously did not lead to satisfying results.

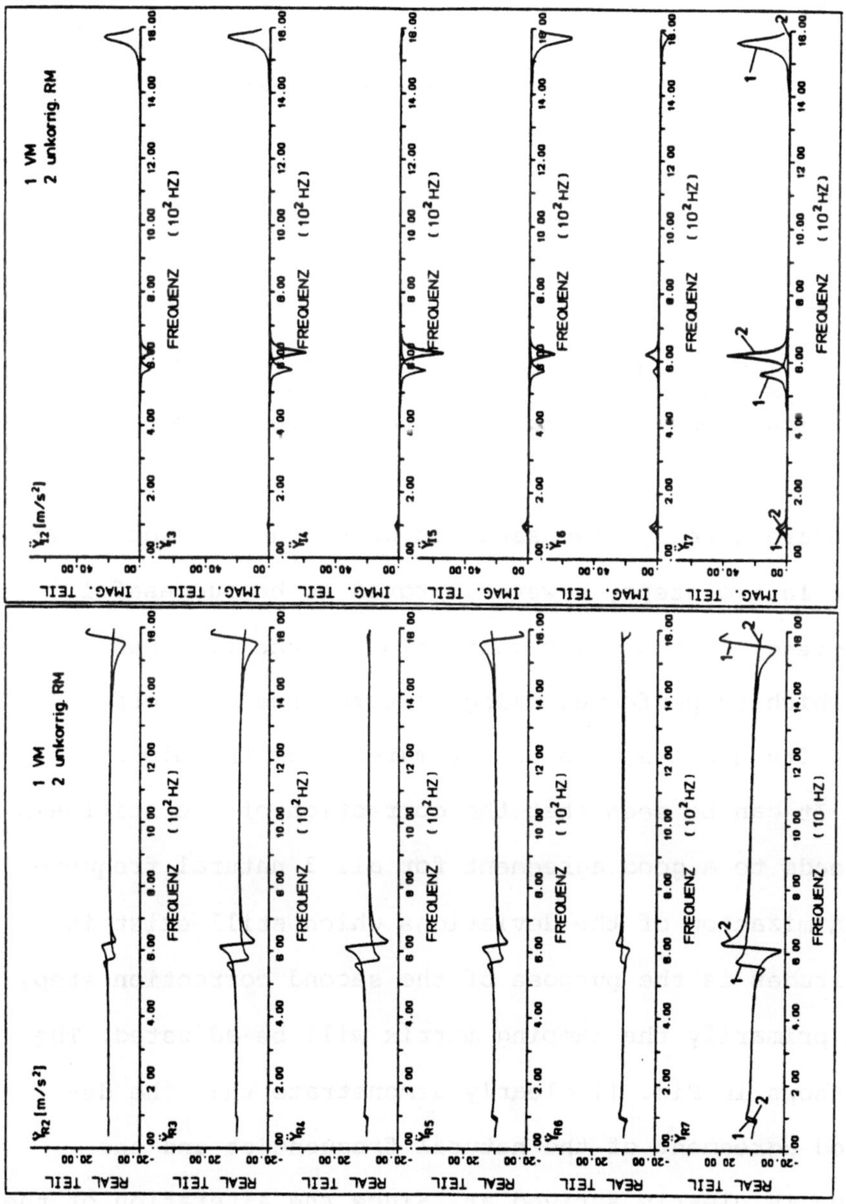

Fig. 9 Comparison of measured and computed (uncorrected analytical model) frequency responses

Natural frequency no.	Deviation of the natural frequency [%]	Deviation of the resonance amplitude [%]
1	11	30
2	11	54
3	11	76

Table 3 Deviations of the natural frequencies and the resonance amplitudes of the uncorrected analytical model from the measured system response

The procedure already discussed, in which the correction is performed in two steps, however, proved to be successful. The efficiency of the prior correction of the stiffness matrix, which is performed using measurements at 8 off-resonance frequencies, can be estimated from the plots of Fig. 10. It can be seen that the correction of the stiffness matrix leads to a good agreement for all 3 natural frequencies. Minimization of the deviations which still exist in the amplitudes is the purpose of the second correction step, in which primarily the damping matrix will be adjusted. The results shown in Fig. 11 clearly demonstrate that the desired good agreement of the natural frequencies and the amplitudes is clearly arrived at. Since the alteration of the stiffness correction coefficients are < 2% in the second

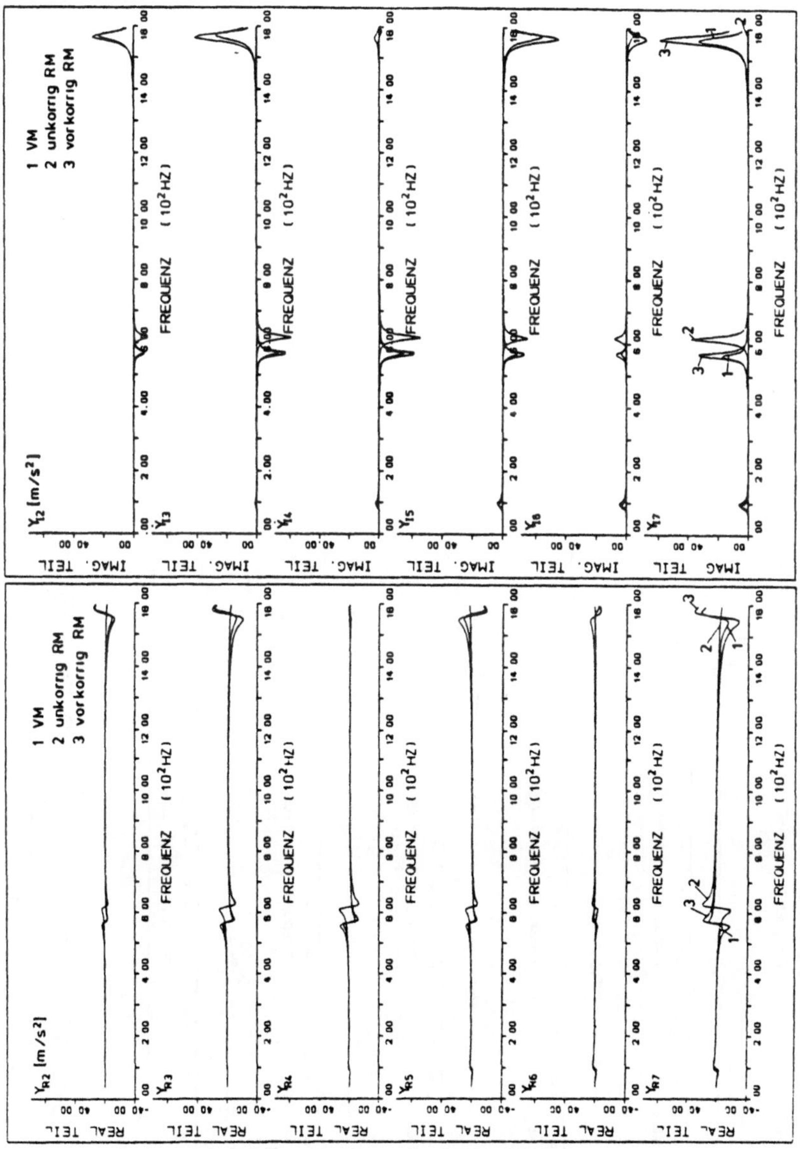

Fig. 10 Result of 1st correction-step (stiffness matrix)

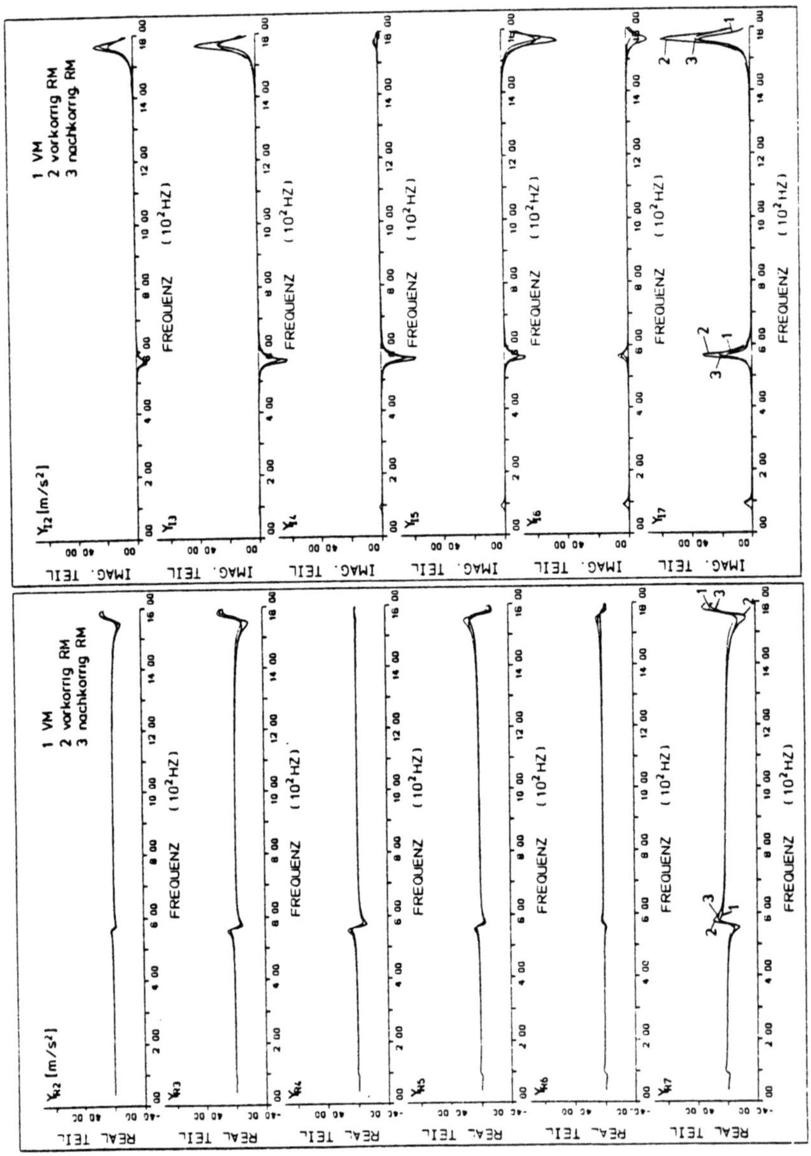

Fig. 11 Result of 2nd correction-step (damping and stiffness matrix)

correction step, it can also be interpreted as the identification of a complete damping matrix. Tab. 4 lists the starting values of the correction coefficients and the final results for all three substructures.

Substructure	Stiffness correction coefficients			Damping correction coefficients		
	$b_{\sigma o}$	$b_{\sigma 4}$	Deviation [%]	$d_{\sigma o}$	$d_{\sigma 4}$	Deviation [%]
$\sigma = 1$	1	.8472	-15,28	1	2.357	135.7
$\sigma = 2$	1	.8544	-14.56	1	1.758	75.8
$\sigma = 3$	1	.7456	-25.44	1	.9255	-7.45

Table 4 Summary of the results for the corrected analytical model. (Correction performed in two steps using frequency responses at 3 points)

The increased reduction to the stiffness towards the free end leads to the conclusion that the assumed fixed boundary condition resulted in an a priori model which is too stiff.

The damping correction coefficients increase proportionally to the distance from the free end of the cantilever. This tendency is possibly caused by the influence of the structural damping.

A method for the correction of stiffness and damping

matrices using measured frequency responses has been developed. The versatility of the method can be judged from the results given for two examples. It is possible to demonstrate its efficiency, in particular by means of the second example, which is of practical importance:

- despite large deviations of the uncorrected model, the correction can be carried out with relatively small computational and experimental effort.
- an expansion of the domain of start-elements leading to convergence towards the absolute minimum is achieved by performing the correction in two steps.
- an interpretation of the correction procedure as method for the identification of complete damping matrices is justified.

References

1 Felgenhauer, H.-P., Korrektur von Rechenmodellen für gedämpfte elastische Systeme mittels gemessener erzwungener Schwingungen, Fortschrittberichte der VDI-Zeitschriften, Reihe 11, Nr. 37, 1981

2 Natke, H.G., Die Korrektur des Rechenmodells eines elastomechanischen Systems mittels gemessener erzwungener Schwingungen, Ing.-Archiv 46, 1977, 169-184

INDIRECT METHODS: CORRECTION OF THE RESULTS OF SYSTEM ANALYSIS BY RESULTS OF IDENTIFICATION - A SURVEY

H.G. Natke
Curt-Risch-Institut
Universität Hannover

1. Introduction

System analysis based on design drawings yields the computational model (Fig. 1). System identification results in the test model (Fig. 2). The question arises: What has to be done with both results? In general, the values of the computational model differ from the corresponding magnitudes of the test model. Therefore a correction is needed to improve the computational model by identification results (Fig. 3). This means that the adjustment of the calculated values to the measured ones has to be introduced in such a manner that the properties of the configurations which are

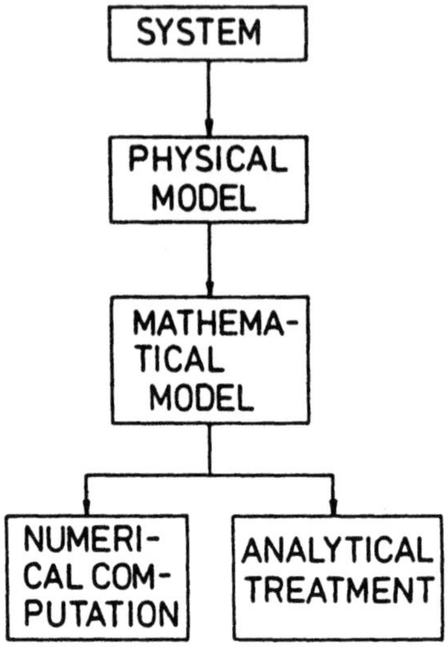

Fig. 1 Computational Model

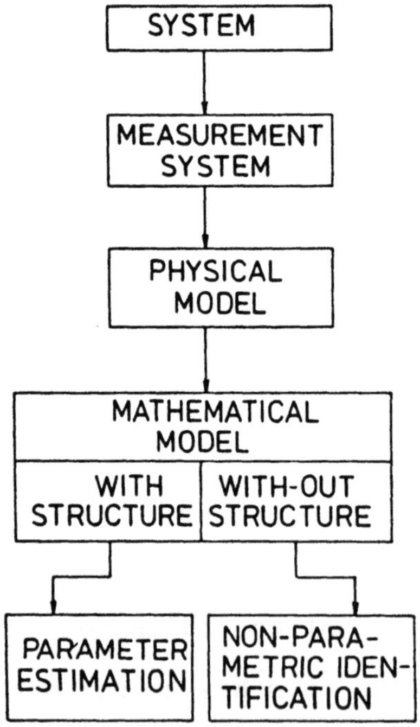

Fig. 2 Test Model

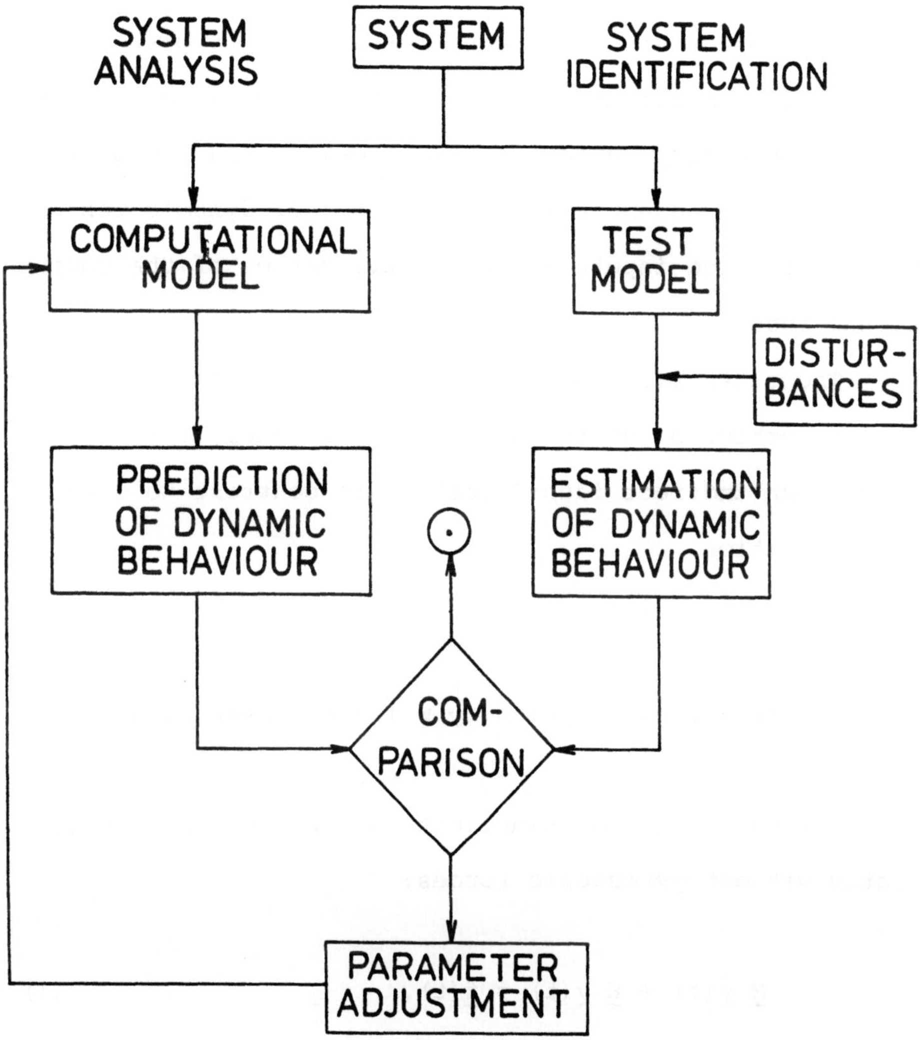

Fig. 3 Comparison of the Computational and Test Model, Parameter Adjustment

not measured will also be improved. If the identified values themselves are not needed for special purposes, the expensive identification, e.g. of eigen-magnitudes, proves to be unnecessary. A more straightforward and inexpensive method is, for example the measurement of dynamic responses to well-defined inputs and the optimization of the computational model.

The classification of parameter adjustment, or generally the correction of the computational model, may be signified as an indirect identification in contrast to direct identification.

2. Mathematical Description by Weighted Least Squares

Let us assume for simplicity a discrete conservative system without gyroscopic forces:

$$\underline{M}\,\ddot{\underline{y}}(t) + \underline{K}\,\underline{y}(t) = \underline{f}(t). \qquad (1)$$

$\underline{M}$: inertia matrix, symmetrical, real elements, positive definite, order n

$\underline{K}$: stiffness matrix, symmetrical, real elements, non-negative definite, order n

$\underline{y}(t)$: vector of the displacements, n components, output vector

$\underline{f}(t)$: vector of external forces, input vector.

An underlined letter means a vector or a matrix. A dot over a letter indicates differentation with respect to time. A superscript T means transposition. The following description is close to that of Natke.[1]

2.1 General Formulation

$V_{\nu n}$, ν, $n \in \mathbb{N}$ describes the real vector space of dimension $n \cdot \nu$ with the inner product $<\ldots>$ and the norm $\|\ldots\|$. Let $\mathcal{L}_{n,n}$ be the space of real symmetrical and positive operators $V_n \rightarrow V_n$, then it is valid with $\underline{T} \in \mathcal{L}_{n,n}$:

$$\bigwedge_{\underline{r},\underline{s} \in V_n} <\underline{T}\ \underline{r},\underline{s}> = <\underline{T}\ \underline{s},\underline{r}>, \quad \bigwedge_{\substack{\underline{r} \in V_n \\ \underline{r} \neq 0}} <\underline{T}\ \underline{r},\underline{r}>> 0. \tag{2}$$

$\mathcal{L}'_{n,n}$ may be the space of real symmetrical operators $V_n \rightarrow V_n$. In the following we will not distinguish between a parameter and the corresponding estimation.

If $\underline{d} \in V_R$, $R \in \mathbb{N}$, is the parameter vector to be estimated and $\underline{G} \in \mathcal{L}_{\nu n, \nu n}$, $\underline{G} \neg \underline{G}(d)$ is the weighting operator, then with the residual $\underline{e}(\underline{d}) \in V_{\nu n}$ which has to be defined later, the cost function is defined by

$$J(\underline{d}) := \langle \underline{G}\,\underline{e}, \underline{e} \rangle = \mathrm{Min}(\underline{d}). \tag{3}$$

The necessary conditions

$$\frac{\partial J(\underline{d})}{\partial d_\sigma} = 2 \langle \underline{G}\,\underline{e}, \frac{\partial \underline{e}}{\partial d_\sigma} \rangle = 0, \quad \sigma = 1(1)R \tag{4a}$$

or

$$\underline{e}^T \underline{G} \frac{\partial \underline{e}}{\partial d_\sigma} = \frac{\partial \underline{e}^T}{\partial d_\sigma} \underline{G}\,\underline{e} = 0 \tag{4b}$$

lead in the case of $\underline{e}$ linear in the parameters with

$$\underline{e} \overset{!}{=} \sum_{\rho=1}^{R} d_\rho \frac{\partial \underline{e}}{\partial d_\rho} - \underline{f}, \quad \underline{f} \in V_{\nu n}, \quad \frac{\partial \underline{e}}{\partial d_\rho} \neg \frac{\partial \underline{e}(\underline{d})}{\partial d_\rho} \tag{5}$$

to a linear system of equations for $\underline{d}$. With the definition, independent of (5),

$$D\underline{e}\ (\underline{d}) := \left[\frac{\partial \underline{e}}{\partial d_1}, \ldots, \frac{\partial \underline{e}}{\partial d_R} \right] \tag{6}$$

results from (5)

$$\underline{e} = D\underline{e}\ \underline{d} - \underline{f}\ , \tag{7}$$

$$\frac{\partial \underline{e}^T}{\partial d_\sigma}\ \underline{G}(D\underline{e}\ \underline{d} - \underline{f}) = 0,\ \sigma = 1(1)R, \tag{8}$$

$$D\underline{e}^T\ \underline{G}\ (D\underline{e}\ \underline{d} - \underline{f}) = \underline{0}. \tag{9}$$

In the non-linear case, when (5) is not valid, a linearization may be introduced e.g. Stoer.[2] Instead of the functional (3) the following functional with given approximation $\tilde{\underline{d}}$ will be considered:

$$< \underline{G}[\underline{e}(\tilde{\underline{d}}) + D\underline{e}(\tilde{\underline{d}})\underline{s}]\ ,\ \underline{e}(\tilde{\underline{d}}) + D\underline{e}(\tilde{\underline{d}})\ \underline{s} > \ = \text{Min}(\underline{s}), \tag{10}$$

$$\underline{s} := \underline{d} - \tilde{\underline{d}}\ , \tag{11}$$

where $\underline{d}$ is a better approximation than $\tilde{\underline{d}}$.

The iteration procedure is:

$$\begin{aligned} &< \underline{G}[\underline{e}(\underline{d})^{(1)}) + D\underline{e}(\underline{d}^{(1)})\underline{s}^{(1)}]\ ,\ \underline{e}(\underline{d}^{(1)}) + D\underline{e}(\underline{d}^{(1)})\underline{s}^{(1)} > = \\ &\qquad = \text{Min}\ (\underline{s}^{(1)}), \\ &\varphi(\tau) := < \underline{G}\ \underline{e}(\underline{d}^{(1)} + \tau\underline{s}^{(1)}),\ \underline{e}(\underline{d}^{(L)} + \tau\underline{s}^{(1)}) > , \end{aligned} \tag{12}$$

m is the smallest integer with

$$\varphi(2^{-2}) < \varphi(0) = < \underline{G}\ \underline{e}(\underline{d}^{(1)}),\ \underline{e}(\underline{d}^{(1)}) >,$$

$$\underline{d}^{(1+1)} := \underline{d}^{(1)} + 2^{-m}\ \underline{s}^{(1)} .$$

The eigen-magnitudes of the system (1) are described by the matrix eigenvalue problem

$$(-\lambda_i\ \underline{M} + \underline{K})\underline{x}_i = \underline{O},\ i = 1(1)n,\ \underline{M},\ \underline{K} \in \mathcal{L}_{n,n} \tag{13}$$

with positive values $\lambda_i := \omega_{oi}^2$ and $\underline{x}_i \in V_n$. The normalization of the eigenvectors may be defined by

$$\|\underline{M}\ \underline{x}_i\| := < \underline{M}\ \underline{x}_i,\ \underline{x}_i > = 1. \tag{14}$$

The equations of motion for sinusoidal excitation follow from eq. (1) to

$$(-\Lambda_r\ \underline{M} + \underline{K})\ \underline{y}_r = \underline{p}_r,\ \underline{f}(t) = \underline{p}\ e^{j\Omega_r t},$$

$$r = 1(1)N,\ N \geq 1,\ \underline{y}_r,\ \underline{p}_r \in V_n. \tag{15}$$

$\Lambda_r := \Omega_r^2 \geq 0$. Equations (13) and (15) describe the computational model of system analysis; the superscript M to the corresponding magnitudes denotes the measured values (test model), and the superscript K denotes the corrected computational model .

The correction is shown only for the stiffness and the flexibility matrix. As shown by Natke,[3] the same formalism can be applied to adjust the inertia matrix if the eigenvalue problem is written for the inverse eigenvalue. For economic and practical reasons the correction is formulated in the form of submatrices (subsystems),

$$\underline{K}^K := \sum_{\varrho=1}^{R} a_\varrho \, \underline{K}_\varrho, \; \underline{K}^K \varepsilon \mathcal{L}_{n,n}, \; \underline{K}_\varrho \varepsilon \mathcal{L}'_{n,n}, \tag{16}$$

$$\underline{C}^K := (\underline{K}^K)^{-1} := \sum_{\varrho=1}^{R} b_\varrho \, \underline{C}_\varrho \, , \; C_\varrho \varepsilon \mathcal{L}'_{n,n}, \tag{17}$$

$\underline{K}_\varrho$ and $\underline{C}_\varrho$ are assumed linear independent. The parameters a_ϱ , b_ϱ are the wanted correction factors. Another possibility may be the indirect adjustment of the stiffness matrix by using the dynamic flexibility matrix

$$\underline{F}_o^K := (-\Lambda_o \, \underline{M} + \underline{K}^K)^{-1} := \sum_{\varrho=1}^{R} c_{o\varrho} \, \underline{F}_{o\varrho} \, , \; \underline{F}_{o\varrho} \varepsilon \mathcal{L}'_{n,n}. \tag{18}$$

Because of linear independence one must choose

$R \leq \mathrm{Min}\ (\nu\, n, \frac{n(n+1)}{2})$ when using eigen-magnitudes

and $R \leq Nn$ when using dynamic responses.

2.2 Mathematical Preparations

With the statements (16) - (18) the system parameter matrices under consideration are functions of several parameters to be estimated. Therefore the corrected eigenvalues λ_i^K and eigenvectors depend on these parameters. The essential equations are briefly listed below:

$$\underline{x}_i^{KT} \underline{M} \underline{x}_i^K = 1, \curvearrowright \underline{x}_i^{KT} \underline{M} \frac{\partial \underline{x}_i^K}{\partial a_\sigma} = \underline{x}_i^{KT} \underline{M} \frac{\partial \underline{x}_i^K}{\partial b_\sigma} = 0, \tag{19}$$

$$(13), (16) \curvearrowright (-\frac{\partial \lambda_i^K}{\partial a_\sigma} \underline{M} + \underline{K}_\sigma) \underline{x}_i^K + (-\lambda_i^K \underline{M} + \underline{K}^K) \frac{\partial \underline{x}_i^K}{\partial a_\sigma} = \underline{0}$$

$$\curvearrowright \frac{\partial \lambda_i^K}{\partial a_\sigma} = \underline{x}_i^{KT} \underline{K}_\sigma \underline{x}_i^K . \tag{20}$$

$$(20) \curvearrowright (-\lambda_i^K \underline{M}+\underline{K}^K) \frac{\partial \underline{x}_i^K}{\partial a_\sigma} = (\underline{x}_i^{KT} \underline{K}_\sigma \underline{x}_i^K \underline{M}-\underline{K}_\sigma) \underline{x}_i^K \tag{21}$$

or

$$\frac{\partial \underline{x}_i^K}{\partial a_\sigma} = \sum_{j=1}^{n} \alpha_{ij} \underline{x}_j^K \tag{22a}$$

$$\underline{x}_r^{KT} (-\lambda_i^K \underline{M}+\underline{K}^K) \alpha_{ir} \underline{x}_r^K = \underline{x}_r^{KT} (\underline{x}_i^{KT} \underline{K}_\sigma \underline{x}_i^K \underline{M}-\underline{K}_\sigma) \underline{x}_i^K,$$

$$\alpha_{ir} = \frac{\underline{x}_r^{KT} \underline{K}_\sigma \underline{x}_i^K}{\lambda_i^K - \lambda_r^K}, \quad i \neq r \; ; \; \alpha_{ii} = 0 . \tag{22b}$$

The derivations in the flexibility formulation are similar.

Finally, the following equations are valid for the dynamic response problem:

$$\left.\begin{aligned}
&\underline{H}_r^K := (-\Lambda_r \underline{M} + \underline{K}^K),\\
&\underline{H}_r^K \underline{y}_r^K = \underline{p}_r, \Lambda_r \neq \lambda_i^K,\\
&\underline{K}_\sigma \underline{y}_r^K + \underline{H}_r^K \frac{\partial \underline{y}_r^K}{\partial a_\sigma} = \underline{o},\\
&\frac{\partial \underline{y}_r^K}{\partial a_\sigma} = -(\underline{H}_r^K)^{-1} \underline{K}_\sigma \underline{y}_r^K.
\end{aligned}\right\} \quad (23)$$

2.3 Procedures and their Comparison

The procedures are listed in Table 1. They contain all used magnitudes including the vector of residuals. For simplicity it is assumed that the number of identified eigenmagnitudes is equal to the number of degrees-of-freedom. This assumption may be dismissed, as shown by Natke et al.[4] The comparison is made in Table 2. With regard to covariance statements and sensitivity investigations see Natke.[1]

If the adjustment procedures lead to a system of linear equations, then a unique solution exists. If the resulting system of equations is non-linear in the parameters,

Specification	1	Correction Using Eigenvalues
Correctionoperator	1	$\underline{K}^K(\underline{a})$
Identified Values	2	λ_i^M
Residual Vector $\underline{e}$	3	$\{\lambda_i^K - \lambda_i^M\} = \{\underline{x}_i^{KT}\underline{K}^K\underline{x}_i^K - \lambda_i^M\}$
$\underline{G}, \nu$	4	$\mathrm{diag}(g_i),\ \nu = 1$
$\frac{\partial \underline{e}}{\partial d_\sigma},\ \sigma = 1(1)R$	5	$\left\{\frac{\partial \lambda_i^K}{\partial a_\sigma}\right\} = \{\underline{x}_i^{KT}\underline{K}_\sigma\underline{x}_i^K\}$
Necessary Condition $\frac{\partial J(\underline{d})}{\partial d_\sigma} = 0$	6	$\sum_{i=1}^{n} g_i \left(\sum_{\varrho=1}^{R} a_\varrho \underline{x}_i^{KT}\underline{K}_\varrho\underline{x}_i^K - \lambda_i^M\right)\underline{x}_i^{KT}\underline{K}_\sigma\underline{x}_i^K = 0$
Essential Additional Equations (Adaptive Method)	7	$(-\lambda_i^K\underline{M} + \underline{K}^K)\underline{x}_i^K = \underline{0},\ \underline{x}_i^{KT}\underline{M}\underline{x}_i^K = 1$
Special Case of "Linear"Method $\underline{f}$	8	$\{\lambda_i^M\}$

Table 1

Specification		2 Correction Using Inverse Eigenvalues
Correctionoperator	1	$\underline{C}^K(\underline{b})$
Identified Values	2	λ_i^M
Residual Vector $\underline{e}$	3	$\left\{ \frac{1}{\lambda_i^K} - \frac{1}{\lambda_i^M} \right\} = \left\{ \underline{x}_i^{KT} \underline{M}\underline{C}^K \underline{M}\underline{x}_i^K - \frac{1}{\lambda_i^M} \right\}$
$\underline{G}, \nu$	4	$\mathrm{diag}\,(g_i)\,\nu = 1$
$\frac{\partial \underline{e}}{\partial d_\sigma}$, $\sigma = 1(1)R$	5	$\left\{ \frac{\partial}{\partial b_\sigma} \left(\frac{1}{\lambda_i^K} \right) \right\} = \left\{ \underline{x}_i^{KT} \underline{M}\underline{C}_\sigma \underline{M}\underline{x}_i^K \right\}$
Necessary Condition $\frac{\partial J(\underline{d})}{\partial d_\sigma} = 0$	6	$\sum_{i=1}^{n} g_i \left(\sum_{\varrho=1}^{R} b_\varrho \underline{x}_i^{KT} \underline{M}\underline{C}_\varrho \underline{M}\underline{x}_i^K - \frac{1}{\lambda_i^M} \right) \underline{x}_i^{KT} \underline{M}\underline{C}_\sigma \underline{M}\underline{x}_i^K = 0$
Essential Additional Equations (Adaptive Method)	7	$\left(\underline{C}^K \underline{M} - \frac{1}{\lambda_i^K} \underline{I} \right) \underline{x}_i^K = \underline{0}, \ \underline{x}_i^{KT} \underline{M}\underline{x}_i^K = 1$
Special Case of "Linear" Method $\underline{f}$	8	$\left\{ \frac{1}{\lambda_i^M} \right\}$

Table 1 (Continued)

Specification		3 Correction Using Eigenvalues and Eigenvectors
Correctionoperator	1	$\underline{K}^{K}(\underline{a})$
Identified Values	2	$\lambda_i^M, \underline{x}_i^M$
Residual Vector $\underline{e}$	3	$\{\underline{u}_i^K - \underline{u}_i^M\}, \underline{u}_i := (\underline{x}_i^T, \lambda_i)^T$
$\underline{G}, \nu$	4	$\mathrm{diag}(\underline{G}_{2i}), \underline{G}_{2i} := \begin{pmatrix} G_{1i}, 0 \\ 0, g_i \end{pmatrix}, \underline{G}_{1i} \in \mathcal{L}_{n,n}, \nu = n+1$
$\frac{\partial \underline{e}}{\partial d_\sigma}, \sigma = 1(1)R$	5	$\left\{\frac{\partial \underline{u}_i^K}{\partial a_\sigma}\right\} = \left\{\left(\frac{\partial \underline{x}_i^K}{\partial a_\sigma}, \frac{\partial \lambda_i^K}{\partial a_\sigma}\right)^T\right\}, \frac{\partial \lambda_i^K}{\partial a_\sigma} = \underline{x}_i^{KT} \underline{K}_\sigma \underline{x}_i^K$ a) $(-\lambda_i^K \underline{M} + \underline{K}^K) \frac{\partial \underline{x}_i^K}{\partial a_\sigma} = (\underline{x}_i^{KT} \underline{K}_\sigma \underline{x}_i^K \underline{M} - \underline{K}_\sigma) \underline{x}_i^K$ b) $\frac{\partial \underline{x}_i^K}{\partial a_\sigma} = \sum_{\substack{j=1 \\ j \neq i}}^{n} \frac{\underline{x}_j^{KT} \underline{K}_\sigma \underline{x}_i^K}{\lambda_i^K - \lambda_j^K} \underline{x}_j^K$
Necessary Condition $\frac{\partial J(\underline{d})}{\partial d_\sigma} = 0$	6	$\sum_{i=1}^{n} \Big[g_i \Big(\sum_{\rho=1}^{R} a_\rho \underline{x}_i^{KT} \underline{K}_\rho \underline{x}_i^K - \lambda_i^M \Big) \underline{x}_i^{KT} \underline{K}_\sigma \underline{x}_i^K +$ $+ (\underline{x}_i^K - \underline{x}_i^M)^T \underline{G}_{1i} \frac{\partial \underline{x}_i^K}{\partial a_\sigma} \Big] = 0$
Essential Additional Equations (Adaptive Method)	7	$(-\lambda_i^K \underline{M} + \underline{K}^K) \underline{x}_i^K = 0, \underline{x}_i^{KT} \underline{M} \underline{x}_i^K = 1$
Special Case of "Linear" Method $\underline{f}$	8	—

Table 1 (Continued)

Specification		4 Correction Using Inverse Eigenvalues and Eigenvectors
Correctionoperator	1	$\underline{C}^K(\underline{b})$
Identified Values	2	$\lambda_i^M,\ \underline{x}_i^M$
Residual Vector $\underline{e}$	3	$\{\underline{\nu}_i^K - \underline{\nu}_i^M\},\ \underline{\nu}_i := \left(\underline{x}_i^T,\ \frac{1}{\lambda_i}\right)^T$
$\underline{G}, \nu$	4	$\mathrm{diag}(\underline{G}_{2i}),\ \underline{G}_{2i} := \begin{pmatrix} \underline{G}_{1i}, 0 \\ 0, g_i \end{pmatrix},\ \underline{G}_{1i} \in \mathcal{L}_{n,n},\ \nu = n+1$
$\frac{\partial \underline{e}}{\partial d_\sigma},\ \sigma = 1(1)R$	5	$\left\{\frac{\partial \underline{\nu}_i^K}{\partial b_\sigma}\right\} = \left\{\left(\frac{\partial \underline{x}_i^{KT}}{\partial b_\sigma},\ \frac{\partial}{\partial b_\sigma}\left(\frac{1}{\lambda_i^K}\right)\right)^T\right\}$ $\frac{\partial}{\partial b_\sigma}\left(\frac{1}{\lambda_i^K}\right) = \underline{x}_i^{KT} \underline{M}\,\underline{C}_\sigma \underline{M}\,\underline{x}_i^K$ a) $\left(\underline{C}^K M - \frac{1}{\lambda_i^K}\underline{I}\right)\frac{\partial \underline{x}_i^K}{\partial b_\sigma} = (\underline{x}_i^{KT} \underline{M}\,\underline{C}_\sigma \underline{M}\,\underline{x}_i^K \underline{I} - \underline{C}_\sigma \underline{M})\,\underline{x}_i^K$ b) $\frac{\partial \underline{x}_i^K}{\partial b_\sigma} = \sum_{\substack{j=1 \\ j\neq i}}^{n} \frac{\underline{x}_j^{KT} \underline{M}\,\underline{C}_\sigma \underline{M}\,\underline{x}_i^K}{1/\lambda_i^K - 1/\lambda_j^K}\,\underline{x}_j^K$
Necessary Condition $\frac{\partial J(\underline{d})}{\partial d_\sigma} = 0$	6	$\sum_{i=1}^{n}\left[g_i\left(\sum_{\varrho=1}^{R} b_\varrho \underline{x}_i^{KT} \underline{M}\,\underline{C}_\varrho \underline{M}\,\underline{x}_i^K - \frac{1}{\lambda_i^M}\right)\underline{x}_i^{KT} \underline{M}\,\underline{C}_\sigma \underline{M}\,\underline{x}_i^K\right.$ $\left. + (\underline{x}_i^K - \underline{x}_i^M)^T \underline{G}_{1i} \frac{\partial \underline{x}_i^K}{\partial b_\sigma}\right] = 0$
Essential Additional Equations (Adaptive Method)	7	$\left(\underline{C}^K \underline{M} - \frac{1}{\lambda_i^K}\underline{I}\right)\underline{x}_i^K = 0,\ \underline{x}_i^{KT} \underline{M}\,\underline{x}_i^K = 1$
Special Case of "Linear" Method $\underline{f}$	8	—

Table 1 (Continued)

Specification	5	Correction Using Generalized Energy
Correctionoperator	1	$\underline{K}^K(\underline{a})$
Identified Values	2	$\lambda_i^M, \underline{x}_i^M$
Residual Vector $\underline{e}$	3	$\{\lambda_i^K \underline{x}_i^{KT} \underline{G}_i \underline{x}_i^K - \lambda_i^M \underline{x}_i^{MT} \underline{G}_i \underline{x}_i^M\}, \underline{G}_i \in \mathcal{L}_{n,n},$
$\underline{G}, \nu$	4	$\mathrm{diag}\,(g_{1i}), \nu = 1$
$\frac{\partial \underline{e}}{\partial d_\sigma}, \sigma = 1(1)R$	5	$\left\{\frac{\partial}{\partial a_\sigma}(\lambda_i^K \underline{x}_i^{KT} \underline{G}_i \underline{x}_i^K)\right\} = \left\{\left(\frac{\partial \lambda_i^K}{\partial a_\sigma} \underline{x}_i^{KT} \underline{G}_i + 2\lambda_i^K \frac{\partial \underline{x}_i^{KT}}{\partial a_\sigma} \underline{G}_i\right) \underline{x}_i^K\right\}$ Partial Derivatives, see Column 3, Row 5
Necessary Condition $\frac{\partial J(\underline{d})}{\partial d_\sigma} = 0$	6	$\sum_{i=1}^{n} g_{1i}(\lambda_i^K \underline{x}_i^{KT} \underline{G}_i \underline{x}_i^K - \lambda_i^M \underline{x}_i^{MT} \underline{G}_i \underline{x}_i^M) \cdot \left(\frac{\partial \lambda_i^K}{\partial a_\sigma} \underline{x}_i^{KT} \underline{G}_i + 2\lambda_i^K \frac{\partial \underline{x}_i^{KT}}{\partial a_\sigma} \underline{G}_i\right) \underline{x}_i^K = 0$
Essential Additional Equations (Adaptive Method)	7	$(-\lambda_i^K \underline{M} - \underline{K}^K) \underline{x}_i^K = \underline{0}, \underline{x}_i^{KT} \underline{M} \underline{x}_i^K = 1$
Special Case of "Linear" Method $\underline{f}$	8	–

Table 1 (Continued)

Specification		6 Correction Using (Force) Zero Vectors of the Eigenvalue Probl.
Correctionoperator	1	$\underline{K}^K(\underline{a})$
Identified Values	2	λ_i^M
Residual Vector $\underline{e}$	3	$\{\varepsilon_i^{(1)}\}, \varepsilon_i^{(1)} := (-\lambda_i^M \underline{M} + \underline{K}^K)\underline{x}_i^K$
$\underline{G}, \nu$	4	$\mathrm{diag}\,(G_i), G_i \in \mathcal{L}_{n,n}, \nu = n$
$\frac{\partial \underline{e}}{\partial d_\sigma}$, $\sigma = 1(1)R$	5	$\left\{\underline{K}_\sigma \underline{x}_i^K + (-\lambda_i^M \underline{M} + \underline{K}^K) \frac{\partial \underline{x}_i^K}{\partial a_\sigma}\right\}$, Partial Derivatives, see Column 3, Row 5
Necessary Condition $\frac{\partial J(\underline{d})}{\partial d_\sigma} = 0$	6	$\sum_{i=1}^{n} \underline{x}_i^{KT}(-\lambda_i^M \underline{M} + \underline{K}^K)\underline{G}_i \cdot \left[\underline{K}_\sigma \underline{x}_i^K + (-\lambda_i^M \underline{M} + \underline{K}^K)\frac{\partial \underline{x}_i^K}{\partial a_\sigma}\right] = 0$
Essential Additional Equations (Adaptive Method)	7	$(-\lambda_i^K \underline{M} + \underline{K}^K)\underline{x}_i^K = 0,\ \underline{x}_i^{KT} \underline{M} \underline{x}_i^K = 1$
Special Case of "Linear" Method $\underline{f}$	8	—

Table 1 (Continued)

Specification		7 Correction Using (Force) Zero Vectors of the Eigenvalue Problem
Correctionoperator	1	$\underline{K}^{K}(\underline{a})$
Identified Values	2	$\lambda_i^M, \underline{x}_i^M$
Residual Vector $\underline{e}$	3	$\{\varepsilon_i^{(2)}\}, \varepsilon_i^{(2)} := (-\lambda_i^M \underline{M} + \underline{K}^K)\underline{x}_i^M$
$\underline{G}, \nu$	4	$\mathrm{diag}(\underline{G}_i), \underline{G}_i \in \mathcal{L}_{n,n}, \nu = n$
$\frac{\partial \underline{e}}{\partial d_\sigma}, \sigma = 1(1)R$	5	$\{\underline{K}_\sigma \underline{x}_i^M\}$
Necessary Condition $\frac{\partial J(\underline{d})}{\partial d_\sigma} = 0$	6	$\sum_{i=1}^{n} \underline{x}_i^{MT}\left(-\lambda_i^M \underline{M} + \sum_{\varrho=1}^{R} a_\varrho \underline{K}_\varrho\right)\underline{G}_i \underline{K}_\sigma \underline{x}_i^M = 0$
Essential Additional Equations (Adaptive Method)	7	–
Special Case of "Linear" Method $\underline{f}$	8	$\{\lambda_i^M \underline{M}\underline{x}_i^M\}$

Table 1 (Continued)

Table 1 (Continued)

Specification		8 Correction Using (Displacement) Zero Vectors of the Eigenvalue Probl.
Correctionoperator	1	$\underline{C}^K(\underline{b})$
Identified Values	2	λ_i^M
Residual Vector $\underline{\varepsilon}$	3	$\{\underline{C}^K\varepsilon_i^{(1)}\}$, $\underline{C}^K\varepsilon_i^{(1)} := (-\lambda_i^M \underline{C}^K \underline{M} + \underline{I})\, \underline{x}_i^K$
$\underline{G}, \nu$	4	$\mathrm{diag}(\underline{G}_i)$, $\underline{G}_i \in \mathcal{L}_{n,n}$, $\nu = n$
$\dfrac{\partial \underline{\varepsilon}}{\partial d_\sigma}$, $\sigma = 1(1)R$	5	$\left\{-\lambda_i^M \underline{C}_\sigma \underline{M}\,\underline{x}_i^K + (-\lambda_i^M \underline{C}^K \underline{M} + \underline{I}) \dfrac{\partial \underline{x}_i^K}{\partial b_\sigma}\right\}$ Partial Derivatives, see Column 4, Row 5
Necessary Condition $\dfrac{\partial J(\underline{d})}{\partial d_\sigma} = 0$	6	$\sum_{i=1}^{n} \underline{x}_i^{KT}(-\lambda_i^M \underline{C}^K \underline{M} - \underline{I})^T \underline{G}_i \left[-\lambda_i^M \underline{C}_\sigma \underline{M}\,\underline{x}_i^K + (-\lambda_i^M \underline{C}^K \underline{M} + \underline{I}) \dfrac{\partial \underline{x}_i^K}{\partial b_\sigma}\right] = 0$
Essential Additional Equations (Adaptive Method)	7	$(-\lambda_i^K \underline{C}^K \underline{M} + \underline{I})\, \underline{x}_i^K = 0$, $\underline{x}_i^{KT} \underline{M}\, \underline{x}_i^K = 1$
Special Case of "Linear" Method $\underline{f}$	8	–

Specification	9	Correction Using (Displacement) Zero Vectors of the Eigenvalue Probl.
Correctionoperator	1	$\underline{C}^{K}(\underline{b})$
Identified Values	2	$\lambda_i^M, \underline{x}_i^M$
Residual Vector $\underline{e}$	3	$\{\underline{C}^K \varepsilon_i^{(2)}\}, \underline{C}^K \varepsilon_i^{(2)} := (-\lambda_i^M \underline{C}^K \underline{M} + \underline{I})\, \underline{x}_i^M$
$\underline{G}, \nu$	4	$\mathrm{diag}(\underline{G}_i), \underline{G}_i \in \mathcal{L}_{n,n},\ \nu = n$
$\frac{\partial \underline{e}}{\partial d_\sigma}, \sigma = 1(1)R$	5	$\{-\lambda_i^M \underline{C}_\sigma \underline{M}\, \underline{x}_i^M\}$
Necessary Condition $\frac{\partial J(\underline{d})}{\partial d_\sigma} = 0$	6	$\sum_{i=1}^{n} \underline{x}_i^{MT} \left[-\lambda_i^M \left(\sum_{\varrho=1}^{R} b_\varrho \underline{C}_\varrho \right) \underline{M} + \underline{I} \right]^T \underline{G}_i \lambda_i^M \underline{C}_\sigma \underline{M}\, \underline{x}_i^M = 0$
Essential Additional Equations (Adaptive Method)	7	-
Special Case of "Linear" Method $\underline{f}$	8	$\{-\underline{x}_i^M\}$

Table 1 (Continued)

Specification		10 Correction Using Force Vectors
Correctionoperator	1	$\underline{K}^K(\underline{a})$
Identified Values	2	$\underline{p}_r^M, \underline{y}_r^M, r = 1(1)N$
Residual Vector $\underline{e}$	3	$\{\underline{p}_r^K - \underline{p}_r^M\}, \underline{p}_r^K = \underline{H}_r^K \underline{y}_r^M$
$\underline{G}, \nu$	4	$\mathrm{diag}(\underline{G}_r), \underline{G}_r \in \mathcal{L}_{n,n}, \nu = N$
$\frac{\partial \underline{e}}{\partial d_\sigma}, \sigma = 1(1)R$	5	$\{\underline{y}_r^M \underline{K}_\sigma\}$
Necessary Condition $\frac{\partial J(\underline{d})}{\partial d_\sigma} = 0$	6	$\sum_{r=1}^{N} \left[\underline{y}_r^{MT}\left(-\Lambda_r \underline{M} + \sum_{\varrho=1}^{R} \underline{a}_\varrho \underline{K}_\varrho\right) - \underline{p}_r^{MT}\right] \underline{G}_r \underline{K}_\sigma \underline{y}_r^M = 0$
Essential Additional Equations (Adaptive Method)	7	–
Special Case of "Linear" Method $\underline{f}$	8	$\{\Lambda_r \underline{M} \underline{y}_r^M + \underline{p}_r^M\}$

Table 1 (Continued)

Specification		11 Correction Using Dynamic Responses
Correctionoperator	1	$\underline{K}^K(\underline{a})$
Identified Values	2	$\underline{y}_r^M, r = 1(1)N$
Residual Vector $\underline{e}$	3	$\{\underline{y}_r^K - \underline{y}_r^M\}$, $\underline{y}_r^K = (\underline{H}_r^K)^{-1}\underline{p}_r$
$\underline{G}, \nu$	4	$\mathrm{diag}(\underline{G}_r)$, $\underline{G}_r \in \mathcal{L}_{n,n}$, $\nu = N$
$\frac{\partial \underline{e}}{\partial d_\sigma}$, $\sigma = 1(1)R$	5	$\{-(-\Lambda_r \underline{M} + \underline{K}^K)^{-1}\underline{K}_\sigma \underline{y}_r^K\}$
Necessary Condition $\frac{\partial J(\underline{d})}{\partial d_\sigma} = 0$	6	$\sum_{r=1}^{N}(\underline{y}_r^K - \underline{y}_r^M)^T \underline{G}_r(-\Lambda_r \underline{M} + \underline{K}^K)^{-1}\underline{K}_\sigma \underline{y}_r^K = 0$
Essential Additional Equations (Adaptive Method)	7	–
Special Case of "Linear" Method $\underline{f}$	8	–

Table 1 (Continued)

Specification		12 Correction Using Dynamic Responses
Correctionoperator	1	$\underline{F}_0^K := (\underline{H}_0^K)^{-1} := \sum_{\varrho=1}^{R} c_{0\varrho} \underline{F}_{0\varrho}$
Identified Values	2	$\underline{y}_r^M, r = 1(1)N$
Residual Vector $\underline{e}$	3	$\{\underline{y}_r^K - \underline{y}_r^M\}, \underline{y}_r^K = \underline{F}_0^K \underline{p}_r$
$\underline{G}, \nu$	4	$\text{diag}(\underline{G}_r), \underline{G}_r \in \mathcal{L}_{n,n}, \nu = N$
$\frac{\partial \underline{e}}{\partial d_\sigma}, \sigma = 1(1)R$	5	$\{\underline{F}_{0\sigma} \underline{p}_r\}$
Necessary Condition $\frac{\partial J(\underline{d})}{\partial d_\sigma} = 0$	6	$\sum_{r=1}^{N} \left[\underline{p}_r^T \left(\sum_{\varrho=1}^{R} c_{0\varrho} \underline{F}_{0\varrho}\right) - \underline{y}_r^{MT}\right] \underline{G}_r \underline{F}_{0\sigma} \underline{p}_r = 0$
Essential Additional Equations (Adaptive Method)	7	–
Special Case of "Linear" Method $\underline{f}$	8	$\{\underline{y}_r^M\}$

Table 1 (Continued)

Table 2: Comparison of the Adjustment Procedures Contained in Table 1

Specification		Method 1	2	3	4	5	6	7	8	9	10	11	12
Test Expenditure:	Exciter Appropriation	(X)	(X)	X	X	X	(X)	X	(X)	X			X
	Equipment Expensive			X	X	X		X		X			
	Great Measurement Duration			X	X	X		X		X			(X)
Measuring Expenditure:	Force Measurement										X		
	Many Measuring points			X	X	X		X		X	(X)	(X)	(X)
	Many Measuring values			X	X	X		X		X	X	(X)	(X)
Computation:	Non-linear Method			X	X	X	X		X			X	
	Iterative Method	X	X	X	X	X	X		X			X	
Computation Expenditure	Additional Matrix Eigenvalue Problem	X		X	X	X	X		X				
	Further Additional Equations x)			X	X	X	X		X			X	
	Adaptive Method	+	+				+		+				

x) Partial Derivations of the Corrected Eigenvectors etc.

one can use the method of steepest descent or, for example, the multidimensional Newton-Raphson method. In that case there may be more than one solution which minimizes the cost function, and it is necessary to find the smallest relative minimum.

The adjustment procedure no.11 (Table 1), which is based on a frequency response residuum, can also be used to correct the stiffness matrix of a system which is actually damped. In that particular case excitation frequencies are selected such that they are off-resonant and thus contain only a slight damping influence. A similar approach is used by Felgenhauer (see the relevant lecture) for the prior correction of the stiffness matrix in a first correction step, during which the damping matrix is held constant.

The correction method no.10 (Table 1) which is based on a force vector residuum leads to a linear simultaneous system of equations in the unknown parameters. The method is thus free from the choice of starting values, but more dependent on the quality of the measurements. Methods 10 and 11 can be performed with non-appropriate constant excitation vectors.

The correction methods 1, 2, 6 and 8 are adaptive (Table 2), which implies that the eigenvectors may also be adjusted despite their absence in the cost function. All of the above-mentioned procedures require an iterative solution technique during which the adaption takes place through the solution of the additional equations (see Table 2).

2.4 Examples

The results of adjusting the flexibility coefficients of a spring-mass system with 6 degrees-of-freedom have been published by Natke et al.[4]

The application to the short-haul aircraft VFW 614 has been made and is reported by Zimmermann et al.[5] Fig. 4 shows the idealization of the structure without control surfaces by a beam model. Fig. 5 contains the convergence of the correction parameters b_ϱ with an input of 20 identified eigenfrequencies (procedure of Table 1, no.2). Similar experience as notified by Zimmermann et al.[5] is made by Demchak et al.[6] The application of the procedure of Table 1, no.11, is demonstrated by Natke[7] for a spring-mass system. The correction of the same system extended through the addition of dampers is described by Felgenhauer.[8,9] The latter reference also contains a correction example performed for a finite element model of a cantilever beam using real measurements. Table 3 contains the iteration results of mass adjustment of the German research platform NORDSEE by means of identified eigenfrequencies by Natke et al.[3]

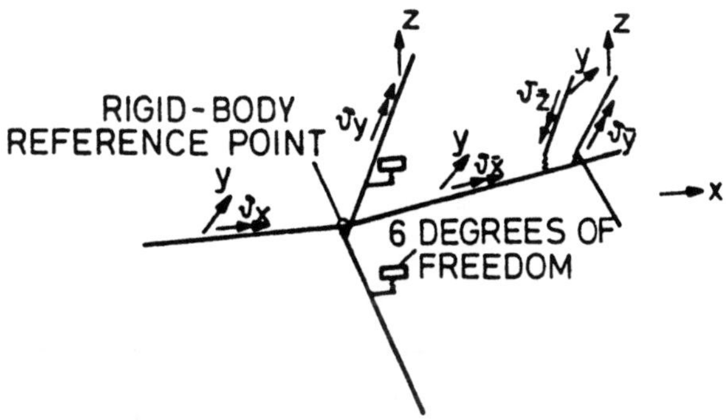

Fig. 4 Idealization of the VFW 614 Structure without Control Surfaces by a Beam Model

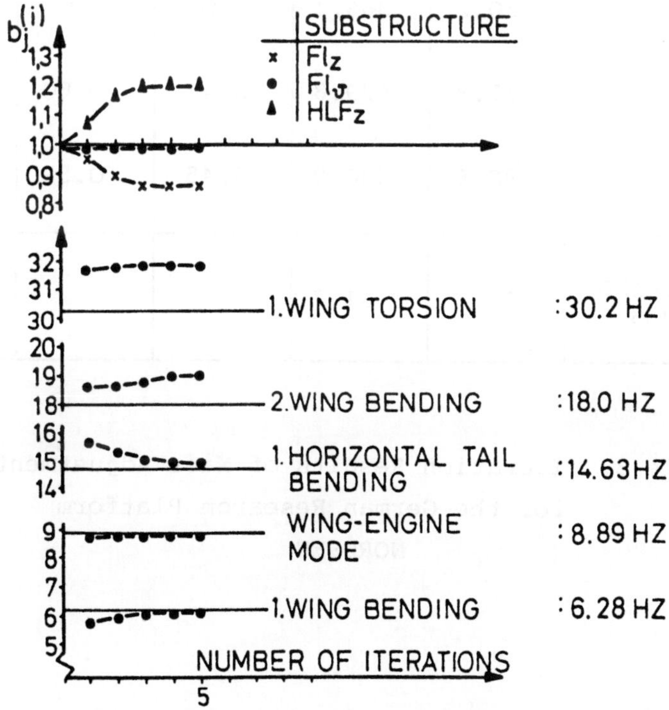

Fig. 5 Convergence of the b_j - Values and Eigenfrequencies against Number of Iterations when Using Measured VFW 614 Eigenfrequencies

NUMBER of ITERATION q	$\lambda_{01}^{K(q)}$	$\lambda_{02}^{K(q)}$	$a_1^{(q)}$	$a_2^{(q)}$	$a_3^{(q)}$
0	150.2	389.1	1.0	1.0	1.0
1	187.8	629.8	3.41	0.60	1.0
2	190.6	630.9	3.45	0.59	1.0
IDENTIFIED VALUES	191.1	631.7			

Table 3: Iteration Results of Mass Adjustment for the German Research Platform NORDSEE

3. Further Methods

A method based on the comparison of calculated and identified eigensolutions and the adjustment of design variables is given in the lecture by Lallement. The lecture by Felgenhauer contains a description and examples of a correction method for damped systems with an analysis-test-criterion based on complex frequency responses. The method is an extension of the adjustment procedure no.11 of Table 1, development by Natke.[7] Collins et al.[10] give the cost function as the variance of the difference between a true vector and the corresponding estimated vector.

Beliveau[11] gives an extension of this procedure to viscous damped systems. The estimated values were determined on the basis of the Maximum-Likelihood and Bayes-criterium.

4. Conclusions

The results of adjustment calculations show:

- o There should be an unambiguous assignment between the degrees-of-freedom of the computational and the test model. If uncertainties exist about the correspondence of certain degrees-of-freedom, these should be omitted.

- If adjustment is based on identified eigenfrequencies, the frequency spectrum of the model must cover all frequencies to be adjusted. A subsystem may be made variable only if an eigenfrequency significant for the subsystem is taken from the test model.
- The simultaneous treatment of subsystems, during which alterations in their flexibilities/stiffnesses lead to different orders of magnitudes in the eigenfrequency shifts, may result in convergence difficulties.
- Major incompatibilities between the physical system of the computational and test models should be avoided, since they lead to convergence problems.

Further investigations are necessary concerning the choice of subsystems and handling of large systems. More experience ought to be gained with real structures.

References

1 Natke, H.G., Vergleich von Algorithmen für die Anpassung des Rechenmodells einer schwingungsfähigen elastomechanischen Struktur an Versuchswerte, ZAMM 59, p. 257-268, 1979

2 Stoer, J., Einführung in die Numerische Mathematik I, Heidelberger Taschenbücher Bd. 105, Springer-Verlag

3 Natke, H.G. and Schulze, H., Parameter Adjustment of a Model of an Offshore Platform from Estimated Eigenfrequencies data, Journal of Sound and Vibration, Vol. 77, No. 2, p. 271 - 285, 1981

4 Natke, H.G., Collmann, D. and Zimmermann, H., Beitrag zur Korrektur des Rechenmodells eines elastomechanischen Systems anhand von Versuchsergebnissen, VDI-Berichte Nr. 221, p. 23-32, 1974

5 Zimmermann, H., Collmann, D. and Natke, H.G., Erfahrungen zur Korrektur des Rechenmodells mit gemessenen Eigenfrequenzen am Beispiel des Verkehrsflugzeuges VFW 614, Z. Flugwiss. Weltraumforsch. 1, Heft 4, p. 278-285, 1977

6 Demchak, L. and Harcrow, H., Analysis of Structural Dynamic Data from Skylab, (1) Technical Discussion, NASA CR 2727, 1976

7 Natke, H.G., Die Korrektur des Rechenmodells eines elastomechanischen Systems mittels gemessener erzwungener Schwingungen, Ing.-Arch. 46, p. 169-184, 1977

8 Felgenhauer, H.-P., Die Korrektur von Rechenmodellen für gedämpfte elastische Systeme mittels gemessener erzwungener Schwingungen, ZAMM 61, Heft 1/5, 1981

9 Felgenhauer, H.-P., Korrektur von Rechenmodellen für gedämpfte elastische Systeme mittels gemessener erzwungener Schwingungen, Fortschritt-Berichte der VDI-Zeitschriften Reihe 11, Nr. 37, 1981

10 Collins, J.D., Hart, G.C., Hasselmann, T.K. and Kennedy, B., Statistical Identification of Structures, AIAA Journal 12, p. 185-190, 1974

11 Beliveau, J.G., Identification of Viscous Damping in Structures from Modal Information, J. Appl. Mech. E 43, p. 335-339, 1976

Application of System Identification with commercially available Hardware and Software

Peter Catchpole, MSc (Eng.)
General Manager
SDRC Engineering Services GmbH
Gustav-Stresemann-Ring 12-16, Wiesbaden

System Analysis via the Building Block Approach.-SABBA

1.0 INTRODUCTION

SDRC SABBA is an interactive system analysis software package available on DEC based processors and the GenRad 2500 series Structural Analysis Systems.

SABBA employs the Building Block Approach to accurately represent complex mechanical systems as either a single or an assembly of components using only a few degrees of freedom. The system natural frequencies and modes are determined together with system response due to external loads.

SABBA is different form other dynamic substructuring programs which employ eigenvalue extraction techniques in that only those degrees of freedom (DOF) that define component connections or enforced displacements/loads need be include in the main solution set. All other DOF data which users may wish to compute (eg. modal vectors, stress etc), may be calculated as a postprocess to the main solution matrix.

Most commonly, a component or entire system is represented in terms of its natural frequencies and modes of vibration. this may be done in two ways:

Experimentally from modal testing - and analysis [1]

or

Analytically from finite element programs, mathematical models or hand calculations.

The system model (or mathematical prototype) can be analyzed to predict ist natural frequencies and modes.

SABBA also determines the system response resulting from dynamic loading: due to Sinusoidal, Random and Transient forces.

Modifications to the model can then be made to predict their effects. By analyzing several design alternatives, the engineer can determine the best design before he builds or alters the hardware prototype.

2. COMPONENT TYPES

A complex system may be divided into discrete sub-assemblies (e.g., tires, frame, suspension, body, etc.) which are separately identified and described. Alternatively, the dynamic characteristics of an entire system may be represented as a single component, as occurs in the case of prototype modal testing. Both sub-assemblies and entire systems are represented in SABBA using the component types described herein.

2.1 Modal Components

Modal components are so named because their input is mode shapes and associated modal parameters (frequency, modal mass, etc.). The modal data may be obtained directly from test or from a mathematical analysis, such as a dynamic FE model solution. Proportional and non-proportional viscous or hysteretic damping formulations may be used.

The complete set of modal parameters required to describe a SABBA modal component is:

1. Natural frequencies (ω_r) real, complex

2. Natural mode shapes (ψ_r) real, complex

3. Modal mass/residue (m_r, $\mathbf{a}_r$)

4. Modal damping ratio (ζ_r, $\mathbf{g}_r$)

A modal representation of a component may be used to define other identical components which are connected to different degrees of freedom without explicitly respecifying the modal characteristics. This is accomplished by defining the subsequent components as Repeated Modal Components. For example, the modal properties of the right front tire of a vehicle may be defined in detail. The characteristics of the other tires may simply be defined as repeated properties of the first tire, which are applied to different degrees of freedom.

SABBA modal components may be considered as three different representations: [1,2,3]

1. dynamic stiffness
2. unconstrained modal method
3. constrained modal method.

2.1.1. Dynamic Stiffness Representation

It is possible to construct directly a 'flexibility' matrix which includes DOF's for component connections. loadings or boundary conditions, using the modal data obtained form experimental or analytical studies

$$(X(j\omega)) = [G(j\omega)] \quad (F(j\omega))$$

where each element of the flexibility matrix represents a frequency response function between input force and output response.

Various representations with proportional or non-proportional damping may be used. [1]

In order to represent the component in the overall system matrix, SABBA inverts the flexibility matrix to obtain the component 'dynamic stiffness' matrix.

$$[G(j\omega)]^{-1} \{X(j\omega)\} = \{F(j\omega)\} \qquad (2.1)$$

or

$$[D(j\omega)] \{X(j\omega)\} = \{F(j\omega)\} \qquad (2.2)$$

where $[D(j\omega)]$ is the component dynamic stiffness matrix. If the user defines more component degrees of freedom than component modes, the flexibility matrix becomes rank-deficient and cannot be directly iverted. In this case, SABBA automatically converts the component to an alternative form as described by the unconstrained modal method.

2.1.2 Unconstrained Modal Method

This method is applicable for individual components analysed with free boundary connection points. The following formulations will be that for a proportional viscous damped system, but can easily be extended to consider the cases of non-proportional viscous or hysteretic damping with residual effects.

Consider an n -degree of freedom system with viscous damping.

$$[M]\ (\ddot{x}) + [C]\ (\dot{x}) + [K]\ (x) = (f) \qquad (2.3)$$

and $[C] = \delta\ [M] + \beta\ [K]$

where δ and β are arbitrary constants.

There exists a set of eigenvalues and eigenvectors which satisty the homogenous equation

$$-\omega^2 r\ [M]\ (\Psi r) + j\omega\ [C]\ (\Psi r) + [K]\ (\Psi r) = 0 \qquad (2.4)$$

The n eigenvectors form a linearly independent set of vectors in n-space. Any vector in n-space can be expressed as an linear combination of these n vectors.

Thus the solution to 2.4 may be written (2.5)

$$(x) = \sum_{r=1}^{n} (\Psi r)\ Qr$$

the following orthogonality conditions apply. (2.6)

$$(\Psi r)^T [C] (\Psi r) = Cr$$

$$(\Psi r)^T [M] (\Psi r) = Mr$$

$$(\Psi r)^T [K] (\Psi r) = Kr$$

using equations 2.5 and 2.6 with 2.4 the equations of motion may be written in modal coordinates as:

$$([\ulcorner K \lrcorner] + j\omega [\ulcorner c \lrcorner] - \omega^2 [\ulcorner m \lrcorner]) (Q) = (\Psi)^T (f) \quad (2.7)$$

$$\text{or} \quad [D] (Q) = (\Psi)^T (f) \quad (2.8)$$

$$\text{where} \quad [D] = [(K + j\omega c - \omega^2 m)]$$

This representation as defined by equations 2.5 and 2.8 can be considered as a form involving modal oscillators and constraint equations.

A modal oscillator is defined for each component mode included as shown in Figure 2.0. SABBA automatically defines the modal degrees of freedom (Qi.j) where i is the component number and j is the mode number. Appropriate scalar springs, masses and dampers are also defined for each modal oscillator. The mode shapes are then converted to constraint equations which connect the modal degrees of freedom to the physical degrees of freedom (X_2).

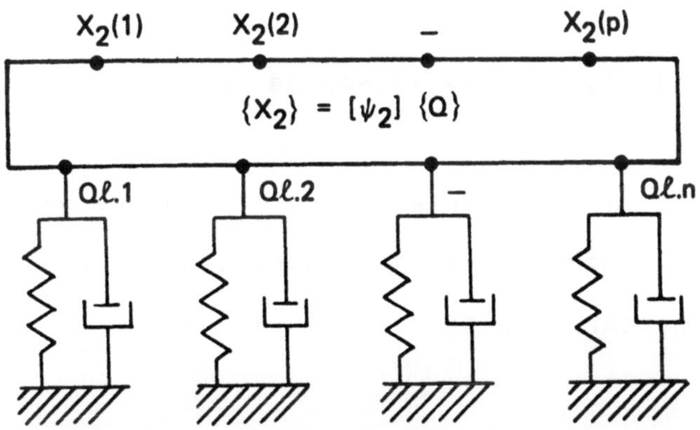

p = Number of free degrees of freedom in $\{X_2\}$
ℓ = Component number
n = Number of modes

Figure 2.0

Schematic Representation of Unconstrained Modal Formulation

2.1.3. Constrained Modal Method

Something it is desirable, or even essential to test a component with one or more connection points restrained to zero motion.

Some reasons for this are:

1. Free-free modes are eliminated avoiding the difficulties in their determination.

2. Fewer modes are required to represent the component since it is tested in a condition that may be similar to its assembled configuration.

3. Damping characteristics can change with frequency. A minimal shift in dominant modes will yield a better damping representation.

Modal testing provides the following data:

$(Xf)=[\Psi]\cdot(Q)$	Displacement mode shapes for the free points.
$(Fr)=[R]\ (Q)$	Force mode shapes for the restrained points.
$[{}^\frown m_\smile]\ ,\ [{}^\frown k_\smile]$	= modal stiffness and mass.

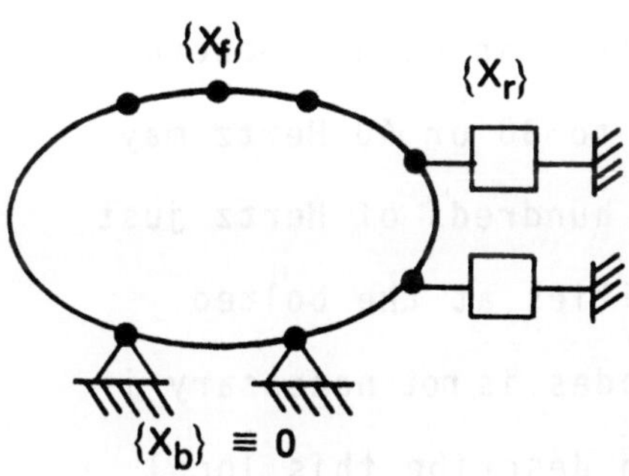

X_f Free degrees of freedom unrestrained ($F_f \equiv 0$) during test/analysis; X_f recorded.

X_r Degrees of freedom restrained ($X_r \equiv 0$) during test/analysis; F_r recorded.

X_b Degrees of freedom restrained ($X_b \equiv 0$) during test/analysis and also restrained boundary conditions in the system model.

Figure 2.1 Restrained Component Degrees of Freedom

2.1.4. Inertia Restraint and Residual Flexibility

Residual effects [1] are one of the most important aspects of the modal component description. For a linear system the physical motion of any point on a structure is governed by the contribution at that point of all the modes of vibration of that system. Errors due to modal truncation can be extreme.

Obtaining sufficient component modes is impractical in cases such as predicting the combined front end sheet metal and firewall interaction in a car. The local residual flexibility at the bolted connections are extremely important in determining the system modes and forced response. If this flexibility is ignored, highly erroneous results can occur. However, to determine system results up to 30 or 40 Hertz may require component modes in the hundreds of Hertz just to describe the local flexibilities at the bolted connections. Obtaining these modes is not necessary if residual flexibility is used to describe this local flexibility.

SABBA provides the capability to develop the flexibility matrix from a finite number of mode shapes within a frequency range ω_1 to ω_2 without loss of accuracy.

2.2. Scalar Components

Scalar components provide the user with a convenient method of specifying the stiffness, mass of damping between any two (2) degrees of freedom in the system model. This elementary version of a matrix component is particularly useful for simple representations of mounts, springs, etc. between components which exhibit symmetric behavior. A scalar component representation should not be used when off-diagonal terms are important or where nonsymmetric behavior exists. Spring and damper scalar components contribute to the dynamical matrix as follows:

$$[D] = a \begin{bmatrix} 1 & -1 \\ -1 & 1 \end{bmatrix} \tag{2.9}$$

where

$a \quad = k$ for a spring

$\quad = j\omega c$ for a viscous damper

$\quad = jg$ for a hysteretic damper

For a lumped mass scalar component (i.e. connected between a physical degree fo freedom and GROUND):

$$[D] = -\omega^2 m. \qquad (2.10)$$

For a mass scalar element connected between two degrees of freedom, the consistent mass matrix for a truss finite element is used:

$$[D] = \frac{-\omega^2 m}{6} \begin{bmatrix} 2 & 1 \\ 1 & 2 \end{bmatrix} \qquad (2.11)$$

Although any two system degrees of freedom can be connected using a scalar element, the physical location of the degrees of freedom should be considered to ensure accurate results. For example, if the line of action of a scalar spring is not coincident with a line between the physical points as shown in Figure 2.2, an unbalanced moment is created which will cause erroneous results. In this case, a matrix component which will account for the noncoincident locations should be used to represent the stiffness between the degrees of freedom.

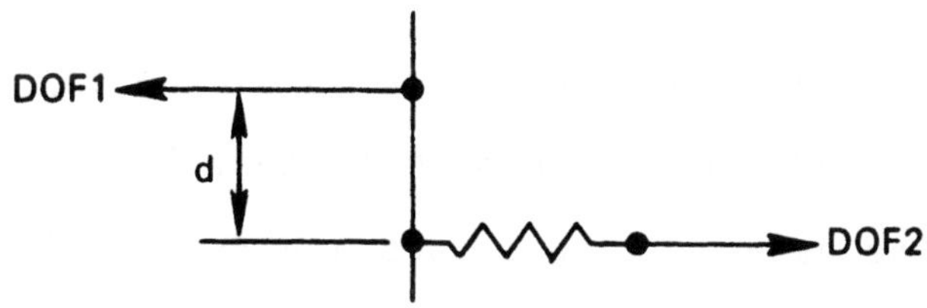

Figure 2.2

Invalid Scalar Spring Connection

2.3. Matrix Components

Matrix components are very useful in describing components having mass, stiffness, and/or damping characteristics which are not easily, or accurately, represented with modal models. Components exhibiting nonsymmetric behavior, such as fluid film bearings or the stiffness and damping properties of a propeller experiencing aerodynamic whirl flutter are normally represented as matrix components.

Another use of a matrix component is the direct introduction into the system model of a single finite element, such as a beam.

Or finite element matrices composed of many different elements which have been generated using guyan reduction techniques. However, care should be employed when using reduced finite elemnt matrices. Since less than 50% of the natural mode shapes and frequencies extracted from a matrix are generally accurate, sufficient matrix size must be specified to ensure the dynamic range accurately spans the frequency range of interest in the problem. If large Guyan reduced matrices are required, it is usually better to employ a modal representation.

Matrix components are described in SABBA by directly specifying mass, stiffness, and/or damping between any degrees of freedom in the system model. The matrices may be symmetric or nonsymmetric, sparce of full and result in a component dynamical matrix of the form.

$$[D] = [K] + j\ [G] + j\omega\ [C] - \omega^2\ [M] \tag{2.12}$$

A matrix definition of a component can be readily repeated for identical components connected to different degrees of freedom without explicitly redefining the matrix.

This is accomplished by defining the subsequent components as Repeated Matrix Components. For example, the matrix properties of a bearing/pedestal subsystem may be defined in detail for connection with a rotor. The characteristics of an other identical subsystem can than be defined as repeated properties applied at different degrees of freedom.

Because of the ability to include non-symmetric matrices, SABBA may be used to determine and study system instabilities.

2.4 Constraint Component

Constraint components allow an arbitrary number of dependent degrees of freedom to be defined in terms of a set of previously defined independent degrees of freedom. Each constraint is described by a single equation that specifies a linear relationship between a dependent degree of freedom and one or more independent degrees of freedom.

$$(X\,d) = [C]\ (Xi) \qquad (2.13)$$

The dependant degrees of freedom may be variables previously defined (e.g. on other components, or new variable names).

Constraint relationships are particularly useful in determining such values as average motion, rigid body motion, relative motion and rigid links.

(2.14)

Average Motion

$$X = aX1 + bX2 + cX3$$

Rigid Body Motion - necessary to characterize a free-free modal component.

$$\begin{bmatrix} X \\ Y \\ Z \\ \theta_x \\ \theta_y \\ \theta_z \end{bmatrix} = \left[\begin{array}{ccc|ccc} 1 & 0 & 0 & 0 & C_z & -C_y \\ 0 & 1 & 0 & -C_z & 0 & C_x \\ 0 & 0 & 1 & C_y & -C_x & 0 \\ \hline 0 & 0 & 0 & 1 & 0 & 0 \\ 0 & 0 & 0 & 0 & 1 & 0 \\ 0 & 0 & 0 & 0 & 0 & 1 \end{array}\right] \begin{bmatrix} X_{cg} \\ Y_{cg} \\ Z_{cg} \\ \theta x_{cg} \\ \theta y_{cg} \\ \theta z_{cg} \end{bmatrix} \qquad (2.15)$$

Relative Motion

$$X = X_2 - X_1 \tag{2.16}$$

Rigid & Links

$$X_1 = X_2 \tag{2.17}$$

For a constraint component, both the dependent and independent degrees of freedom can act as connections to other components. Forces may be applied to both sets of degrees of freedom; however, displacement, velocity, or acceleration cannot be specified at dependent degrees of freedom.

2.5 Auxiliary Component

Auxiliary components allows the user to extract system response at degrees of freedom which are not included directly in the system solution. This capability allows large problems to be solved quickly and economically, since the central solution is not affected by the number of auxiliary equations which are evaluated.

The values determined in auxiliary components may be dependent upon any set of degrees of freedom in the system model. The displacement at a physical degree of freedom in a modal component, for example, can be evaluated by an auxiliary equation descibing the relationship of the physical motion to the modal coordinates Q, and the mode shapes, ψ :

$$X_i = \sum_{j=1}^{n} \psi_{ij} Q_j \qquad (2.18)$$

Dynamic stresses or strains may similarly be evaluated if modal stresses or strains have been obtained from dynamic component tests or analyses.

$$\sigma_i = \sum_{j=1}^{n} \psi^{\sigma}_{ij} Q_j \qquad (2.19)$$

where ψ^{σ}_{ij} = Modal stress coefficients

Evaluating the average motion of several degrees of freedom, such as a vehicle operator seat, may also simply be obtained from the motion at each seat corner, X_i,

$$X_{ave} = \frac{1}{n} \sum_{j=1}^{n} X_i \qquad (2.20)$$

An auxiliary component is very much like a constraint component with one major difference; the dependent degrees of freedom are not part of the solution. This means that the dependent degrees of freedom cannot be used as connections between components nor can loads be applied to them. In fact, the auxiliary dependent degrees of freedom can be specified only once.

2.6. SABBA Beam Component

SABBA provides a general beam element which is an essential modeling capability in many design and troubleshooting problems where scalar elements simply cannot provide for the inclusion of beam effects. The beam must be able to operate over the entire frequency range (not just in the range of interest), cope with arbitrary orientations, offsets and eccentricities and must allow for multiple-spans in one component.

Two formulations are provided in SABBA

1. a finite element formulation

2. a timoshenko formulation (accurate at all frequencies)

Both formulations provide the following

1. Coupled vibrations of Torsion, Bending (axial, flexural)

2. Rigid link offsets

3. Eccentricity of shear ratio

4. Two forms of beam orientation input

5. End releases

6. Shear ratio area

7. Consistent mass formulation

SABBA Beam Matrix Reduction

The beam component matrix $[K-\omega^2 M]$ is initially formulated to include all degrees of freedom for all spans. The matrix is then reduced to the retained degrees of freedom via matrix condensation as follows:

Let X_1 = retained DOF

X_2 = eliminated (reduced out) DOF

$[D] = [K - \omega^2 M]$

$[D](X) = (F)$

Then

$$\begin{bmatrix} D_{11} & D_{12} \\ D_{21} & D_{22} \end{bmatrix} \cdot \begin{bmatrix} X_1 \\ X_2 \end{bmatrix} = \begin{bmatrix} F \\ 0 \end{bmatrix} \qquad (2.21)$$

Expanding

$$D_{11}X_1 + D_{12}X_2 = F$$

$$D_{21}X_1 + D_{22}X_2 = 0$$

$$X_2 = -D_{22}^{-1}D_{21}X_1$$

Therefore

$$D_{11}X_1 - D_{12}D_{22}^{-1}D_{21}X_1 = F$$

$$\text{or } \left[D_{11} - D_{12}D_{22}^{-1}D_{21}\right] X_1 = F$$

The reduced $[D]$, called $[D]^r$ is:

$$[D]^r = \left[D_{11} - D_{12}D_{22}^{-1}D_{21}\right] = \left[K-\omega^2 . M\right]^R \quad (2.22)$$

An interesting advantage of this technique is that it may be possible to solve some very large beam problems, as each element can be 'long' and unused DOF are reduced out.

3.0. LOADS FOR SABBA

SABBA can solve directly the final system equations to obtain eigenvalues and eigenvectors.
In many applications the response of a system due to a multiple input situation is required.

For forced response, SABBA solves for one class of problems steady state dynamics of linear systems for which a Fourier Transform of the steady state input exists. Regardless of whether the input is SINUSOIDAL, RANDOM or TRANSIENT, internally SABBA obtains a series of solutions spanning the appropriate frequency range.

3.1. Sinusoidal or Random Analysis

For these types analysis, the user specifies the Fourier transform of his input and SABBA works entirely in the frequency domain. Since it never uses its FFT algorithm it never needs to envoke the limitations imposed by Shannons theorem.

SABBA allows the specification of force or displacement (velocity or acceleration) at any physical DOF in the system.

Sinusoidal $P(t) = A \sin \Omega t$ $P(s) = A\delta(S - j\Omega)$

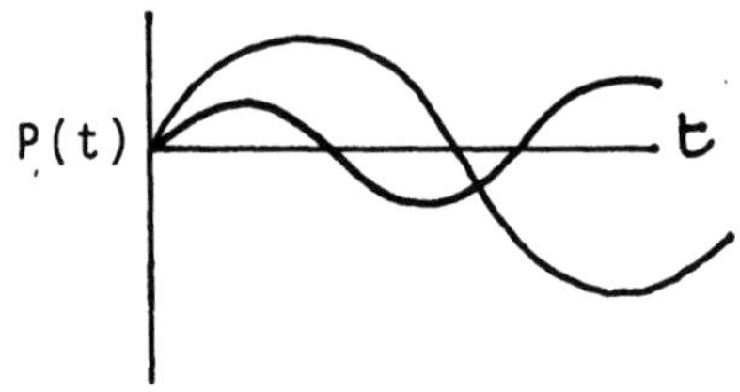

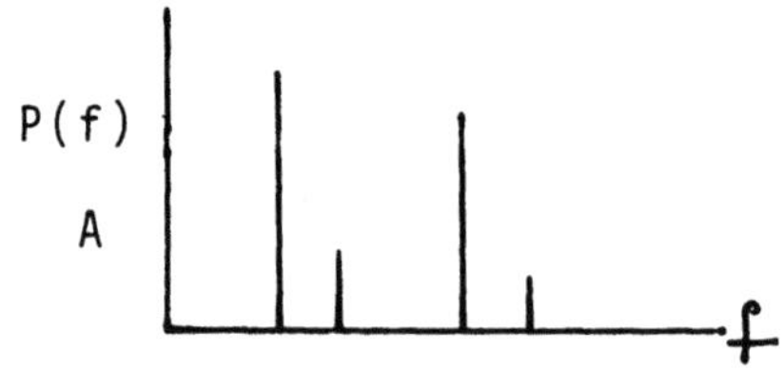

The user may vary the frequency fi (Hertz) arbitrarily; amplitude and phase remain constant.

Unbalance $P(t) = me\,\Omega^2 \sin \Omega t$ $P(s) = me\,S^2 \delta(S - j\Omega)$

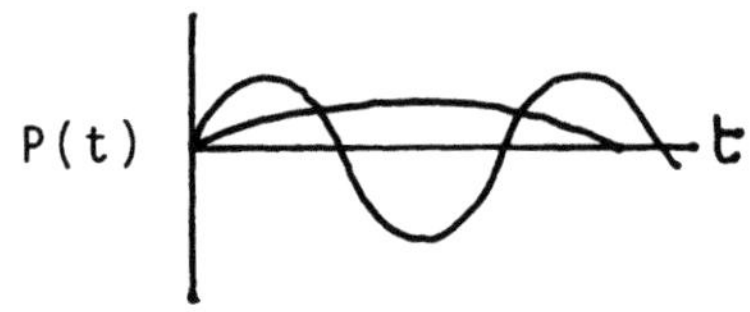

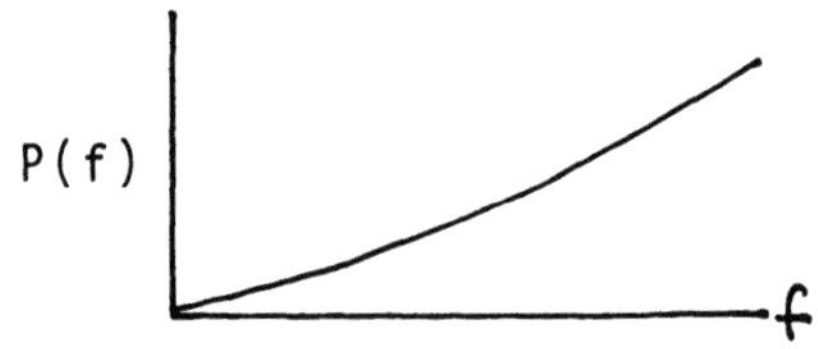

Rotating unbalance forces are specified in terms of the unbalance and its phase angle.

Random $P(t) = A(\Omega) \sin \Omega t$ $P(s) = A(\Omega)\delta(S - j\Omega)$

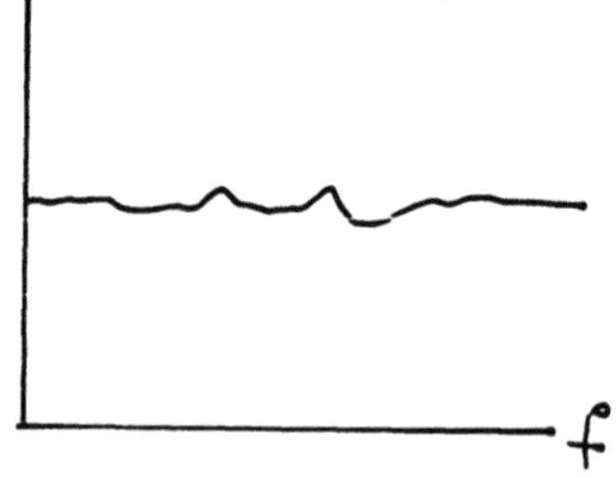

Here the variation of amplitude and phase can be arbitrarily specified with frequency varied to suit the problem.

SABBA interpolates to find intermediate values at each frequency where an solution is obtained.

3.2. Transient

Transient loads can be applied in the form of an arbitrary amplitude versus time history, L(t). The time history of the input is first transformed into the frequency domain by the finite Fourier transform.

$$L(f) = \int_0^T L(t)e^{\frac{-2\pi jft}{T}} dt = F[L(t)] \qquad 3.1$$

SABBA uses a Discrete Fourier Transform (Cooley-Tukey FFT) whose basic relations for application are:

$T = N \cdot \Delta t$ N= number of pts recorded (e.g. 1024)

Shannons theorem: $f\ max \leqslant \frac{1}{2\Delta t}$

and N time points are required for N/2 complex quantities in the frequency domain

Therefore $\frac{fmax}{N/2} \leqslant \Delta f$ -- resolution

Hence $\Delta f \leqslant {}^{1}/_{N}\ \Delta t \leqslant {}^{\cdot 1}/_{T}$

The concept is illustrated in Figure 3.1.

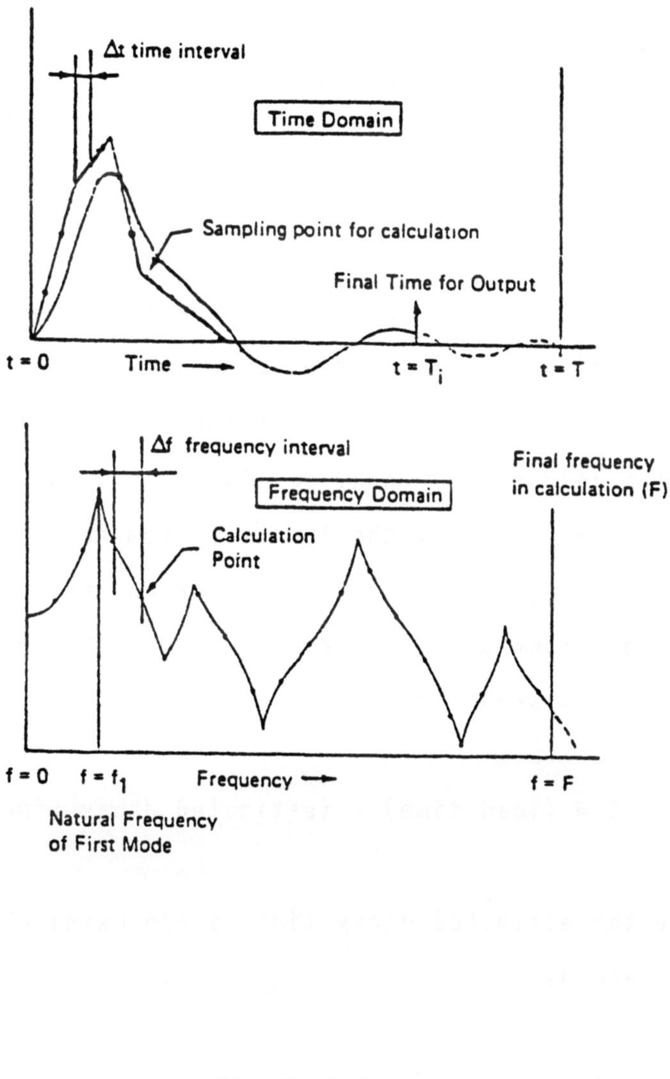

Figure 3.1

Discrete Fourier Transform

To perform this operation SABBA must determine an appropriate period, T, and sample rate, Δt. Since the transient response is assumed to be periodic in the period, T, the response must decay to a small value at the end of the period. Hence

$$L(o) = L(T) = o$$

Enough time must elapse after the termination of the input load to allow the response to damp out. SABBA automatically determines the time required for the response to decay to 10% of its peak value using an estimate of the lowest system natural frequency, ω_1, supplied by the user and an assumed damping ration of 0.1. The total period of the transient response as shown in Figure 3.1 is

$$T = (\text{load time}) + (\text{estimated decay time})$$

where the estimated decay time is approximately $(3.66/\omega_1)$ seconds.

The sample rate, Δt, is determined by SABBA from an estimate of the highest frequency resolution, f_{max}, desired using Shannon's Theorem.

$$\Delta t = \frac{1}{2 \cdot f_{max}}$$

The total number of time samples is then

$$N = \frac{T}{\Delta t}$$

The user must exercise care in estimating the lowest and highest frequencies. An unrealistically low estimate of the lowest system natural frequency will produce an unnecessarily long period. Similarly, a high estimate of the highest frequency of resolution will cause an unnecessarily fine sample rate. Both effects could produce a very large number of solution steps, resulting in long analysis times.

If, for example, the system model represents the system dynamics up to 70 Hz, then a maximum frequency resolution of 75 Hz is normally sufficient. However, the user should also perform a Fourier analysis of the transient load to ensure that all significant input is within the frequency range represented by the system. If for this example the transient input has significant frequency content above 100 Hz, then realistic dynamic response can generally not be predicted without additional system definition.

4.0. SABBA MATRIX ASSEMBLY

4.1. Equilibrium Considerations

Compatability between component DOF's is essentially defined at the component level stage (i.e. using scalar, constraint, beam, components etc.).

Equilibrium is automatically satisfied by adding the component matrices into the system stiffness matrix.

4.2. System Equations (SABBA MATRIX PARTITIONS)

SABBA divides the system matrix into six partitions.

$$\begin{bmatrix} F_1 \\ F_2 \\ F_3 \\ 0 \\ 0 \\ F_6 \end{bmatrix} = \begin{bmatrix} D_{11} & D_{12} & \cdots & D_{16} \\ D_{21} & D_{22} & & \\ \vdots & & & \\ \vdots & & & \\ D_{61} & D_{62} & - - - & D_{66} \end{bmatrix} \cdot \begin{bmatrix} X_1 \\ X_2 \\ X_3 \\ X_4 \\ X_5 \\ 0 \end{bmatrix} \qquad (4.1)$$

X_1 = Independent DOF in a constraint

X_2 = Independent DOF not in a constraint

X_3 = Dependent DOF in a constraint

X_4 = Specified Motion Independent DOF not in a constraint

X_5 = Specified Motion Independent DOF in a constraint

X_6 = Boundary Conditions (ground DOF).

Only the first two (2) partitions are needed in the direct solution.

Partition 3 can be determined from the constraint relationship relating dependent DOF's to independent DOF's.

Constraint Equations

$$X_3 = \left[C_1, C_5\right] \cdot \begin{bmatrix} X_1 \\ \\ X_5 \end{bmatrix} \tag{4.2}$$

Partitions 4 and 5 contain in the solution vector X the nonzero displacements, velocities and accelerations. These are user-defined.

Specified Degrees of Freedom

$$\begin{bmatrix} X_4 \\ X_5 \end{bmatrix} = \begin{bmatrix} X_4(f) \\ X_5(f) \end{bmatrix} \tag{4.3}$$

Degrees of freedom defined as ground plus any zero defined are shown in Partition 6.

<u>Boundary Conditions</u>

$$(X_6) = 0 \tag{4.4}$$

Applied loads are displayed in partitions 1 through 3 in the force vector, F.

Loads

External Forces

$$\begin{bmatrix} F_1 \\ F_2 \\ F_3 \end{bmatrix} = \text{Force moment, unbalance, etc.} \qquad (4.5)$$

In assembling the partitioned matrix SABBA checks for many input data errors. For example an error will be reported if a degree of freedom is defined such that it appears in two different partitions or if both a force and a boundary are defined for the same degree of freedom.

SABBA will optionally display a system summary such as that shown in Figure 4.1. which provides a complete description of the system model. This summary is useful for model debugging and documentation. The density of nonzero terms in the system matrix and the constraint equations (if any) are displayed together with a number of degrees of freedom in each matrix partition.

The input data file names, the load types and the solution frequency range are also documented with an estimate of the fraction of total memory used during problem set up and solution.

When the system matrix is being developed, memory usage depends primarily on the component size. If memory usage approaches a value of 1., the sizes of the components may not be further expanded, although the number of components may be increased.

During execution the memory usage depends on the size of the first two (2) partitions plus the constraint equations, in addition to the maximum component size. If execution memory utilization approaches a value of 1, no components can be added to the system which will either increase the number of degrees of freedom in the first two partitions or add constraint equations.

The system summary includes a component summary which lists the component number, component type, component matrix size and force calculation option (if selected) for each component.

The degrees of freedom associated with each row in the system matrix (N1-N6) are also listed. If during execution a singular system matrix is encountered, SABBA will report the row number of the singularity so the user may determine the corresponding degree of freedom name using the above list.

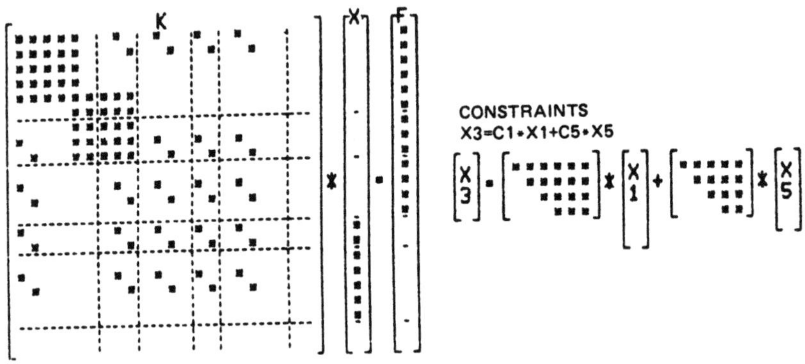

MATRIX PARTITIONS

NO	SIZE	DESCRIPTION
1	6	INDEP. CONSTRAINT DOF
2	3	DOF NOT IN A CONSTRAINT
3	4	DEPENDENT CONSTRAINT DOF
4	2	SPEC. DOF NOT CONSTR.
5	5	SPEC. INDEP. CONSTR. DOF
6	2	DOF = 0.0 (B.C.)

COMPONENT FILE = COMP.DAT
LOAD FILE = LOAD.DAT
EXECUTION FILE = EXECF.DAT
OUTPUT FILE = RESULT.DAT
MEMORY USAGE – INPUT=.50, EXEC=.5
ANALYSIS=SINUSOIDAL, FREQ. STEPS=2
F0(HZ)=1.0E+0 , FN(HZ)=3.0E+0

• COMPONENTS •

ID	TYPE	SIZE	F												
1	DMI	5	Y	X1	X2	X3	X4	X5	X6	X7	X8	X9	X10	X11	Y12
2	DMI	5	Y	Y13	Y14	Z15	Z16	Z17	Z18	Z19	Z20	Z21	GROUND		
3	DMI	5													
4	DMI	5	Y												

Fig. (4.1.) System Summary Display

5. FINAL SYSTEM EQUATIONS AND SOLUTION

By matrix manipulation, the reduced set of system equations may be presented as:

$$\begin{bmatrix} K_{11}+K_{13}\psi_1+\psi_1^{\,t}K_{31}+\psi_1^{\,t}K_{33}\psi_1, & K_{12}+\psi_1^{\,t}K_{32} \\ K_{21}+K_{23}\psi_1 & , \quad K_{22} \end{bmatrix} \cdot \begin{bmatrix} X_1 \\ X_2 \end{bmatrix}$$

$$= \begin{bmatrix} F_1+\psi_1^{\,t}F_3 \\ F_2 \end{bmatrix} - \begin{bmatrix} K_{14}+\psi_1^{\,t}K_{34}, & K_{15}+\psi_1^{\,t}K_{35}+K_{13}\psi_5+\psi_1^{\,t}K_{33}\psi_5 \\ K_{24} \quad , & K_{25}+K_{23}\psi_5 \end{bmatrix} \times \begin{bmatrix} X_4 \\ X_5 \end{bmatrix} \tag{5.1}$$

$$\begin{bmatrix} F_1^* \\ F_2^* \end{bmatrix} = \begin{bmatrix} D_{11}^* & D_{12}^* \\ D_{21}^* & D_{22}^* \end{bmatrix} = \begin{bmatrix} X_1 \\ X_2 \end{bmatrix}$$

Where the (D*) elements include the component equilibrium equations modified by the constraint eqautions and (F*) terms include the effects of constraint and specified displacements.

5.1. Eigenvalue Solution

SABBA will calculate directly the frequencies, damping and mode shapes. In this case specification of forces is not required. A schematic of this solution is given in Fig. 5.1. A detailed theoretical explanations of the algorithm and techniques used is given in Appendix A.

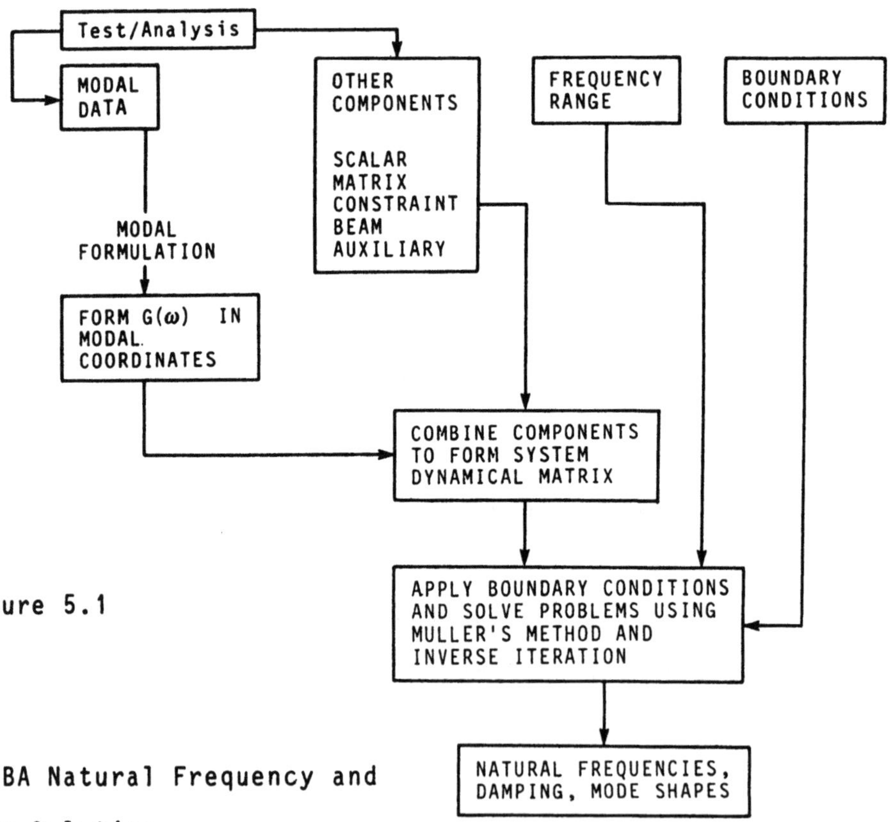

Figure 5.1

SABBA Natural Frequency and Mode Solution

5.2 Forced Response Solution

The forced response solution in SABBA is not determined from say a modal data based created from the eigenvalue solution, but calculated directly from the data describing each component. A process of forward and backward substitution is used which assures a numerically stable solution for symmetric and non-symmetric matrices. The forced response schematic is shown in Figure 5.2.

SABBA provides a graphic solution monitor which displays the solution sequences as shown in Figure 5.3. The logarithm of the system matrix determinant is disdisplayed as each step in the solution sequence is performed.

At each system resonance the natural frequency is extracted and reported at the right side of the display together with the number of accurate significant digits.

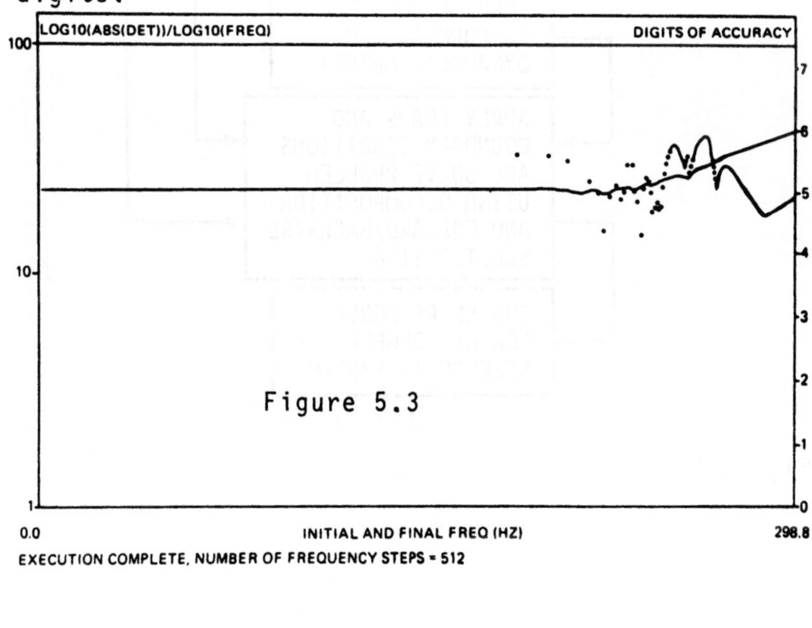

Figure 5.3

SABBA Forced Response Solution

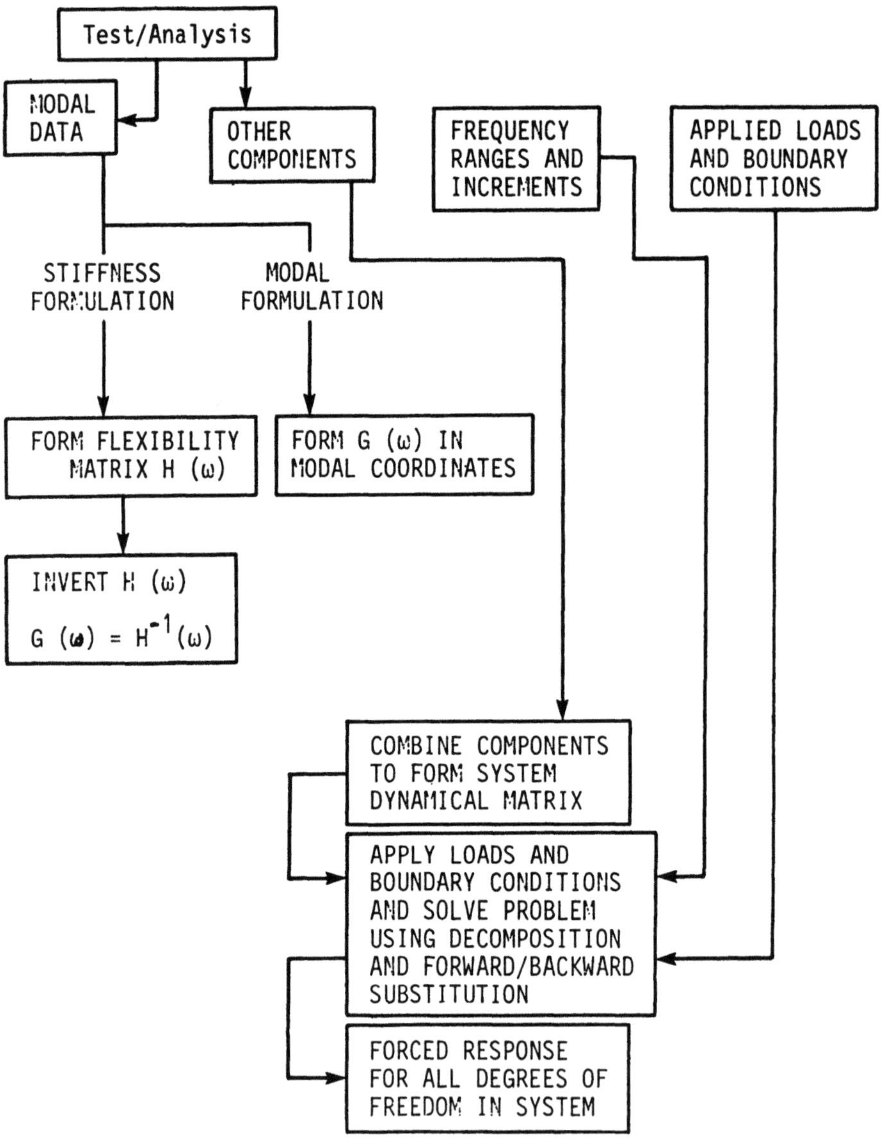

Figure 5.2

6.0. POST PROCESSING OPERATIONS

For an eigenvalue solution, SABBA computes the mode shapes at all DOF selected as input (dependent and independent DOF's). This information can be directly read into the SDRC MODAL-PLUS program [4] for display purposes. One example is given in Figure 6.1

For forced response solution, SABBA output includes the response at all system degrees of freedom and auxiliary equation vaiables plus component loads selected by the user. This output is placed on the user-supplied OUTPUT FILE for subsequent post-processing by the SABPOST program. SABPOST mathematically manipulates and graphically display any of the SABBA output data on a cathode ray tube (CRT) terminal. Frequency response functions generated by MODAL-PLUS can also be processed by SABPOST.

The SABPOST command structure and capabilities have been designed for maximum user convenience and ease of use. An extensive HELP capability provides on-line documentation of all commands for interactive user assistance.

SABBA and MODAL-PLUS frequency response functions can be mathematically manipulated in SABPOST to provide other forms of the data for graphical display. The user can generate autocorrelation, crosscorrelation, auto spectra, cross spectra, coherence, and transfer functions. The user may also add, subtract, multiply, divide, differentiate, integrate, and Fourier transform functions using SABPOST.

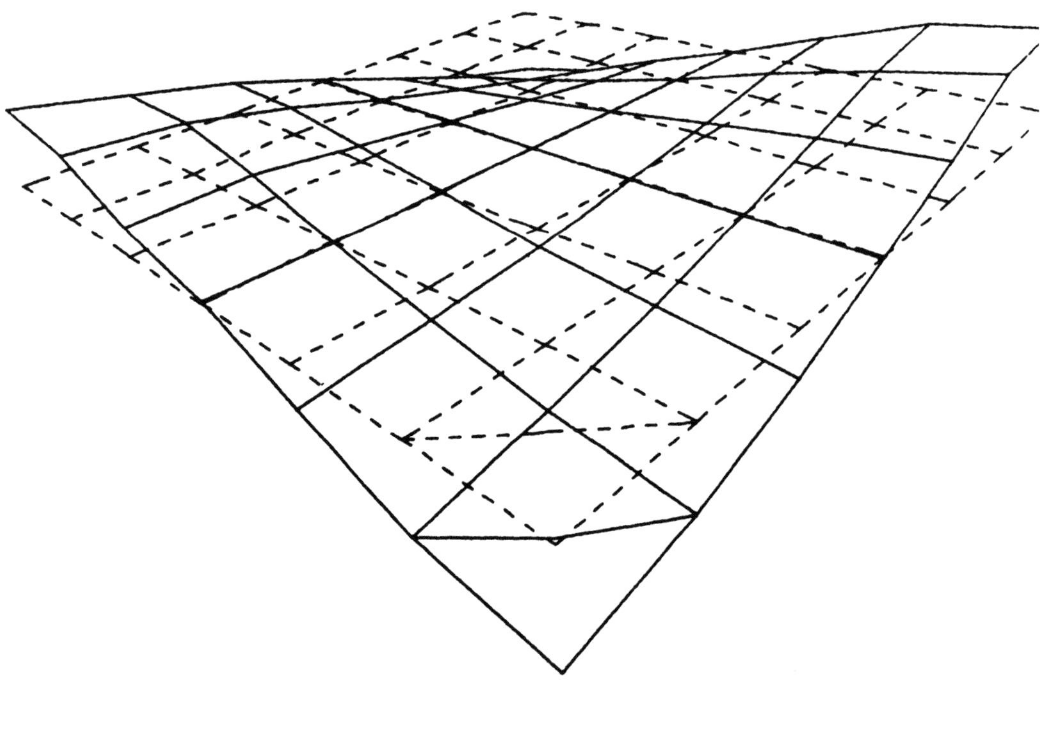

Figure 6.1.

Example of modal display

Extensive file and data base management has also been provided. The ability to query data files and the SABPOST data base allows the user to determine the status of the SABPOST session at any given time. Data changes and mathematical manipulation of data can be saved via a checkpoint feature. Output can also be listed at the terminal or written to a file for subsequent printing on a line printer or other device.

Graphic Displays

Data can be displayed in many forms on linear, logarithmic, or decibel axes. This allows for common display forms as well as many uncommon forms which the user may desire. Available forms include transient response, Bode, real part vs. frequency, imaginary part vs. frequency, phase vs. frequency, argand, and modulus plots. Examples of some common plot forms are shown in Figures 6.2 through 6.4.

Mathematical Capabilities and Operations

The mathematical capabilities in SABPOST plus the constraint and auxiliary equation features in SABBA allow the user to determine such quantities as displacement, velocity, acceleration, force (i.e. component loads), moment, relative motion, and stress.

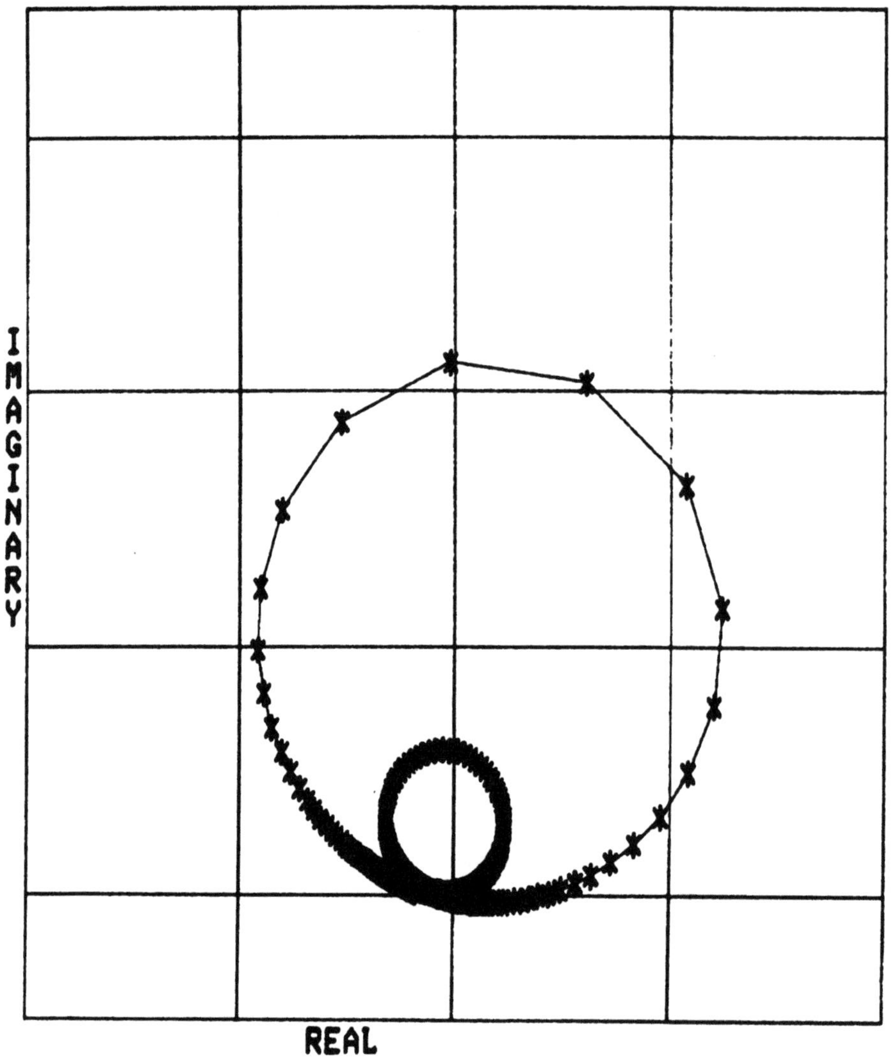

Figure 6.2.

Argand or Nyquist display

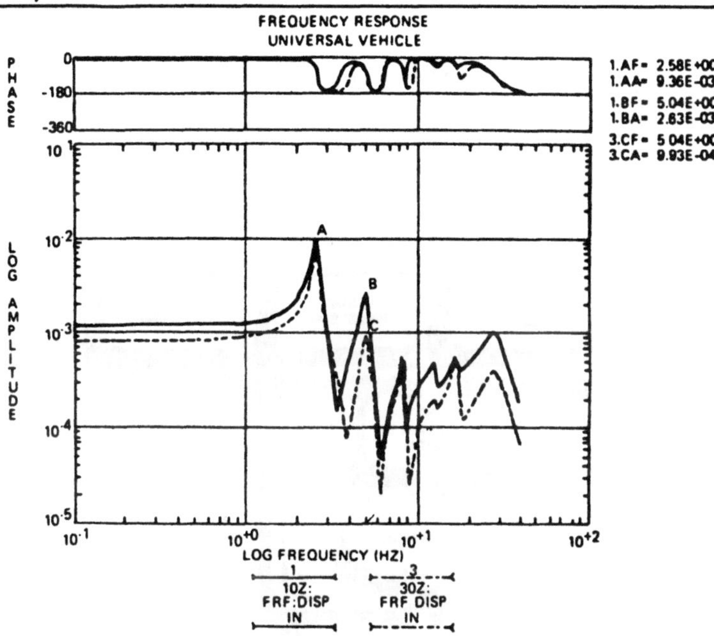

Figure 6.3. Example of Bode Plot

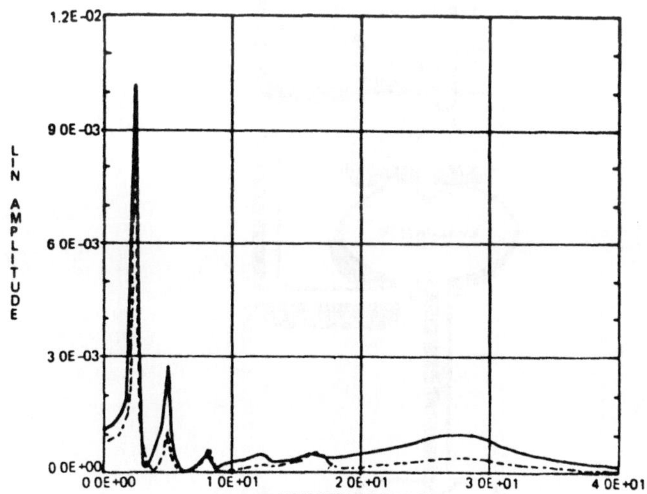

Figure 6.4.

Example of Modulus Plot

7.0. SYSTEM-MODELING APPLICATIONS

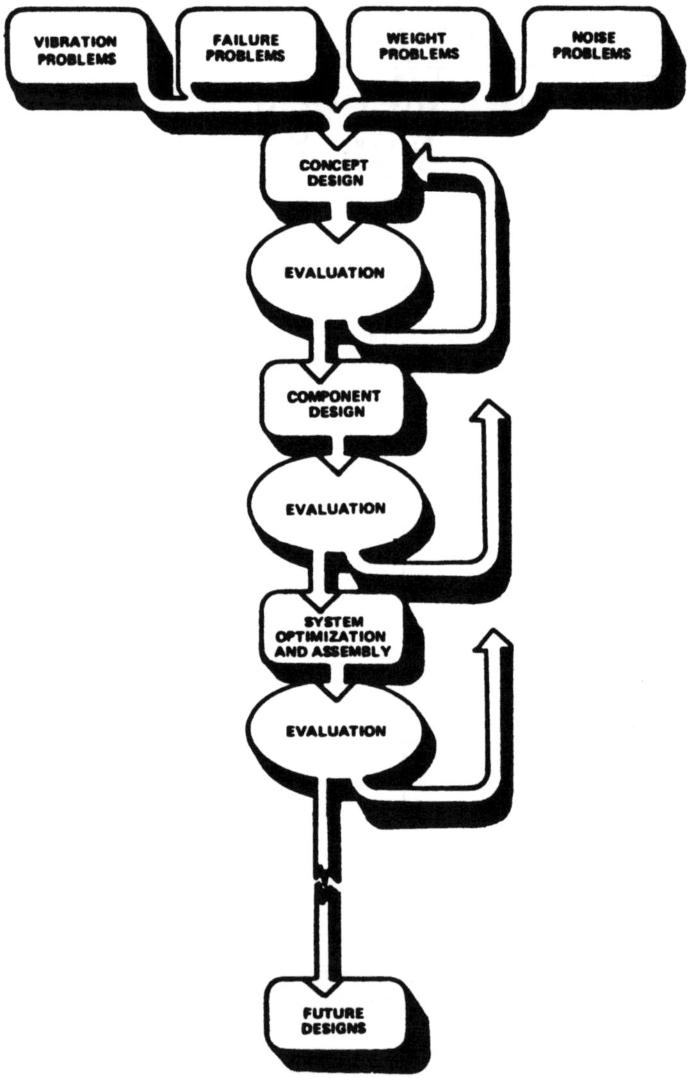

7.1. Concept Design

For concept design, approximate descriptions of major components are assembled in a SABBA system model. Analysis of the system helps evaluate different concepts and guides new component design. For example, you may want to evaluate the effect of changing the location, stiffness, and damping of connections between components.

After selecting a satisfactory design concept, the tasks of modifying or designing new components may be delegated to several people. Your initial system model guides component design by describing loads and by suggesting suitable dynamic characteristics for individual components. It should be emphasized that apparently well-designed components do not necessarily combine to create an optimal system. Interactions between components should be considered as early as possible in the design process.

7.2. Component Design

Component design for structural dynamics usually involves either finite element analysis or modal testing.

Both approaches have a place. For example, modal tests provide accurate results at reasonable cost, if the component already exists or can be prototyped inexpensively. For certain complex and non-linear structures, testing may also be the best way to get accurate results. Finite element analysis is generally preferred when a new component may require several design iterations.

Regardless of whether finite element models or test results are used for component design, SABBA can incorporate the component characteristics in a system model. This is accomplished by the MODAL-PLUS and NIP programs without manual transcription of data. Thus, it is relatively easy to upgrade the model and evaluate system dynamics by including new data from either modal tests or finite element analysis. This information guides component design toward desired system performance.

7.3. System Optimization

After component designs are complete, a prototype of the entire system is usually assembled. This prototype may be entirely satisfactory or it may require some adjustment to achieve the desired dynamic performance. So far, the role of system modeling has been to guide concept and component design, so this first prototype will be nearly optimal. Now, the role of system modeling is to avoid additional prototypes, if it is necessary to further optimize system performance.

This is accomplished by testing the entire system (MODAL-PLUS) and incorporating the experimental results in a SABBA model. This experimentally-derived system model can be used with confidence to evaluate evolutionary design modifications, instead of building a series of prototypes.

7.4. Future Designs

The application of SABBA extends to future designs. System modeling provides insight into component interactions which is not achieved by trial-and-error methods. The systematic understanding gained by modeling can be applied to future designs which involve similar components or modified systems.

The goal is to design a high performance product before it is manufactured. However, many people have first applied SABBA as a troubleshooting tool. It is very easy to test an existing product and incorporate the results in a system model, to understand the cause of a problem and evaluate possible solutions.

REFERENCES

(1) Klosterman, A.
"On the Experimental Determination and use of Modal Representations of Dynamic Characteristics".
Ph. D. Dissertation, University of Cincinnati.
1971.

(2) Klosterman, A. , McClelland,W.A.
"Combining Experimental and Analytical Techniques for Dynamic System Analysis".
Tokyo Seminar on Finite Element Analysis
November, 1973

(3) McClelland, W. Hay
"Frame Design Analysis under Complete Vehicle Boundary Conditions".
SAE Paper 741142, Earthmoving
Industry Conference Peoria, USA 1974

(4) Modal Plus Users Handbook and Theoretical Manual
SDRC Engineering Services. 1980

APPENDIX A

SABBA Eigenvalue Extraction

1.0. FUNDAMENTALS OF THE DETERMINANT METHOD

The basic notion employed in the determinant method of eigenvalue extraction isvery simple. If the elements in a matrix [A] are polynomial functions of the operator p, then the determinant of [A] can be expressed as

$$D(A) = (p - p_1)(p - p_2)(p - p_3) \ldots (p - p_n) \quad (1)$$

where $p_1, p_2, p_3 \ldots p_n$ are the eigenvalues of the matrix. The value of the determinant vanishes for $p = p_i$, $i = 1,2,3 \ldots n$.

In the determinant method, the determinant is evaluated for trial values of p, selected according to some iterative procedure, and a criterion is established to determine when D(A) is sufficiently small or when p is sufficiently close to an eigenvalue. The eigenvector is then found by solution of the equation.

$$[A]\,(u) = 0 \qquad (2)$$

with one of the elements of (u) preset.

For evaluating the determinant of a matrix SABBA employs the triangular decomposition

$$[A] = [L] . [U] \tag{3}$$

where [L] is a lower unit triangular matrix (unit values on the diagonal) and [U] is an upper triangular matrix. The determinant of [A] is equal to the product of the diagonal terms of [U].

(4)

One version of the triangular decomposition operation is provided. In this version row interchanges are used in order to improve numerical stability.

(5)

The matrix [A] may be expressed as

$$[A] = p^2 [M] + p [C] + [K + iB]$$

for complex eigenvalue problems.

The main strength of the determinant method is its insensitivity to the functional form of the elements of the [A] matrix which could, for example, contain poles as well as zeroes, or be transcendental functions of p. These effects occur frequently in hydroelastic and aeroelastic problems. It is also necessary for the general problem which could have nonsymmetric matrices, frequency dependent component, residual effects, complex modes, etc.

1.1. Iteration Algorithm

Wilkinson's recent, but already standard, treatise (1) includes an authoritative discussion of polynomial curve-fitting schemes for tracking the roots of a determinant. He shows that little is to be gained by using polynomials higher than the second degree. Accordingly, Muller's quadratic method (Wilkinson, p.435) is used in SABBA. The form of the algorithm in our notation is as follows

Consider a series of determinants, D_{k-2}, D_{k-1}, D_k, evaluated for trial values of the eigenvalue, $p = p_{k-2}, p_{k-1}, p_k$. A better approximation to the eigenvalue is obtained from the following calculations. Let

$$h_k = p_k - p_{k-1} \quad , \qquad (6)$$

$$\lambda k = \frac{h_k}{h_{k-1}} \quad , \qquad (7)$$

$$\delta k = 1 + \lambda_k \quad , \qquad (8)$$

(1) Wilkinson, J.H., "The Algebraic Eigenvalue Problem", Clarendon Press, Oxford, 1965.

Then

$$h_{k+1} = \lambda_{k+1} h_k \quad , \qquad (9)$$

$$p_{k+1} = p_k + h_{k+1} \quad , \qquad (10)$$

where

$$\lambda_{k+1} = \frac{-2D_k \delta_k}{g_k \mp \left[g_k^2 - 4D_k \delta_k \lambda_k \left(D_{k-2} \lambda_k - D_{k-1} \delta_k + D_k \right) \right]^{1/2}} \qquad (11)$$

in which

$$g_k = D_{k-2} \; {}_k^2 - D_{k-1} \delta_k^2 + D_k \left(\lambda_k + \delta_k \right) \qquad (12)$$

The (+) or (-) sign in Equation (11) is selected to minimize the absolute value of λ_{k+1}. In the case when p_k, p_{k-1}, and p_{k-2} are all arbitrarily selected initial values (starting points), the starting points are arranged such that

$$D_k \leqslant D_{k-1} \leqslant D_{k-2} \qquad (13)$$

and the (+) or (-) sign in Equation 11 is selected to minimize from p_{k+1} to the closest starting point rather than to p_k.

1.2. Scaling

In calculating the determinant of [A] (Equation 3), some form of scaling must be employed, because the accumulated product will rapidly overflow or underflow the floating point size of a digital computer. Accordingly, the accumulated product af the diagonal terms of [U] is calculated and stored as a scaled number

$$D = d \times 10^n \qquad (14)$$

where

$$\frac{1}{10} \leqslant d \leqslant 1 \qquad (15)$$

The arithmetic operations indicated in Equations 11 and 12 are calculated in scaled arithmetic. λ_{k+1} is then reserted to unscaled form.

1.3. Sweeping of Previously Extracted Eigenvalues

Once an eigenvalue has been found to satisfactory accuracy, a return to that eigenvalue by the iteration algorithm can be prevented by dividing the determinant by the factor $(p - p_i{}')$, where $p_i{}'$ is the accepted approximation to p_i, in all subsequent calculations.

Thus

$$D^{(1)}(A) = \frac{D(A)}{p-p_i'} \tag{16}$$

should be used in place of D(A) after the first eigenvalue has been found. In general, the reduced determinant used for finding the i+1st eigenvalue is

$$D^{(i)}(A) = \frac{D^{(i-1)}(A)}{p - p_i'} = \frac{D(A)}{(P-p_1')(p-p_2')\ldots(p-p_i')} \tag{17}$$

Wilkinson states that this sweeping procedure is quite satisfactory provided that all p_i' have been calculated to an accuracy that is limited only by round-off error.

1.4. Search Procedures

Three initial values of p (starting points) are neededin order to start the iteration algorithm. The determinant mehod is essentially a root-tracking method that finds nearby roots with ease and remote roots with difficulty. Thus it is not advisable to use the three starting points for all eigenvalues because the eigenvalues are usually distributed throughout a region of the p-plane and at least some of .....

... the eigenvalues will be remote from any given set of three starting points.

The sets of starting points are uniformly distributed in an interval of p. The user specifies the lowest and highest eigenvalues of interest, R_{min} and R_{max}. SABBA estimates the number of roots in the range N_E, using the Sturm sequence property. This property states that the number of roots below a frequency is equal to the number of non-positiue diagonal elements in U of equation (3). The $3N_E + 1$ starting points are then located such that the first coincides with R_{min}, the last coincides with R_{max} and the rest are place at uniform intervals between R_{min} and R_{max}. The process of extracting eigenvalues is initiated from the three smallest starting points, p_{s1}, p_{s2} and p_{s3}. After all eigenvalues within the "range" of these starting points have been extracted, p_{s1} is dropped, p_{s4} is added, and the search is repeated.

When searching for roots from any set of starting points the search is terminated if a root is predicted outside the local search region. The local search region extends to 20% below the current search region.

This improves the chance of finding eigenvalues that were initially by-passed.

The search for eigenvalues is finally terminated when no additional roots are to be found.

1.5. Convergence Criteria

The convergence criteria are based on successive values of the estimated eigenvalue. No tests on the magnitude of the determinant or on any of the diagonal terms of the triangular decomposition are necessary or desirable.

Wilkinson shows that, for h_k sufficiently small, the magnitude of h_k is approximately squared for each successive iteration when using Muller's method to find isolated roots. This is an extremely rapid rate of convergence. In a very few iterations the "zone of indeterminancy" is reached within which h_k remains small but exhibits random behavior due to round-off error.

Wilkinson states that, if it is desired to calculate the root to the greatest possible precision, the convergence criterion for accepting p_k as a root should be

$$|h_{k+1}| > |h_k| \tag{18}$$

We accept his advice, tempered by practical considerations. The first of these is that Equation 18 may be satisfied during the first few iterations while the root tracking algorithm is picking up the "scent". Thus it must, in addition, be required that $|h_k|$, $|h_{k-1}|$ and $|h_{k-2}|$ be reasonably small. The second practical consideration is that several iterations may be wasted within the zone of indeterminacy while waiting for Equation 18 to be satisfied. This can be avoided by accepting p_k if $|h_k|$ is sufficiently small. Finally, if the number of iterations becomes excessively large without satisfying a convergence criterion, it is best to give up and proceed to a new set of starting points.

In light of this, SABBA accepts a root when the estimate is the same as the previous estimate to within six significant digits for both the real and imaginary parts.

To avoid excessive iterations in the case of very light or near zero damped modes, damping less than 1% of critical is considered zero.

Experience with the determinant method indicates that, on average, convergence to isolated roots is achieved in about six iterations. The number of iterations indreases for problems with multiple roots, many close roots, or roots clustered just beyond the ends of the desired region.

1.6. Recovery of Eigenvectors

Once an approximate eigenvalue, p_j, has been accepted, the eigenvector is determined by back substitution (i.e. inverse iteration) into the previously computed triangular decomposition of $[A\,(p_j)]$ where

$$[A(p_j)](u) = [L(p_j)][U(p_j)](u) = 0 \quad , \qquad (19)$$

At the root, the decomposed matrix will have the form

$$A=LU, U = \begin{bmatrix} U_{11} & & & & & & \\ & U_{22} & & & & & \\ & & \varepsilon_1 & & & & \\ & & & U_{44} & & & \\ & & & & \varepsilon_2 & & \\ & & & & & U_{66} & \\ & & & & & & \text{etc} \end{bmatrix}$$

The appearance of multiple ε's (on the diagonal) denotes multiple (or very close) roots, which although Hermitian (complex conjugate transpose) will have a double root. Thus we must be able to extract multiple vectors at the same frequency.

First the theory for a general matrix (complex and nonsymmetric) A(p), p= complex root.

right vectors: $A\Phi = \lambda_i \Phi_i$ at i-th root

left vectors: $\psi_i^t A = \lambda_i \psi_i^t$

or $A^t \psi_i = \lambda_i \psi_i$

Note that we find ϕ_i by inverse iteration,

$$A\Phi_i = (L \cdot U)\,\Phi_i = \lambda_i \Phi_i = F_i$$

Thus we simply assume some arbitrary vector F_i and perform a forward and backward substitution, since A is nearly singular (if a diagonal is =0, set it to ε).

The solution is ϕi, or it converges to ϕi in only a few steps.

To find additional vectors at the same root requires that we make them orthogonal to the previous vectors at the same root. This is due to the nature of iverse iteration as the trial vector F_i will converge to the strongest singularity. There are two choices, either make the vectors orthogonal, or remove the vector from the matrix. The former is an algebraic operation on the RHS, while the latter requires modification of the A-matrix and dimensional changes. Thus we choose the orthogonality approach due to simplicity of programming.

MODAL SYNTHESIS
MODAL CORRECTION - MODAL COUPLING

Dr. E. Breitbach
Deutsche Forschungs- und Versuchsanstalt
für Luft- und Raumfahrt e.V.
Institut für Aeroelastik
Bunsenstr. 10, D-3400 Göttingen
W. Germany

1. INTRODUCTION

Modal synthesis is a technical term closely related to the development of the *finite element technique*. Especially in dealing with large structural systems with some thousand physical degrees of freedom, modal synthesis has proved to be an efficient means to reduce computational errors and to minimize computer costs by decomposing the system into several subsystems, each of which is analyzed separately. Based on the results of these component analyses the subsystems can be brought together again to the complete system by means of modal synthesis methods.

Since the mid-fifties much fundamental work has been devoted to this technique (also termed: modal coupling, component mode synthesis, modal combination, branch mode analysis, etc.), particularly in the USA and UK where some key publications related to modal synthesis originated.[1,2,3,4]

In the last decade, much effort has been put into elaborating modal synthesis approaches, allowing a broader range of application. Thus, the capability of modal synthesis techniques may be enhanced considerably in combination with modal survey testing (MST).[5,6] The mathematical models of the different substructures can be established numerically on the basis of finite element calculations or experimentally by means of modal survey tests.

Some typical problems which can be solved by means of modal synthesis are:

- *Modification of mass, stiffness and damping properties of aircraft control systems*
- *Modifications of mass inertia properties in the usually large number of various wing-with-stores configurations of fighter aircraft (usually some hundred in the lifetime of an aircraft)*
- *Attachment of additonal boosters to a launcher system*
- *Payload alterations of space systems for instance by exchanging the entire payload module of a satellite*
- *Coupling of subsystems designed and manufactured within the scope of multi-lateral cooperative projects.*

In the following sections some modal coupling concepts as well as a modal correction method are described. In particular the following topics will be dealt with:

- *Convergence problems due to frequency range truncation and incomplete load conditioning in the coupling points*
- *Flexible, rigid, mixed coupling*
- *Statically determinate and overdeterminate coupling conditions.*

2. FUNDAMENTAL RELATIONS

On the simplified condition of linearity and viscous damping (later on replaced by so-called structural damping defined as the imaginary part of a complex stiffness) the equations of motion of an elastodynamic system can be formulated in the time domain in terms of the physical deflections $u(t)$ as follows:

$$A\,\ddot{u}(t) + B\,\dot{u}(t) + C\,u(t) = P(t) \tag{1}$$

where

A	mass matrix
B	damping matrix
C	stiffness matrix
$P(t)$	column matrix of external forces
$u(t)$	column matrix of the dynamic response where $\dot{u}(t)$ and $\ddot{u}(t)$ are first and second order differentials with respect to time t.

Applying to Eq.(1) the modal transformation

$$u(t) = \Phi\,q(t) \tag{2}$$

where

Φ	modal matrix with the normal mode shapes Φ_r $(r = 1,2,\ldots,n)$ as columns
$q(t)$	column matrix of the generalized coordinates

and premultiplication by Φ^T changes Eq.(1) to

$$M\,\ddot{q}(t) + D\,\dot{q}(t) + K\,q(t) = Q(t) \tag{3}$$

where

$M = \Phi^T A\,\Phi$	diagonal matrix of the generalized masses M_r
$D = \Phi^T B\,\Phi$	generalized damping matrix, not necessarily diagonal, with the elements D_{rs}
$K = \Phi^T C\,\Phi$	diagonal matrix of the generalized stiffnesses $K_r = M_r\,\omega_r^2$ with ω_r denoting the circular normal frequency
$Q = \Phi^T P\,\Phi$	column matrix of the generalized forces Q_r .

In the case of a harmonic excitation

$$Q(t) = Q\,e^{j\omega t} \quad , \quad j = \sqrt{-1} \tag{4}$$

the structure responds harmonically, too

$$q(t) = q\,e^{j\omega t} \tag{5}$$

where ω designates the circular excitation frequency. Insertion of Eqs.(4) and (5) into Eq.(3) and replacing the viscous-type damping definition by the above-mentioned "complex stiffness" definition leads to

$$(-\omega^2 M + jD + K)\,q = Q \quad . \tag{6}$$

Another fundamental relation repeatedly used in the following sections is a simplified form of *Lagrange's* equations

$$\frac{d}{dt}\left(\frac{\partial E_k}{\partial \dot{q}_r}\right) + \frac{\partial E_p}{\partial q_r} = 0 \quad , \quad r = 1,2,\ldots,n \tag{7}$$

where

E_k kinetic energy

E_p potential or stiffness energy .

3. MODAL CORRECTION METHOD

The so-called modal correction method,[5,7,8] has been successfully applied for more than a decade to problems with relatively small changes of mass, damping and stiffness. The following list of some specific problems may give an idea of the application range of this method:

- *Changes of mass, mass moments of inertia and location of the center of gravity of aircraft external stores*

- *Changes of mass distribution, hinge stiffness and hinge damping of aircraft controls*
- *Payload alterations of satellites.*

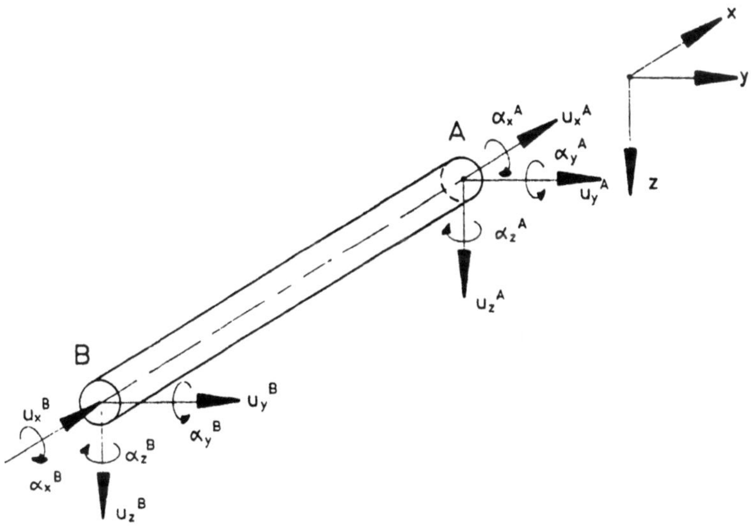

Figure 1: Flexible beam-type element.

Considering for example a payload suspension device which can be represented by a simple beamlike element as sketched in Figure 1, changes of the stiffness properties can be expressed by a change of the potential energy

$$\Delta E_p = \frac{1}{2} u_f^T \Delta C u_f \tag{8}$$

where u_f can be formulated in accordance with Figure 1 as follows:

$$u_f = \begin{Bmatrix} u_f^A \\ \hline u_f^B \end{Bmatrix} \tag{9}$$

and

$$u_f^A = \left(u_x^A,\ u_y^A,\ u_z^A,\ \alpha_x^A,\ \alpha_y^A,\ \alpha_z^A\right)^T \tag{10a}$$

$$u_f^B = \left(u_x^B,\ u_y^B,\ u_z^B,\ \alpha_x^B,\ \alpha_y^B,\ \alpha_z^B\right)^T \quad . \tag{10b}$$

The stiffness changes are represented by the 12×12 matrix ΔC to be added to the stiffness matrix C of the basic unchanged system.

By applying the modal transformation

$$u_f = \Phi_f\, q \tag{11}$$

to Eq.(8), ΔE_p converts to

$$\Delta E_p = \frac{1}{2}\, q^T\, \Phi_f^T\, \Delta C\, \Phi_f\, q \tag{12}$$

where the matrix

$$\Phi_f = \begin{bmatrix} \Phi_{f1}^A, \Phi_{f2}^A, \ldots, \Phi_{fr}^A, \ldots, \Phi_{fn}^A \\ \hline \Phi_{f1}^B, \Phi_{f2}^B, \ldots, \Phi_{fr}^B, \ldots, \Phi_{fn}^B \end{bmatrix} \tag{13}$$

contains the normal mode deflections at the endpoints A and B common to the stiffness element and the structural system of which the stiffness element is part. Thus, for instance

$$\Phi_{fr}^A = \left(u_{xr}^A, u_{yr}^A, u_{zr}^A, \alpha_{xr}^A, \alpha_{yr}^A, \alpha_{zr}^A\right)^T \quad . \tag{14}$$

Applying Lagrange's operation (7) to Eq.(12) leads to the modal stiffness correction matrix

$$\Delta K = \Phi_f^T\, \Delta C\, \Phi_f \tag{15}$$

which has to be added to matrix K in Eq.(6).

To illustrate the appearance of matrix ΔC in the simple case of a one-dimensional element, let the beam-type element shown in Figure 1 degenerate into a torsion spring with a stiffness change Δc_t, then

$$\Delta C = \Delta c_t \begin{bmatrix} 1 & -1 \\ -1 & 1 \end{bmatrix} . \tag{16}$$

The simplest way to take into account damping changes is to consider ΔC as a complex stiffness matrix with the imaginary part representing the damping alteration.

Changes of mass, mass moments of inertia and center of gravity location occur especially in the course of dynamic and aeroelastic qualification of fighter aircraft with sometimes extremely large numbers of different external stores configurations. To keep test effort and cost as low as possible only a relatively small number of representative external stores configurations are analyzed in ground vibration tests whereas the modal characteristics of the much more numerous remaining configurations can be determined in a purely numerical way by means of the modal correction approach based on the mathematical models of the test configurations.

Let the difference between the mass inertia properties of two different types of interchangeable external stores be represented by changes of mass m, mass moments of inertia Θ_x, Θ_y, Θ_z and the location of the center of gravity (x_{CG}, y_{CG}, z_{CG}), then the difference between the kinetic energies of the two external stores can be expressed by

$$\Delta E_k = E_{k2} - E_{k1} \tag{17}$$

where the subscripts $_1$ and $_2$ indicate the basic configuration 1 and the changed configuration 2 . The energy terms E_{k1} and E_{k2} can be written as follows:

$$E_{k1} = \frac{1}{2} u_1^T I_1 u_1 \tag{18a}$$

$$E_{k2} = \frac{1}{2} u_2^T I_2 u_2 \quad . \tag{18b}$$

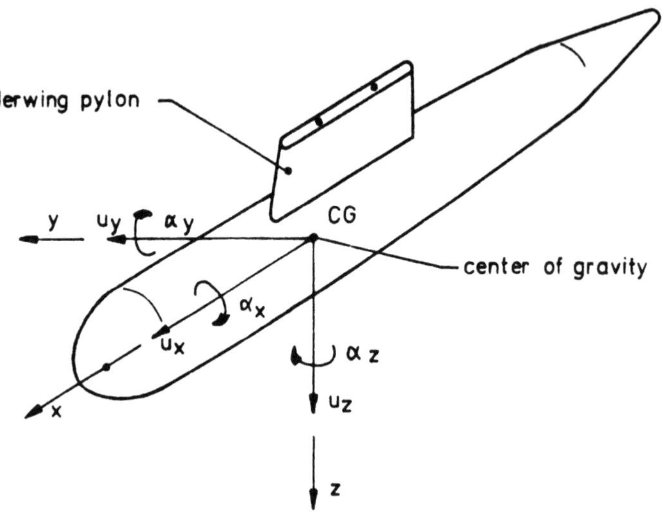

Figure 2: Aircraft underwing store with pylon.

According to Figure 2 the displacement vectors

$$u_1 = \left(u_x^1, u_y^1, u_z^1, \alpha_x^1, \alpha_y^1, \alpha_z^1\right)^T \tag{19a}$$

$$u_2 = \left(u_x^2, u_y^2, u_z^2, \alpha_x^2, \alpha_y^2, \alpha_z^2\right)^T \tag{19b}$$

and the mass inertia matrices

$$I_1 = \begin{bmatrix} m_1 & & & & & \\ & m_1 & & & & \\ & & m_1 & & & \\ & & & \Theta_{x1} & & \\ & & & & \Theta_{y1} & \\ & & & & & \Theta_{z1} \end{bmatrix} \tag{20a}$$

$$I_2 = \begin{bmatrix} m_2 & & & & & \\ & m_2 & & & & \\ & & m_2 & & & \\ & & & \Theta_{x2} & & \\ & & & & \Theta_{y2} & \\ & & & & & \Theta_{z2} \end{bmatrix} \tag{20b}$$

are formulated with respect to the centers of gravity (x_1, y_1, z_1) and (x_2, y_2, z_2), where $x_2 = x_1 + \Delta x$, $y_2 = y_1 + \Delta y$, $z_2 = z_1 + \Delta z$.

Expressing vector u_2 by vector u_1 leads to the transformation

$$u_2 = T\, u_1 \tag{21}$$

where

$$T = \left[\begin{array}{ccc|ccc} 1 & & & 0 & \Delta z & -\Delta y \\ & 1 & & -\Delta z & 0 & \Delta x \\ & & 1 & \Delta y & -\Delta x & 0 \\ \hline & & & 1 & & \\ & 0 & & & 1 & \\ & & & & & 1 \end{array}\right] . \tag{22}$$

Then, the change of the kinetic energy is

$$\Delta E_k = \frac{1}{2}\, \dot{u}_1^T (T^T I_2 T - I_1)\, \dot{u}_1 \quad . \tag{23}$$

Applying the modal transformation

$$u_1 = \Phi_1\, q \tag{24}$$

and subsequently Lagrange's operation (7) to Eq.(23) results in the modal correction matrix:

$$\Delta M = \Phi_1^T (T^T I_2 T - I_1)\, \Phi_1 \tag{25}$$

where the modal matrix Φ_1 contains the measured modal displacements as the column vectors

$$\Phi_{1r} = \left(u_{xr}, u_{yr}, u_{zr}, \alpha_{xr}, \alpha_{yr}, \alpha_{zr}\right)^T \quad . \tag{26}$$

The addition of matrix ΔM to M in Eq.(6) results in the equations of motion of the changed system.

In the simplified case of a lumped mass changed by Δm, the matrix ΔM (25) degenerates into

$$\Delta M = \Phi_1^T \, \Delta I \, \Phi_1 \tag{27}$$

where

$$\Delta I = \Delta m \begin{bmatrix} 1 & & \\ & 1 & \\ & & 1 \end{bmatrix} \quad . \tag{28}$$

The capability of the modal correction method has been proved successfully for a considerable number of applications. It is worth mentioning that this method works very well even if problems with larger structural changes are to be attacked.

Normal Mode	Normal frequencies f_r		
	Basic Config.	Changed Config.	
	measured	measured	calculated
r = 1	3.14	3.32	3.32
2	4.38	4.34	4.39
3	4.65	5.03	4.97
4	6.18	6.76	6.74
5	6.75	7.99	7.45
6	15.4	15.06	15.4

Table 1: Application of the Modal Correction Method to a sweepable wing with the wing pivot stiffness c_α = 646.7 N-m/rad increased by Δc_α = 661.5 N-m/rad .

Table 1 shows the results of an application of the modal correction method to a wing model as shown in Fig.3 with the wing pivot stiffness c_α (sweepable wing configuration) changed by Δc_α . Two other examples are described in Ref.[9] dealing with changes of the root bending stiffness of a sailplane wing and in Ref.[10] treating the problem of an aircraft with many interchangeable external stores configurations.

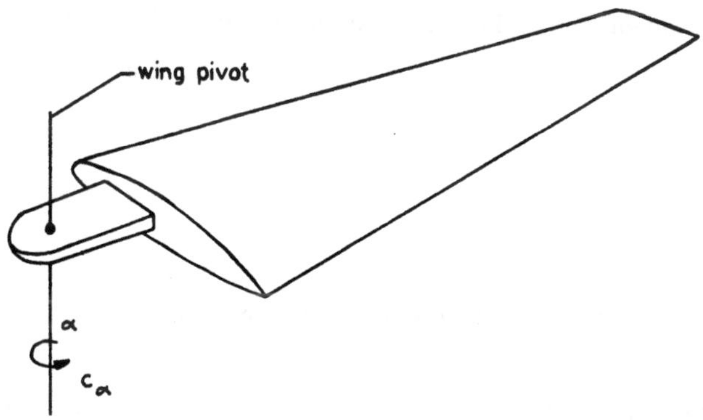

Figure 3: Sweepable wing with pivot stiffness c_α .

A computer software package for general application of the modal correction approach has been elaborated in the DFVLR Institute of Aeroelasticity and published in Ref.[11].

4. MODAL COUPLING

One can basically distinguish between three different types of coupling conditions:

- *Rigid coupling*
  In the case of coupling two substructures with interfaces which can be considered approximately rigid, the displacements at the coupling points have to fulfill the compatibility condition.
- *Flexible coupling*
  In the case of coupling substructures by means of flexible elements,

a special coupling approach can be used if elastic properties of the flexible elements are known.

- *Mixed coupling*
  It often occurs that a coupling element has to be considered rigid with respect to some of its degrees of freedom whereas the others may be treated as flexible.

By neglecting external forces and structural damping for the sake of simplifying the derivation, the elastodynamic equations of motion of the complete system with the coupling elements removed can be written as follows:

$$(-\omega^2 M + K)\, q = 0 \tag{29}$$

where for the case of two substructures A and B

$$M = \left[\begin{array}{c|c} \diagdown M^A & 0 \\ \hline 0 & M^B \diagdown \end{array}\right], \qquad K = \left[\begin{array}{c|c} \diagdown K^A & 0 \\ \hline 0 & K^B \diagdown \end{array}\right],$$

$$q = \left((q^A)^T,\ (q^B)^T\right)^T \; : \tag{30}$$

In what way Eq.(29) has to be changed due to rigid, flexible or mixed coupling is shown in the following sections. A computer software package comprising these three types of substructure coupling has been worked out by the Institute of Aeroelasticity and published in Ref.[12].

4.1 Flexible coupling

This approach,[5,6,8] starts at the formulation of the stiffness or potential energy of a coupling element between the substructures A and B which can be expressed quite similar to Eq.(8) by

$$E_p = \frac{1}{2} u_f^T C^{AB} u_f \tag{31}$$

where C^{AB} denotes the stiffness matrix of the coupling element and

$$u_f = \begin{Bmatrix} u_f^A \\ \hline u_f^B \end{Bmatrix} \quad . \tag{32}$$

The column vectors u_f^A and u_f^B contain the arbitrary displacements at the coupling points of the substructures A and B , respectively. By developing vector u_f into the modal series expansion

$$u_f = \Phi_f \, q \tag{33}$$

where

$$\Phi_f = \left[\begin{array}{c|c} \Phi_f^A & 0 \\ \hline 0 & \Phi_f^B \end{array}\right] \tag{34}$$

Eq.(31) can be transformed into

$$E_p = \frac{1}{2} q^T \Phi_f^T C^{AB} \Phi_f \, q \quad . \tag{35}$$

The submatrices Φ_f^A and Φ_f^B contain the normal mode displacements at the coupling points corresponding to the normal modes of the uncoupled substructures A and B , respectively.

Application of Lagrange's operation (7) to Eq.(35) results in the modal coupling matrix

$$\Delta K = \Phi_f^T C^{AB} \Phi_f \quad . \tag{36}$$

Addition of ΔK to matrix K in Eq.(29) leads to the equations of motion of two substructures A and B coupled by a flexible spring element, the stiffness matrix of which is denoted by C^{AB} .

4.2 Rigid coupling approach

Rigid coupling of two substructures A and B which is described for instance in Refs.[4] and [5] requires the fulfillment of the compatibility condition in the coupling points

$$u_R^A - u_R^B = 0 \tag{37}$$

where the column matrices u_R^A and u_R^B contain the displacements at the coupling points of the substructures A and B , respectively. Eq.(37) can be expanded again into a series of the normal mode displacements at the coupling points such as

$$Gq = 0 \tag{38a}$$

where

$$G = \left[\Phi_R^A \;\vdots\; -\Phi_R^B \right] \quad . \tag{38b}$$

By partitioning matrix G into an invertable square matrix $\bar{\Phi}$ and a matrix Φ^* Eq.(38) can be reordered as follows:

$$Gq = 0 \quad \rightarrow \quad \overline{Gq} = 0 \tag{39}$$

where

$$\overline{Gq} = \left[\Phi^* \;\vdots\; \bar{\Phi} \right] \begin{Bmatrix} p^* \\ \hline \bar{p} \end{Bmatrix} \quad . \tag{40}$$

By using relation (40) the column matrix $\bar{p}$ may be expressed by the column matrix p^* as follows:

$$\bar{p} = - \bar{\Phi}^{-1} \Phi^* p^* \qquad (41)$$

so that column vector $\bar{q}$ can be expressed by the relation

$$\bar{q} = T p^* \qquad (42)$$

where

$$T = \begin{bmatrix} 1 \\ \hline - \bar{\Phi}^{-1} \quad \Phi^* \end{bmatrix} \qquad (43)$$

meaning that matrix $\bar{\Phi}$ must not be singular. Application of the reordering scheme (39) to Eq.(29) and application of transformation (42) leads to the equations of motion of the coupled system

$$T^T (- \omega^2 \bar{M} + \bar{K}) T p^* = 0 \quad . \qquad (44)$$

As to the order of the column matrices $\bar{q}$, $\bar{p}$ and p^* it has to be mentioned that if n^A and n^B are the numbers of generalized degrees of freedom of the uncoupled substructures A and B and if n_C is the number of constraints, then

$\bar{q}$ is of order $(n^A + n^B) \times 1$

$\bar{p}$ is of order $n_C \times 1$

p^* is of order $(n^A + n^B - n_C) \times 1$.

4.2.1 Some remarks to the problem of statically overdeterminate coupling. In contrast to the flexible coupling approach rigid coupling entails in any case a loss of degrees of freedom due to the compatibility condition at the coupling points. Thus, fulfillment of the compatibility condition, as usually expressed by a number of n_C constraints, leads to a reduction of the generalized coordinates by the number of constraints. No disadvantage results unless a case of statically overdeterminate coupling is considered. The most extreme situation imaginable may be characterized by

$$n^A + n^B = n_C \tag{45}$$

which no longer makes sense because of a total loss of degrees of freedom. Thus, problems with statically overdeterminate coupling conditions must be treated delicately. The problem may be eased somewhat by

- *Using the flexible coupling or mixed coupling approach as far as possible even though the coupling joints are relatively stiff*
- *Using more normal modes, i.e. generalized coordinates of the substructures in question, which is not always possible especially if the modal data are determined experimentally.*

4.2.2 An approach to establish independent coordinates. The determination of the transformation matrix T, Eq.(43), can be seriously hampered by the requirement that a square matrix $\bar{\Phi}$ be found which can be inverted. This process can be facilitated considerably by means of the so-called "zero eigenvalue theorem" published in Ref.[13], where a comprehensive proof is given. The mathematical background of this approach is also described for instance in Ref.[14] especially aiming at a method to determine the rank of a matrix by considering the well-known Gram's determinant.

With special view to the rigid coupling method described above, the mathematical problem can be characterized as follows. A transformation matrix T has to be determined so that the transformation (42) relates the dependent coordinates $\bar{q}$ to a set of independent coordinates p^*. The following steps must be performed to attain this aim:

- *Calculate a matrix*

  $$E = G^T G$$

  then E *will be symmetric of order* $(n^A + n^B) \times (n^A + n^B)$ *and positive semi-definite.*

- *Determine the eigenvalues* λ_i $(i = 1,2,\ldots,n^A + n^B)$ *of matrix* E *and the corresponding modal matrix* H *the columns of which represent the eigenvectors* H_i *of matrix* E .
- *Identify the columns of* H *which correspond to the zero eigenvalues of matrix* E . *This step requires special attention as to calculation roundoffs which can affect the rigorous distinction between the finite eigenvalues and the zero eigenvalues to be identified.*
- *Arrange the eigenvectors* H_i *corresponding to the zero eigenvalues in a matrix which is the required transformation matrix* T *of order* $(n^A + n^B) \times (n^A + n^B - n_C)$. *The number of constraints* n_C *is equivalent to the number of positive eigenvalues.*

4.3 Mixed coupling

As is well known, coupling conditions sometimes occur which cannot clearly be categorized as pure rigid coupling or pure flexible coupling. This special type of coupling is termed mixed coupling which can best be explained with the example of a wing-aileron system:

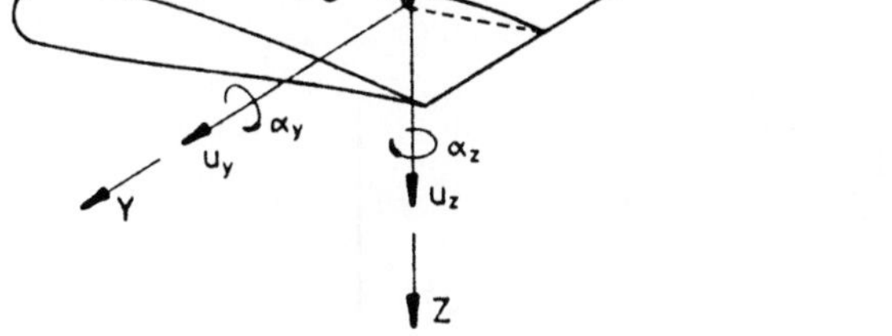

Figure 4: Wing-aileron system.

As is evident from Figure 4, the only flexible coupling condition is the one related to the hinge rotation $\alpha_y = \alpha_y^A - \alpha_y^B$. The coupling stiffness related to the other degrees of freedom are much higher by more than one order of magnitude, so that these connections can be considered rigid. Due to this fact, coupling between aileron and wing can be carried out in good approximation by satisfying the compatibility condition for the displacements $u_x^i, u_y^i, u_z^i, \alpha_x^i, \alpha_z^i$, (i = A,B) , and by assuming the hinge connection to be flexible.

In such a case of mixed coupling conditions the equations of motion can be formulated as follows:

$$T^T(-\omega^2\ \overline{M} + \overline{K} + \Delta\overline{K})T\ q = 0 \tag{47}$$

where matrix T accounts for the rigid coupling condition. Matrix $\Delta\overline{K}$ relating to the flexible hinge connection is basically equivalent to Eq.(36) but reordered in accordance with Eq.(39).

4.4 Exemplary application of the mixed coupling approach

To give an impression of how the modal coupling concepts derived above work, a simple system consisting of some lumped masses and helical stiffnesses is dealt with as an example. The structure of this system shown in Figure 5 is characterized by 9 masses, 9 helical springs and 2 rigid connections. The system is assumed to be subdivided into 3 substructures A , B and C , the modal characteristics of which are known. They are:

$$M^A = \begin{bmatrix} 2.76 & & \\ & 2 & \\ & & 2.76 \end{bmatrix} \tag{48a}$$

$$M^B = M^C = \begin{bmatrix} 5 & & \\ & 2.5 & \\ & & 2.5 \end{bmatrix} \tag{48b}$$

$$K^A = 10^4 \begin{bmatrix} 0.528 & & \\ & 1 & \\ & & 3.62 \end{bmatrix} \tag{49a}$$

$$K^B = K^C = 10^4 \begin{bmatrix} 0 & & \\ & 0.864 & \\ & & 2.261 \end{bmatrix} \tag{49b}$$

$$\Phi^A = \begin{bmatrix} 1 & 1 & -0.618 \\ 0.618 & 0 & 1 \\ 1 & -1 & -0.618 \end{bmatrix} \tag{50a}$$

$$\Phi^B = \Phi^C = \begin{bmatrix} 1 & -0.809 & 0.309 \\ 1 & 1 & 1 \\ 1 & 0.309 & -0.809 \end{bmatrix} . \tag{50b}$$

Connecting subsystem A with the subsystems B and C by two flexible links as shown in Figure 5 leads to a coupling stiffness matrix

$$C^{ABC} = 10^4 \begin{bmatrix} 1 & 0 & -1 & 0 \\ 0 & 1 & 0 & -1 \\ -1 & 0 & 1 & 0 \\ 0 & -1 & 0 & 1 \end{bmatrix} \tag{51}$$

and correspondingly to matrix

$$\Phi_f = \begin{bmatrix} 1 & 1 & -0.618 & 0 & 0 & 0 & 0 & 0 & 0 \\ 1 & -1 & -0.618 & 0 & 0 & 0 & 0 & 0 & 0 \\ 0 & 0 & 0 & 1 & -0.809 & 0.309 & 0 & 0 & 0 \\ 0 & 0 & 0 & 0 & 0 & 0 & 1 & -0.809 & 0.309 \end{bmatrix} \tag{52}$$

comprising the modal deflections at the coupling points.

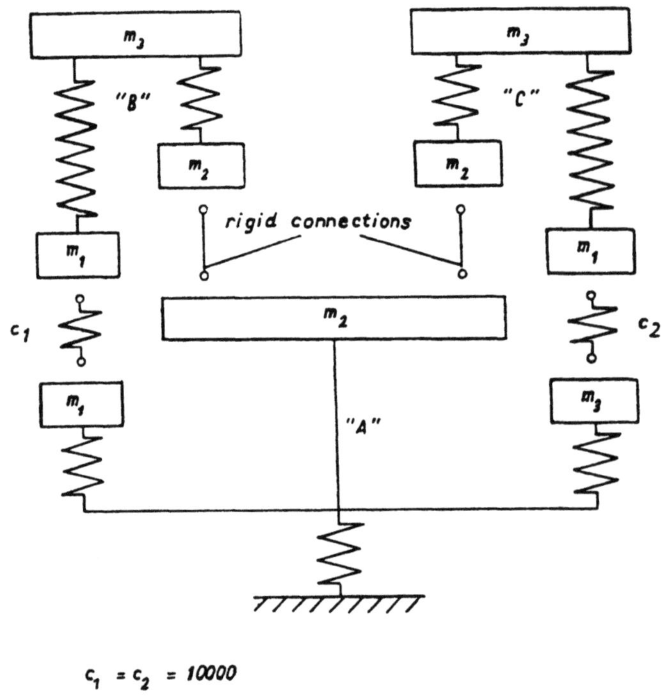

Figure 5: Application of the mixed coupling approach to a simple mass-spring system.

According to Eq.(36) the modal coupling matrix ΔK can be determined from Eqs.(51) and (52) as follows

$$\Delta K = \Phi_f^T \, C^{ABC} \, \Phi_f \quad . \tag{53}$$

Connecting the subsystem A rigidly by two rigid links as shown in Fig.5 with the subsystems B and C results in the compatibility matrix

$$G = \begin{bmatrix} 0.618 & 0 & 1 & -1 & -1 & -1 & 0 & 0 & 0 \\ 0.618 & 0 & 1 & 0 & 0 & 0 & -1 & -1 & -1 \end{bmatrix} \quad . \tag{54}$$

Determination of the eigenvectors H_i of matrix $E = G^T G$ results in the modal matrix

$$H = \begin{bmatrix} .364 & 0 & .311 & .00488 & .124 & .00457 & .178 & 0 & .851 \\ 0 & 0 & 0 & 0 & 0 & 0 & 0 & 1 & 0 \\ .589 & 0 & .503 & .00790 & .201 & .00740 & .288 & 0 & -.526 \\ -.295 & .408 & .243 & -.399 & .0970 & -.706 & .140 & 0 & 0 \\ -.295 & -408 & .299 & -.410 & .0914 & .708 & .131 & 0 & 0 \\ -.295 & .408 & .223 & .820 & .0889 & .00913 & .127 & 0 & 0 \\ -.295 & -.408 & .699 & .00314 & -.291 & .00301 & -.417 & 0 & 0 \\ -.295 & -.408 & -.00173 & .00405 & -.291 & .00374 & .812 & 0 & 0 \\ -.295 & -.408 & -.00159 & .00374 & .864 & .00348 & -.0029 & 0 & 0 \end{bmatrix} \tag{55}$$

and in the matrix

$$\Lambda = \begin{bmatrix} 5.763 & & & & & & & & \\ & 3 & & & & & & & \\ & & 0 & & & & & & \\ & & & 0 & & & & & \\ & & & & 0 & & & & \\ & & & & & 0 & & & \\ & & & & & & 0 & & \\ & & & & & & & 0 & \\ & & & & & & & & 0 \end{bmatrix} \tag{56}$$

containing the eigenvalues λ_i ,

The first two eigenvalues are finite and the last seven have the value zero. Therefore the last seven columns of the modal matrix constitute a suitable transformation matrix T . With matrix ΔK as formulated in Eq.(53), with

$$M = \left[\begin{array}{c|c|c} \ddots M^A & & \\ \hline & \ddots M^B \ddots & \\ \hline & & M^C \ddots \end{array}\right] \tag{57}$$

and

$$K = \left[\begin{array}{c|c|c} \ddots K^A & & \\ \hline & \ddots K^B \ddots & \\ \hline & & K^C \ddots \end{array}\right] \tag{58}$$

and with the transformation matrix T , the elastodynamic equations of motion of the coupled system can be established as formulated in Eq.(47).

4.5 Convergence problem

Whether a modal coupling calculation results in a satisfactory description of the elastodynamic behavior of the complete system depends on the conditions under which the different substructures have been tested.

First of all, for every substructure only a limited number of normal modes can be measured. This truncation of the frequency range may result in convergence problems. Even more difficult to cope with is the problem arising from the requirement that the dynamic load distribution in the substructure modal survey test must be closely representative of the dynamic load distribution in the complete system. In other words, the normal modes of the substructures must contain dynamic load distributions, a superposition of which rapidly converges to the dynamic load distribution in the complete system.

This *load conditioning* problem may best be illustrated by the example of a modularly designed communication satellite as schematically sketched in Figure 6.

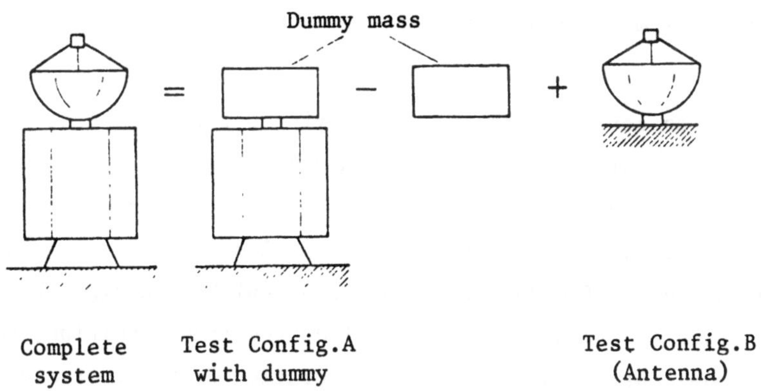

Figure 6: Coupling scheme of a satellite structure - Test configurations.

In the case of different interchangeable antenna systems with the rest of the structure unchanged, it proves convenient to decompose the satellite into two different substructures conditioned in such a way that they can serve as the basic MST configurations A and B . Configuration A consists of the satellite with the antenna system replaced by a nearly rigid dummy, approximately representing the mass inertia properties of the antenna system. Moreover, configuration A is fixed at its base to a heavy seismic block to simulate the connection between satellite and rocket launcher. Test configuration B consists of the antenna fixed to the seismic block to approximate actual load conditions at the interface between antenna base and satellite structure more realistically than possible with a free unloaded interface.

According to Figure 6 the main steps to be performed in establishing the mathematical model of the complete system may be listed as follows:

1. MST on subsystem A (basic satellite structure plus mass dummy)
2. MST on subsystem B (antenna)
3. Removal of the dummy mass from subsystem A by means of a *modal correction calculation*
4. Coupling of subsystem B to the basic satellite structure (subsystem A minus mass dummy) by means of the *rigid coupling approach.*

There are no serious problems in carrying out the steps 1, 2 and 3. However, step 4 entails a special problem deserving closer consideration. It is obvious that coupling subsystem B rigidly to the satellite structure has to involve not only a set of elastic normal modes, but also the six rigid body degrees of freedom of the antenna. Unlike the rigid body degrees of freedom which are related to a so-called free-free condition, the elastic normal modes correspond to test configuration B. Due to these various boundary conditions the rigid body modes Φ_{Or} and the elastic normal modes Φ_{Er} are not mutually orthogonal. This inconsistency generally results in coupling off-diagonal elements (due to the positive definite mass matrix A^B of the antenna) in the generalized mass matrix

$$M^B = (\Phi^B)^T A^B \Phi^B = \left[\begin{array}{c|c} \ddots M_O^B \ddots & M_{OE}^B \\ \hline M_{EO}^B & \ddots M_E^B \ddots \end{array}\right] \tag{59}$$

where

$$\Phi^B = \left[\Phi_0^B \quad \Phi_E^B \right] \tag{60}$$

and

$M_0^B = (\Phi_0^B)^T A^B \Phi_0^B$ diagonal matrix of the generalized masses M_{Or}^B corresponding to the rigid body degrees of freedom Φ_{Or}^B

$M_E^B = (\Phi_E^B)^T A^B \Phi_E^B$ diagonal matrix of the generalized masses M_{Er}^B corresponding to the elastic normal modes Φ_{Er}^B

$M^B_{OE} = (\Phi^B_0)^T A^B \Phi^B_E$ coupling matrix due to the non-orthogonality of the normal modes Φ^B_{Or} and Φ^B_{Er}, $M^B_{OE} = (M^B_{EO})^T$.

As experience has shown, a mass matrix A^B stemming for instance from a finite element analysis is fairly correct and usually much closer to the physical reality than the complementary analytical stiffness matrix C^B can be. Thus, the matrix $M^B_{OE} = (M^B_{EO})^T$ can easily be determined on the basis of the measured normal modes Φ^B_{Er} and Φ^B_{Or} and the mass matrix A^B.

The generalized stiffness matrix can be written as follows

$$K^B = (\Phi^B)^T C^B \Phi^B = \begin{bmatrix} \ddots K^B_0 \ddots & K^B_{OE} \\ K^B_{EO} & \ddots K^B_E \ddots \end{bmatrix} \quad (61)$$

where, because $(\Phi^B_0)^T C^B = 0$ with C^B semi-definite

$$K^B_0 = (\Phi^B_0)^T C^B \Phi^B_0 = 0 \quad (62)$$

$$K^B_{OE} = (\Phi^B_0)^T C^B \Phi^B_E = 0 \quad (63)$$

$$K^B_{EO} = (K^B_{OE})^T \quad . \quad (64)$$

Furthermore

$$K^B_E = (\Phi^B_E)^T C^B \Phi^B_E \quad (65)$$

is a diagonal matrix comprising the generalized stiffnesses $K^B_{Er} = (\omega^B_{Er})^2 M^B_{Er}$ of the elastic normal modes of configuration B with ω^B_{Er} denoting the r-th circular normal frequency.

Thus, Eqs.(59) through (63) show that the mathematical model of configuration B can be established despite the inconsistency of the non-orthogonal sets of normal modes Φ^B_{Or} and Φ^B_{Er} in a relatively simple way without any information on the stiffness distribution of the antenna. The only antenna data required are

- *a reliable mass matrix* A^B *(stemming for instance from a finite element analysis) to calculate matrix* $M^B_{OE} = (M^B_{EO})^T$
- *a complete set of modal data related to the rigid body degrees of freedom (diagonal matrix* M^B_0 *and normal mode shapes* Φ^B_{0r} *).*
- *a complete set of modal data measured in an MST in configuration B (diagonal matrix* M^B_E *, normal mode shapes* Φ^B_{Er} *and normal frequencies* ω^B_{Er} *).*

Setting up the mathematical model of the complete satellite system requires also the knowledge of the elastic normal modes and the related data of configuration A and, in the case of rigid coupling, the transformation matrix T .

It should be mentioned that the modal coupling approach can also be applied to numerous other problems such as those occuring when wing/external stores configurations or wing/control surface combinations are treated by means of substructure coupling.

5. CONCLUDING REMARKS

The modal coupling methods as well as the modal correction approach constitute an efficient tool to treat a special range of structural dynamics problems. As is shown, all these methods can be employed without any information on the stiffness distribution. Only a reliable mass model attained for instance from a finite element analysis, the properties of the coupling interface elements and the normal mode data (generalized masses, normal frequencies and normal mode shapes) of the different subsystems are necessary to set up the equations of motion of the coupled system.

Aside from this still underdeveloped treatment of problems arising from frequency range truncation and incomplete interface load conditioning, there remain numerous open questions on which special emphasis should be placed in the future. Probably the three most severe problems yet to be solved can be described as follows:

- *Consideration of structural nonlinearities, especially those occuring in the coupling interfaces*
- *Establishment and solving of the nonlinear equations of motion*
- *Development of methods to deal with rigid coupling problems with special regard to continuously connecting interfaces (connection along a coupling line or in a coupling plane) with statically over-determined coupling conditions.*

6. REFERENCES

1. Hurty, W.C., Dynamic analysis of structural systems by component mode synthesis, *Techn. Memorandum No.32-530*, Jet Propulsion Laboratory, Pasadena, Calif., 1964.

2. Gladwell, G.M.L., Branch mode analysis of vibrating systems, *Journ. Sound Vibration*, 1, 41, 1960.

3. Goldman, R.L., Vibration analysis by dynamic partitioning, *AIAA Journ.*, 7 (6), 1152, 1969.

4. Hu, S.N., Review of modal synthesis techniques and new approach, *Shock and Vibr. Bul.*, 40 (4), 25, 1969.

5. Breitbach, E., Investigation of spacecraft vibrations by means of the modal synthesis approach, *ESA-SP-121*, 1976.

6. Bertram, A., Dynamic qualification of large space structures by means of modal coupling techniques, *Proc. XXX. Inter'l Astronaut. Congress*, Munich, W.Germany, 1979.

7. Küssner, H.G., Breitbach, E., Bestimmung der Korrekturglieder der Bewegungsgleichungen bei Änderung eines elastomechanischen Systems, *AVA-Report 69 J 101*, 1969.

8. Breitbach, E., Treatment of the control mechanisms of light airplanes in the flutter clearance process, *NASA CP 2085, Part II*, 437, 1979.

9. Niedbal, N., Untersuchung des aeroelastischen Verhaltens des Segelflugzeugs ASW-17 bei Änderung der Tragflügelbiegesteifigkeit, *DFVLR-Report IB 253-73 C 12*, 1973.

10. Breitbach, E., Niedbal, N., Berücksichtigung von Modifikationen der in Standschwingungsversuchen untersuchten Außenlastkonfigurationen des Alpha-Jets mittels modaler Korrekturen auf Basis der Versuchsergebnisse, *DFVLR-Report IB 253-74 C 18*, 1974.

11. Freymann, R., Development of modal techniques using experimental modal data - Work package 1, Small modifications of structural configuration - Users' and Programmers' manual for the computer program DIMOC, *DFVLR-Report IB 253-78 C 02*, 1978.

12. Bertram, A., Development of modal techniques using experimental modal data - Work package 2, Structural dynamic synthesis - Users' and Programmers' manual for the computer program MODAC, *DFVLR-Report IB 253-79 C 07*, 1979.

13. Walton, W.C., Steeves, E.C., A new matrix theorem and its application for establishing independent coordinates for complex dynamical systems with constraints, *NASA TR R-326*, 1969.

14. Courant, R., Hilbert, D., *Methods of Mathematical Physics, Vol.1*, Interscience Publishers, Inc., New York, 1953.

LINEARIZED ERROR ANALYSIS IN SYNTHESIS TECHNIQUES

H.G. Natke
Curt-Risch-Institut
Universität Hannover

1. Introduction

Large and complex structures need system analysis based on drawings and also system identification for investigation of the dynamic behaviour and qualification of the structure. This is realized by substructure handling and synthesis techniques. This means the partitioning of the structure into several substructures, analysis and/or testing the substructures, and coupling them to the whole structure using substructure results. Advantages and disadvantages are to be found in the previous lecture or in Craig and Chang.[1]

The procedure can be described by the following steps:

- o Defining substructures
- o Calculations or testing substructures
- o Assembling the equations of motions of substructures: description of the whole structure
- o Solving the equations for the whole structure using substructure results

Figs. 1 and 2 contain some examples of the structures indicated, for which synthesis methods are suitable.

Off-shore platforms sited on relatively weak soil need the substructure soil for dynamic investigations. The research and development of new fast railway systems can be carried out by steps, that is, by investigations of the shown substructures and by combining the results with reference to the entire structure.

2. Basic equations and problem formulation

The basic equations of linearized error analysis result firstly from synthesis formalism. We consider two substructures A and B as shown below:

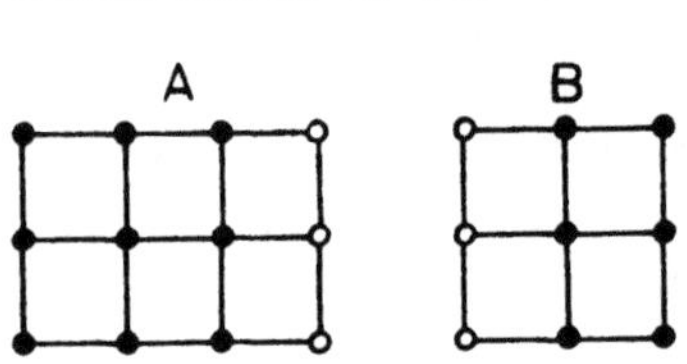

● Interior points, assembled in the vector $\underline{u}_{SS}(t)$

o Junction, interface points, assembled in the vector $\underline{u}_{SV}(t)$

SUBSTRUCTURE 1: PLATFORM CONSTRUCTION
SUBSTRUCTURE 2: SOIL

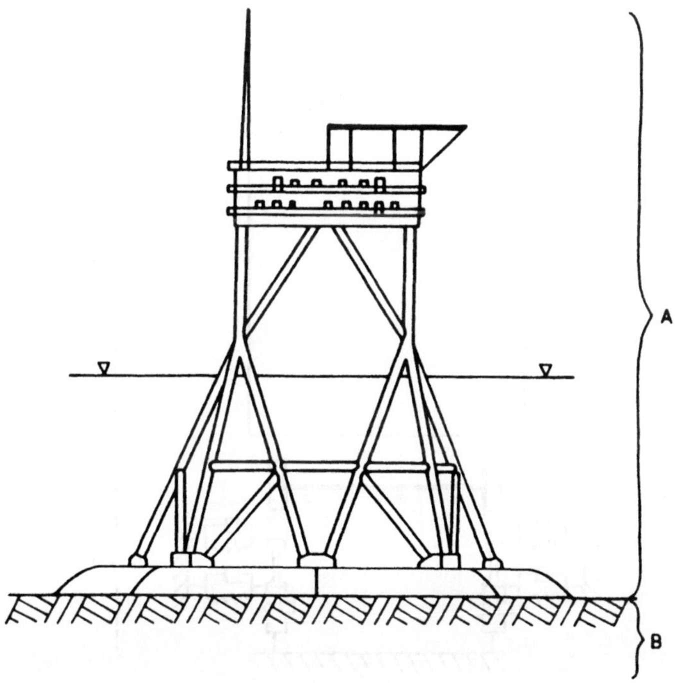

Fig. 1 Offshore platform

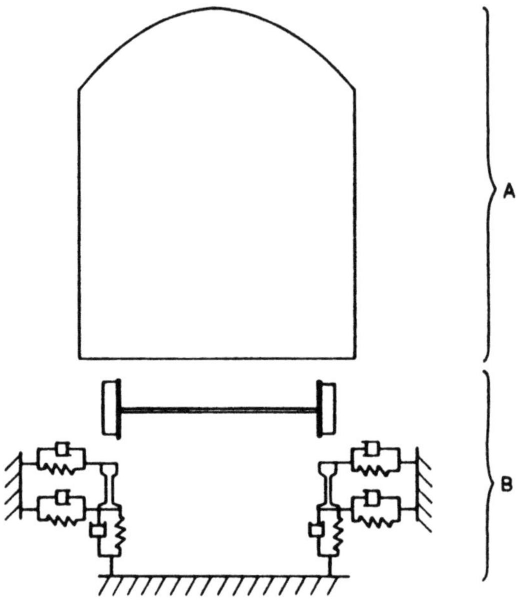

Fig. 2 Fast railway system

$$\underline{u}_S(t) = \begin{pmatrix} \underline{u}_{SS}(t) \\ \underline{u}_{SV}(t) \end{pmatrix} \begin{matrix} \}n_S - n_V \\ \}n_V \end{matrix}, \quad \underline{H}_S = \begin{pmatrix} \underline{H}_{SS} & \underline{H}_{SV} \\ \underline{H}_{VS} & \underline{H}_{VV} \end{pmatrix}, \underline{f}_S(t) = \begin{pmatrix} \underline{f}_{SS}(t) \\ \underline{f}_{SV}(t) \end{pmatrix}. \quad (1)$$

The structure points are divided into interior and interface points, where the latter are defined as those points which are joints to adjacent subsystems. The position concerns the dynamic operator $\underline{H}_S$, the displacement vector $\underline{u}_S(t)$ and the force vector $\underline{f}_S(t)$, where $\underline{f}_{SV}$ is the vector of intersection forces.

The equations of motion are given as:

$$\underline{H}_S(p)\ \underline{u}_S(t) = \underline{f}_S(t) \quad , \quad S = A, B \quad (2)$$

with the dynamic operator $\underline{H}_S$ depending on the differential operator p.

$\underline{u}_S(t)$ and $\underline{f}_S(t)$ are of the same dimension n_S.

As an example, the equations of motion of a linear viscous damped discrete system are given as:

$$\begin{aligned} \underline{H}_S(p)\ \underline{u}_S(t) &:= (p^2\underline{M}_S + p\ \underline{B}_S + \underline{K}_S)\ \underline{u}_S(t) \\ &= \underline{M}_S\ \ddot{\underline{u}}_S(t) + \underline{B}_S\ \dot{\underline{u}}_S(t) + \underline{K}_S\ \underline{u}_S(t). \end{aligned} \quad (3)$$

The assumptions made are:

- o linearity
- o multi-degree-of-freedom system

o and rigid connected subsystems.

The coupling is carried out firstly by the kinematic compatibility condition

$$\underline{u}_{AV} = \underline{u}_{BV} =: \underline{u}_V \tag{4}$$

so that the junction coordinates do not gape, and secondly, by the force equilibrium condition

$$\underline{f}_{AV} = \underline{f}_{BV} =: \underline{f}_V \; . \tag{5}$$

The resulting equations, fully described and briefly formulated as $\underline{H}$ applied to $\underline{u}$ equal to $\underline{f}$, are:

$$\begin{pmatrix} \underline{H}_{AA} & \underline{O} & \underline{H}_{AV} \\ \underline{O} & \underline{H}_{BB} & \underline{H}_{BV} \\ \underline{H}_{VA} & \underline{H}_{VB} & \underline{H}_{VV} \end{pmatrix} \begin{pmatrix} \underline{u}_{AA} \\ \underline{u}_{BB} \\ \underline{u}_V \end{pmatrix} = \begin{pmatrix} \underline{f}_{AA} \\ \underline{f}_{BB} \\ \underline{f}_V \end{pmatrix} \tag{6}$$

$$\underline{H}_{VV} := \underline{H}^A_{VV} + \underline{H}^B_{VV}$$

$$\underline{H}(p)\ \underline{u}(t) = \underline{f}(t) \;, \quad n = n_A + n_B + n_V \; .$$

Besides the synthesis formalism, a transformation in generalized coordinates is used:

$$\underline{u}(t) = \underline{T}\,\underline{q}(t)\ ,\ \det \underline{T} \neq 0\ ,\quad \underline{q}(t) \in C_n$$
$$\underline{V},\quad \det \underline{V} \neq 0\ ,\quad \text{order } n\ , \tag{7}$$

$\underline{V}$ and $\underline{T}$ are regular transformation matrices of the order n, which contain the eigenvectors of the substructures as column vectors in a certain arrangement. $\underline{V}$ is the modal matrix of left eigenvectors, $\underline{T}$ the modal matrix of right eigenvectors.

The transformed equations of motion are:

$$\underline{V}^T\,\underline{H}(p)\,\underline{T}\,\underline{q}(t) = \underline{V}^T\,\underline{f}(t)\ . \tag{8}$$

The upper index T means the transposition.

Partition of the generalized coordinates into primary (retained) coordinates $\underline{q}_W(t)$ and secondary (reduced out) coordinates $\underline{q}_R(t)$, corresponding to the transformation matrices in known parts $\underline{V}_W$, $\underline{T}_W$ and unknown matrices $\underline{V}_R$, $\underline{T}_R$, yields the additive decomposition of the displacement vector:

$$\underline{q}(t) = \begin{pmatrix} \underline{q}_W(t) \\ \underline{q}_R(t) \end{pmatrix} \begin{matrix} \}\ n-m \\ \}\ m \end{matrix}$$

$$\underline{V} = (\underline{V}_W, \underline{V}_R)\ ,\quad \underline{T} = (\underline{T}_W, \underline{T}_R),\quad \underline{V}_W, \underline{T}_W \text{ known}$$
$$\underline{u}(t) = \underline{T}_W\,\underline{q}_W(t) + \underline{T}_R\,\underline{q}_R(t)\ . \tag{9}$$

The resulting problem can be described as:

Given only the submatrices $\underline{V}_W$, $\underline{T}_W$ as parts of the substructure modal matrices, this leads to an approximate modal synthesis. The approximation errors are required. The tool used for solution is dynamic condensation.

Partition of the generalized coordinates describing the motion of the whole structure in primary and secondary coordinates yields in the second row of

$$\begin{pmatrix} \underline{V}_W^T \, \underline{H} \, \underline{T}_W \, , & \underline{V}_W^T \, \underline{H} \, \underline{T}_R \\ \underline{V}_R^T \, \underline{H} \, \underline{T}_W \, , & \underline{V}_R^T \, \underline{H} \, \underline{T}_R \end{pmatrix} \begin{pmatrix} \underline{q}_W(t) \\ \underline{q}_R(t) \end{pmatrix} = \begin{pmatrix} \underline{V}_W^T \, \underline{f}(t) \\ \underline{V}_R^T \, \underline{f}(t) \end{pmatrix} , \tag{10}$$

the secondary coordinates depending on the vector of external forces and the vector of primary coordinates in:

$$\underline{q}_R(t) = (\underline{V}_R^T \, \underline{H} \, \underline{T}_R)^{-1} \, \underline{V}_R^T \left[\underline{f}(t) - \underline{H} \, \underline{T}_W \, \underline{q}_W(t) \right] . \tag{11}$$

Substitution of eq. (11) in the first row of eq. (10) results in:

$$\underline{H}_W(p) \, \underline{q}_W(t) = \underline{f}_W(t) \tag{12}$$

with

$$\begin{aligned} \underline{H}_W(p) &:= \underline{V}_W^T \, \underline{H} \left[\underline{I} - \underline{T}_R \, (\underline{V}_R^T \, \underline{H} \, \underline{T}_R)^{-1} \, \underline{V}_R^T \, \underline{H} \right] \underline{T}_W \, , \\ \underline{f}_W(t) &:= \underline{V}_W^T \left[\underline{I} - \underline{H} \, \underline{T}_R (\underline{V}_R^T \, \underline{H} \, \underline{T}_R)^{-1} \, \underline{V}_R^T \right] \underline{f}(t) \, . \end{aligned} \tag{13}$$

This is a system of equations of the reduced order n-m in the primary coordinates. It contains the unknown modal submatrices $\underline{V}_R$ and $\underline{T}_R$ and an inverse matrix, which, in the case of eigenvibrations, depends on the unknown eigenvalues.

Considering that only the submatrices $\underline{V}_W$ and $\underline{T}_W$ are known,and in order to avoid calculating with the eigen - value - dependent inverse, approximations are necessary for the modal synthesis.

In the following only eigenvibrations are considered which are described by the matrix eigenvalue problem.

The approximate formulation is signed by the upper index L:

$$\underline{H}_W^{(L)}(\lambda^{(L)})\hat{\underline{q}}_W^{(L)} = \underline{0} \quad \text{instead of } \underline{H}_W(\lambda)\ \hat{\underline{q}}_W = \underline{0}\ . \tag{14}$$

The most simple and most far-reaching approximation consists of truncating the secondary coordinates (L = O):

$$\underline{H}_W^{(O)} := \underline{V}_W^T\ \underline{H}\ \underline{T}_W\ . \tag{15}$$

The dynamic operator contains only the known modal sub-matrices without an inverse. Further approximations use the static condensation, in the case L = 1 all dynamic coupling terms between the primary and secondary coordinates are, consequently, neglected:

$$\underline{H}_W^{(1)} := \underline{H}_W^{(0)} + \left[\underline{H}_W(p) - \underline{H}_W^{(0)}(p)\right]\Big/_{p=0}$$
$$= \underline{V}_W^T \left[\underline{H} - \underline{K}\,\underline{T}_R\,(\underline{V}_R^T\,\underline{H}\,\underline{T}_R)^{-1}\,\underline{V}_R^T\,\underline{K}\right]\underline{T}_W \,. \tag{16}$$

In the case L = 2, the dynamic operator of the secondary coordinates is substituted by the corresponding stiffness matrix:

$$\underline{H}_W^{(2)} := \underline{V}_W^T\,\underline{H}\left[\underline{I} - \underline{T}_R\,(\underline{V}_R^T\,\underline{H}\,\underline{T}_R)^{-1}\,\underline{V}_R^T\,\underline{K}\right]\underline{T}_W \,, \tag{17}$$

Other approximations are possible, H.G. Natke.[2]

The above approximations signed by $L \geq 1$ contain the unknown submatrices $\underline{V}_R$, $\underline{T}_R$ and an inveuse independent of the unknown eigenvalues. Because the only condition used with regard to transformation in generalized coordinates is the regularity of the transformation mations one can therefore choose arbitrary linear independent vectors for $\underline{V}_R$, $\underline{T}_R$ so that the completed matrices $(\underline{V}_W, \underline{V}_R)$ and $(\underline{T}_W, \underline{T}_R)$ are regular. This choice does not influence the eigensolutions of the exact problem, but those of the approximated problems with $L \geq 1$. If $\underline{V}_W$, $\underline{T}_W$ are chosen as a nearly complete vector system, the influence of the choice of $\underline{V}_R$, $\underline{T}_R$ is small.

3. Error Analysis

The error definitions of the approximated eigenvalues, the dynamic operator and the eigenvectors are

$$\left.\begin{aligned} \Delta\lambda^{(L)} &:= \lambda - \lambda^{(L)} \\ \Delta \underline{H}_W^{(L)}(\lambda^{(L)}, \Delta\lambda^{(L)}) &:= \underline{H}_W(\lambda) - \underline{H}_W^{(L)}(\lambda^{(L)}) \\ \Delta \underline{q}_W^{(L)} &:= \hat{\underline{q}}_W - \hat{\underline{q}}_W^{(L)}. \end{aligned}\right\} \tag{18}$$

With these definitions introduced into eq. (14), there follows a system of (n-m) equations for each degree of freedom:

$$\Delta \underline{H}_W^{(L)}(\lambda^{(L)}, \Delta\lambda^{(L)})\, \hat{\underline{q}}_W^{(L)} + \left[\underline{H}_W^{(L)}(\lambda^{(L)}) + \Delta \underline{H}_W^{(L)}(\lambda^{(L)}, \Delta\lambda^{(L)})\right] \cdot \Delta \hat{\underline{q}}_W^{(L)} = \underline{0} \tag{19}$$

With the abbreviations:

$$\left.\begin{aligned} \underline{H}_{1W}(\lambda^{(L)}) &:= \frac{\partial}{\partial\lambda} \underline{H}_W(\lambda) \Big/_{\lambda = \lambda^{(L)}} \\ \underline{H}_{2W}(\lambda^{(L)}) &:= \frac{1}{2} \frac{\partial}{\partial\lambda} \underline{H}_{1W}(\lambda) \Big/_{\lambda = \lambda^{(L)}} \end{aligned}\right\} \tag{20}$$

and with the expansion of the error of the dynamic operator:

$$\Delta \underline{H}_W^{(L)}(\lambda^{(L)}, \Delta\lambda^{(L)}) = \underline{H}_W(\lambda^{(L)}) - \underline{H}_W^{(L)}(\lambda^{(L)}) + \underline{H}_{1W}(\lambda^{(L)})\, \Delta\lambda^{(L)} + \underline{H}_{2W}(\lambda^{(L)})\, (\Delta\lambda^{(L)})^2 + \ldots\ldots \tag{21}$$

the equations of the linearized errors can be achieved by linearization of eq. (19) and (21):

$$\Delta\lambda_{lin}^{(L)}\,\underline{H}_{1W}(\lambda^{(L)})\hat{\underline{q}}_W^{(L)}+\underline{H}_W^{(L)}(\lambda^{(L)})\Delta\hat{\underline{q}}_{W1in}+\underline{H}_W(\lambda^{(L)})\hat{\underline{q}}_W^{(L)}=\underline{0}. \quad (22)$$

The result is a linear system of equations (22) in the unknowns of the n-m-1 components of the vector $\Delta\hat{\underline{q}}_W^{(L)}$ (because of normalization only in n-m-1 components) and the eigenvalue error.

With the left eigenvector of the approximate problem:

$$\hat{\underline{r}}_W^{(L)^T}\,\underline{H}_W^{(L)}(\lambda^{(L)}) = \underline{0} \quad (23)$$

one obtains - through left multiplication with the transposed left eigenvector - a scalar equation for $\Delta\lambda_{lin}^{(L)}$ with known right side for each degree of freedom

$$\Delta\lambda_{lin}^{(L)} = -\frac{\hat{\underline{r}}_W^{(L)^T}\,\underline{H}_W(\lambda^{(L)})\,\hat{\underline{q}}_W^{(L)}}{\hat{\underline{r}}_W^{(L)^T}\,\underline{H}_{W1}(\lambda^{(L)})\,\hat{\underline{q}}_W^{(L)}}\;. \quad (24)$$

The value resulting from (24) substituted in (22) gives an inhomogeneous linear system of equations of the order n-m-1 which allow calculation of the linearized error vector with regard to the approximated eigenvector for each degree of freedom. As can be seen, the expense of calculation is not very great.

There is no need for additional calculation or identification of substructural modal magnitudes.

The question remains of the inaccuracy of the linearized errors. For this reason the errors in second approximation are useful.

The error analysis of second approximation results in a matrix eigenvalue problem of second degree

$$\left[(\Delta\lambda_{quad}^{(L)})^2 \underline{H}_{2W}(\lambda^{(L)}) + \Delta\lambda_{lin}^{(L)} \underline{H}_{1W}(\lambda^{(L)} + \underline{H}_{W}(\lambda^{(L)})\right] \cdot$$
$$\cdot (\underline{\hat{q}}_{W}^{(L)} + \Delta\underline{\hat{q}}_{Wquad}^{(L)}) = \underline{0} \quad . \qquad (25)$$

The eigenvalue of (25) is the eigenvalue error being sought and the eigenvector of (25) is the corrected approximate eigenvector in generalized coordinates. The smallest absolute eigenvalue is relevant for each degree-of-freedom.

The error analysis therefore includes

first : supplement of $\underline{V}_W$, $\underline{T}_W$ with arbitrary vectors to regular transformation matrices $\underline{V}$, $\underline{T}$ in the modified problems $L \geq 1$

second: solving the approximate modal synthesis equations

third: calculation of the linearized errors: that means for each degree-of-freedom calculation of a scalar equation and solving a system of linear equations of the order n-m-1

fourth : examination of the linearized error values with the bilinear form resulting from the quadratic eigenvalue problem simplified with the approximate eigenvector, that means neglecting $\Delta \hat{\underline{q}}_W^{(L)}$.

The expense of the error analysis has to be compared with the expense when repeating the whole calculation with a greater number n-m including the calculation or identification of additional eigenmagnitudes of the substructures, or it has to be compared with the expense when solving iteratively the dynamic condensated problem. Discussion of the expense in system identification is omitted in most cases.

4. Range of Application with Regard to Modal Transformations

Synthesis methods are summarized and discussed by R.R. Craig Jr. and C.-J. Chang.[1,3,4] With regard to the reduction of the number of degrees-of-freedom, they are dealt with by H. Röhrle.[5] The author[6] has produced a summary including a linearized error calculation for the modal transformation using fixed interface normal modes of the substructures.

Some essential modal transformations are listed below (determining the arrangement of the subsystem eigenvectors in the matrices $\underline{V}$ and $\underline{T}$):

T1: Fixed interface normal modes

T2: Fixed interface normal modes and constraint modes (for convergence improvement).

These transformations are simple, but the boundary conditions for the substructures can be realized only approximatively.

T3: Free interface normal modes

T4: Free interface normal modes and attachment modes (case distinction: regularity of the stiffness matrix)

T5: Loaded interface normal modes, in order to obtain boundary conditions which are suitably loaded.

5. Examples

The chosen system is a spring-mass system, as shown below,consisting of 9 masses equal to 1, and 9 springs with stiffness constants equal to 10^3. The system is divided into 2 subsystems, each described by 5 equations and each including the junction coordinate.

SPRING-MASS-SYSTEM

9 DOF WITH

$k_1 = \dots = k_9 = 10^3$

$m_1 = \dots = m_9 = 1$

$n_A - n_V = 4, \; n_V = 1$

$n_B - n_V = 4$

MODAL TRANSFORMATION T1:

FIXED INTERFACE NORMAL MODES:

$$\underline{V} = \underline{T} = \begin{pmatrix} \hat{\underline{U}}_{AA} & \underline{0} & \underline{0} \\ \underline{0} & \hat{\underline{U}}_{BB} & \underline{0} \\ \underline{0} & \underline{0} & 1 \end{pmatrix}$$

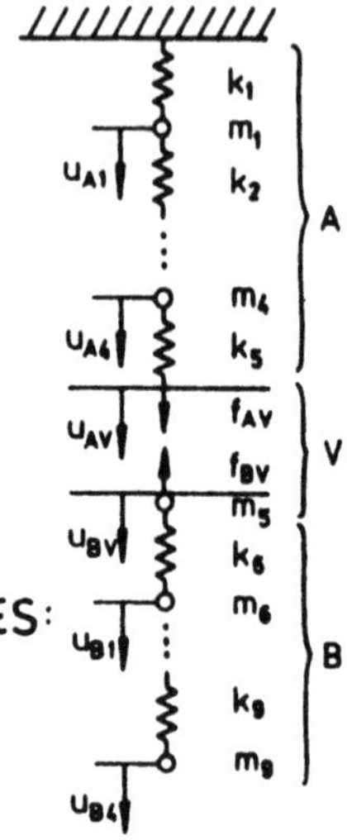

Fig. 3 Example of a system divided into subsystems

In order to obtain large errors, a non-appropriate transformation was chosen: T1 with fixed interface normal modes, that means the substructures are fixed at point V. The corresponding eigenfrequencies are shown in Table 1, and it becomes clear that the fundamental degree-of-freedom would be difficult to approximate.

f_0 [Hz]	0,83	2,47	4,04	5,51	6,82	7,94	8,85	9,52	9,93
f_{0A} [Hz]		3,11		5,92		8,14		9,57	
f_{0B} [Hz]	1,75		5,03		7,71		9,46		

Table 1 Eigenfrequencies of the undamped systems

We consider first the undamped structure and neglect the last degree-of-freedom of substructure A (m_A = 1) and the last two degrees-of-freedom of substructure B (m_B = 2). The approximate modal synthesis method chosen is L = 0, which means truncation of all secondary coordinates. Table 2 shows the exact (see the left-hand column) and linearized relative errors (see the right-hand columns) in p.c.. The first row shows the errors of the eigenfrequencies, the other rows show the errors of the eigenvector components in physical coordinates. The 4 columns represent the first four degrees-of-freedom of the entire structure. As can be seen, the linearized errors are good approximations, although the values are relatively large.

Other calculations, for example with L = 1, give the same good approximations of the linearized errors, but the errors are smaller, as can be seen in Table 3.

Tables 4 and 5 contain the exact, linearized error and the errors in second approximation for the eigenfrequencies

DOF i	1		2		3		4	
$\frac{\Delta f_{oi}^{(o)}}{f_{oi}}$ %, $\frac{\Delta f_{oi\,lin}^{(o)}}{f_{oi}}$ %	-38,7	-29,7	- 4,6	- 4,2	- 4,1	- 3,9	- 1,0	- 1,0
$\frac{\Delta \hat{u}_{oi,k}^{(o)}}{\hat{u}_{oi,k}}$ %, $\frac{\Delta \hat{u}_{oi,k\,lin}^{(o)}}{\hat{u}_{oi,k}}$ %								
k = 1	- 1,7	- 1,6	-20,6	-18,9	13,5	14,1	- 8,6	- 7,3
2	24,3	24,8	- 9,6	- 8,0	5,6	6,0	-14,0	-12,3
3	7,4	6,5	-15,1	-13,3	34,1	32,0	-36,8	-35,5
4	21,3	19,8	- 3,3	- 1,5	-40,6	-25,9	- 8,5	- 7,0
5	22,7	22,9	68,7	77,0	14,0	15,5	-17,4	-17,2
6	-13,2	-12,3	59,4	53,2	-61,5	-58,7	23,6	23,8
7	-15,7	-13,8	28,2	25,7	42,6	43,6	66,4	67,0
8	- 7,5	- 4,9	11,2	9,7	-22,2	-20,1	24,0	23,9
9	21,1	18,1	6,5	8,4	2,2	5,4	2,8	4,9

Table 2 The undamped system neglecting the last dof of substructure A (m_A = 1) and the last two dof of substructure B (m_B = 2).
Approximation L = O: Truncation of sec. coordinates

DOF i	1		2		3		4	
$\frac{\Delta f_{oi}^{(1)}}{f_{oi}}\%$, $\frac{\Delta f_{oi\,lin}^{(1)}}{f_{oi}}\%$	- 0,6	- 0,6	- 0,5	- 0,5	- 1,0	- 1,0	- 0,5	- 0,5
$\frac{\Delta \hat{u}_{oi,k}^{(1)}}{\hat{u}_{oi,k}}\%$, $\frac{\Delta \hat{u}_{oi,k\,lin}^{(1)}}{\hat{u}_{oi,k}}\%$								
k = 1	-17,8	-17,0	- 7,3	- 6,8	6,3	7,6	- 3,1	- 2,2
2	13,7	14,2	2,9	3,4	- 4,5	- 3,1	-10,6	- 9,5
3	-10,1	- 9,4	- 5,8	- 5,2	15,7	16,7	-25,1	-24,1
4	4,2	4,8	1,5	2,0	4,2	6,1	- 3,3	- 2,3
5	20,5	21,0	93,8	93,9	23,1	24,1	-27,5	-26,5
6	-12,3	-11,7	32,5	32,8	-41,5	-39,7	17,1	17,8
7	- 7,8	- 7,1	10,3	10,7	36,3	37,2	65,1	65,5
8	6,2	6,7	- 5,4	- 4,8	-16,0	-14,4	16,9	17,6
9	- 0,4	0,3	- 0,1	0,4	- 1,1	0,5	0,3	1,5

Table 3 The undamped system neglecting the last dof of substructure A (m_A = 1) and the last two dof of substructure B (m_B = 2).
Approximation L = 1: Static condensation

DOF i	1	2	3	4	5	6	7
$\frac{\Delta f_{Di}^{(o)}}{f_{Di}}$ %	- 16,27	- 1,56	- 1,25	-0,27	-0,91	-0,13	-3,80
$\frac{\Delta f_{Di\,lin}^{(o)}}{f_{Di}}$ %	- 14,91	- 1,53	- 1,25	-0,27	-0,90	-0,13	-2,56
$\frac{\Delta f_{Di\,quad}^{(o)}}{f_{Di}}$ %	- 16,27	- 1,56	.- 1,25	-0,27	-0,91	-0,13	-1,52
$\frac{\Delta \gamma_i^{(o)}}{\gamma_i}$ %	- 1,26	0,66	-0,50	0,73	0,03	-0,83	0,26
$\frac{\Delta \gamma_{i\,lin}^{(o)}}{\gamma_i}$ %	- 1,28	0,66	-0,51	0,73	0,05	-0,80	0,61
$\frac{\Delta \gamma_{i\,quad}^{(o)}}{\gamma_i}$ %	- 1,26	0,66	-0,50	0,73	0,03	-0,83	0,50

Table 4 The hysteretic damped system with non-modal damping.

$m_A = 1$, $m_B = 1$, $L = 0$
$d_1 = d_3 = \ldots = d_9 = 30$, $d_2 = d_4 = \ldots = d_8 = 20$

DOF i	1	2	3	4	5	6
$\frac{\Delta f_{Di}^{(o)}}{f_{Di}}$ %	-38,71	- 4,61	- 4,08	- 1,00	- 7,83	-12,15
$\frac{\Delta f_{Di\,lin}^{(o)}}{f_{Di}}$ %	-29,01	- 4,24	- 3,89	- 0,96	- 4,42	- 0,70
$\frac{\Delta f_{Di\,quad}^{(o)}}{f_{Di}}$ %	-38,70	- 4,61	- 4,07	- 1,00	- 3,14	- 0,67
$\frac{\Delta \gamma_i^{(o)}}{\gamma_i}$ %	-26,11	- 3,18	- 6,75	1,30	-22,06	4,85
$\frac{\Delta \gamma_{i\,lin}^{(o)}}{\gamma_i}$ %	- 8,48	- 2,26	- 6,06	1,36	6,41	2,53
$\frac{\Delta \gamma_{i\,quad}^{(o)}}{\gamma_i}$ %	-26,07	- 3,18	- 6,76	1,30	8,55	2,32

Table 5 The hysteretic damped system with non-modal damping.

$m_A = 1$, $m_B = 2$, $L = 0$

$d_1 = d_3 = \ldots d_9 = 30$, $d_2 = d_4 = \ldots = d_8 = 20$

in the upper part, and the damping ratios in the lower part of the hysteretic non-modal damped structure in the same configurations as before. Now the results of all degrees-of-freedom are presented. The linearized errors are approximations of the exact values for the first four degrees-of-freedom, and for these the errors in second approximation are equal to the exact values.

The viscous damped system with non-modal damping coefficients yields approximate errors similar to these. The comparison provides no further information.

6. Conclusions

Approximate synthesis formulations can be used with known modal submatrices $\underline{V}_W$, $\underline{T}_W$ and - if wanted - with supplemented arbitrary matrices $\underline{V}_R$, $\underline{T}_R$ so that the transformation matrices are regular. Linearized error analysis requires the calculation of one scalar equation and the solution of one system of linear equations of reduced order for each degree-of-freedom. In order to check these values, one can take bilinear forms resulting from the quadratic eigenvalue problem with regard to the errors in second approximation, while using the approximate eigensolutions of the modal synthesis or the solutions corrected by the linearized errors. If necessary, one can determine the errors in second approximation by solving the quadratic eigenvalue problem.

If the interface boundary conditions of the substructures are adequate to the real loading, and therefore the related modal transforms are adequate to the dynamic behaviour of the structure, and if the chosen approximation method signed by L is carried out in dependence on the number of secondary coordinates and the frequency range, then the linearized errors are useful for correcting the approximate solutions.

Extentions can be made with regard to non-rigid coupling in the interface points taking into account the energy parts of the junction elements, Breitbach.[7]

References

1 Craig, Jr. R.R. and Chang, C.-J., Substructure coupling for dynamic analysis and testing, NASA CR - 2781, Washington, D.C., 1977

2 Natke, H.G., Angenäherte Fehlerermittlung für Modalsynthese-Ergebnisse innerhalb der Systemanalyse und Systemidentifikation, ZAMM vol. 61, part 1, p. 41-58, 1981

3 Craig, Jr. R.R. and Chang, C.-J., A review of substructure coupling methods for dynamic analysis, NASA CP - 2001, vol. 2, p. 393-408, 1976

4 Craig, Jr. R.R., Methods of component mode synthesis, The Shock and Vibration Digest 9, vol. 11, p. 3-10, 1977

5 Röhrle, H., Reduktion von Freiheitsgraden bei Strukturdynamik-Aufgaben, Habilitationsschrift Universität Hannover 1979

6 Natke, H.G., Substrukturmethoden bei dynamischen Untersuchungen, CRI-K 1/78, Mitteilung des Curt-Risch-Institutes der Technischen Universität Hannover, p. 17-46, 1978

7 Breitbach, E., Bestimmung der Kopplungsglieder und der Bewegungsgleichungen einer aus mehreren elastomechanischen Teilsystemen zusammengesetzten Flugkonstruktion, Z. Flugwiss., vol. 18, part 9/10, 1970

IDENTIFICATION OF NONLINEAR SYSTEMS

Dr. E. Breitbach
Deutsche Forschungs- und Versuchsanstalt
für Luft- und Raumfahrt e.V.
Institut für Aeroelastik
Bunsenstr.10, D-3400 Göttingen
W. Germany

1. INTRODUCTION

At first glance few problems arise in the area of structural dynamics which cannot be solved by means of today's experimental and analytical tools. Thus, the elastodynamic characteristics can be determined by using common experimental or analytical methods if structural linearity can be assumed to be a proper approximation. However, as experience has shown, the world of structural dynamics is not as linear as desired, which becomes evident when dealing with such problems as control mechanisms and wing/external stores connections of airplanes as well as related problems in spacecraft engineering and many other areas. Consequently, setting up dependable mathematical models requires special attention to nonlinear effects, including modifications of linearized standard procedures.

In the first part of this lecture some of the most frequently occurring types of control system nonlinearities are described as typical examples of nonlinearities. Later on it is shown in detail how the elastodynamic equations of motion of structural systems with localized nonlinearities may be formulated by using various suitably modified ground vibration test (GVT) or modal survey test (MST) procedures all based on the well-known modal synthesis approach, as dealt with in a previous lecture. Moreover, emphasis is placed on the elaboration of an experimental-analytical method to determine the nonlinear characteristics of special mechanical joints and coupling elements.

2. GENERAL REMARKS

2.1 Sources of localized structural nonlinearities

Structural dynamics and aeroelastic investigations are usually carried out on the basis of simplified linearized mathematical models. In many cases this approach has been adequate to ensure sufficient information on a structure's behavior. However, in the last few years, it has become evident that disregarding nonlinear phenomena can lead to hazardously misleading results. For example, it is shown in Ref. [1] that so-called concentrated or localized nonlinearities in airplane control systems have a significant effect on flutter behavior. Nonlinearities of this kind may be produced by such things as

(1) Backlash in joints and linkage elements
(2) Solid friction in control cable and pushrod ducts as well as in the hinge bearings
(3) Kinematic limitation of the control surface stroke
(4) Application of special spring tab systems provided for pilot handling relief.

A structural dynamics or aeroelastic investigation may become even more complicated if it is necessary to account for items such as the following:

(1) Preload changes, for instance due to maneuver loads and specially trimmed flight attitudes
(2) Changes in friction and backlash over the lifetime of a structure
(3) Additional mass, stiffness and damping forces randomly activated during operation of a system.

Coping with all these difficulties requires special measures throughout all investigations in the course of the development of a structural system. First of all, the experimental methods (GVT or MST) used to determine the elastodynamic characteristics of a system have to be modified so that a consistent and superpositionable set of orthogonal, or well-defined nonorthogonal, normal modes can be measured.

Of course the simplest solution would be to build mechanical joints and linkage elements without either friction or backlash. However, aside from a considerable increase in manufacturing costs, there is no guarantee that such an ideal condition could be kept unchanged for the lifetime of a structure. Moreover, nonfriction is not necessarily equivalent to better system qualities, because friction can be of tremendous advantage, for instance in solving fatigue and dynamic response problems.

Because of the author's professional background it seems sensible to deal with the subject of this lecture from the standpoint of an aeroelastician. In doing so, the general problem concerning the identification of nonlinear structures can best be illustrated and made clear by investigating the special problems of control surface nonlinearities. It is no problem to transfer the results of these special airplane-related investigations to other areas of structural dynamics.

2.2 Illustrative examples of control system nonlinearities

To get a realistic impression of control mechanism nonlinearities, the measured force deflection diagrams $F(\beta)$ of the rudder and aileron system (antisymmetric and symmetric cases) of a sailplane are shown in Figures 1,3 and 5. Using the principle of the *energetic equivalence*,[1,2] the stiffness and damping properties of a nonlinear force deflection diagram can be approximated by the so-called equivalent complex stiffness:

$$C_e(B) = C_e'(B) + jC_e''(B) \quad , \quad (j = \sqrt{-1}\,) \tag{1}$$

where B denotes the amplitude of an harmonic deflection

$$\beta = B \cos \varphi \quad . \tag{1a}$$

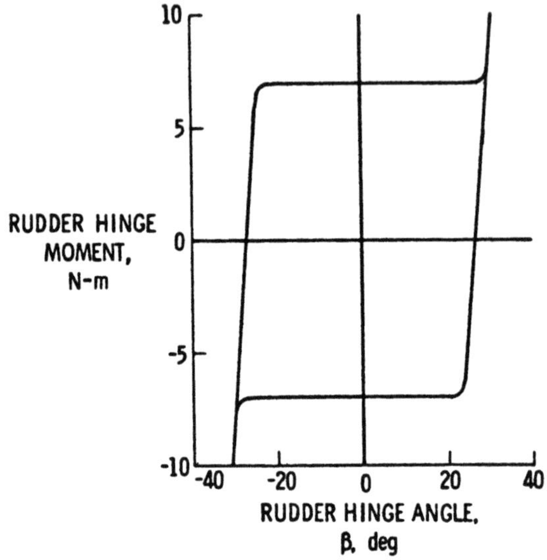

Figure 1: Force deflection diagram - rudder system.

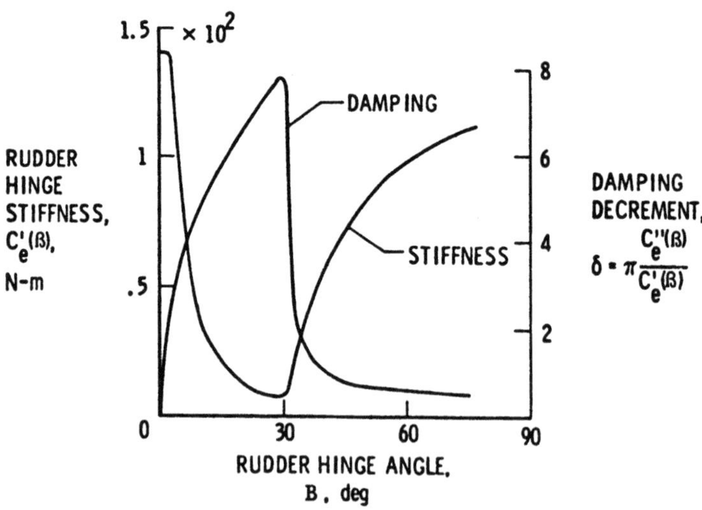

Figure 2: Hinge stiffness and damping versus hinge angle - rudder system.

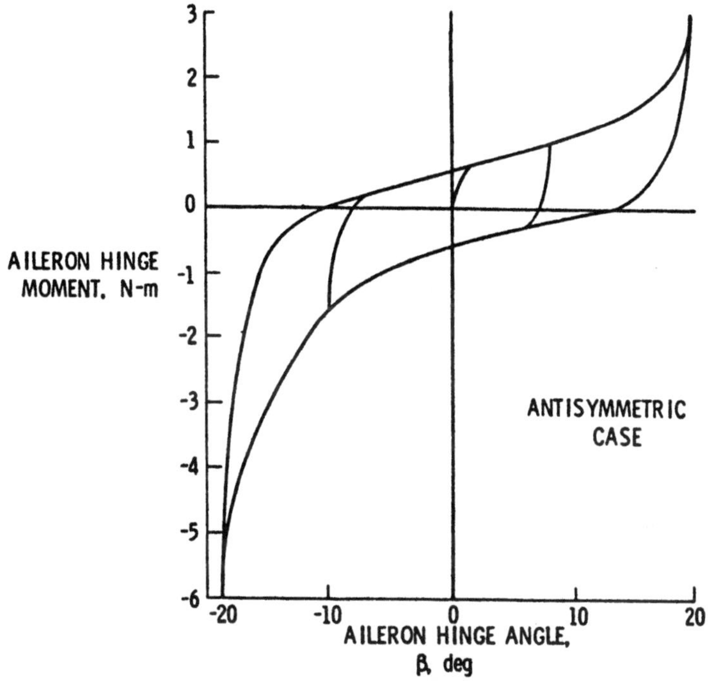

Figure 3: Force deflection diagram - aileron system, antisymmetric case.

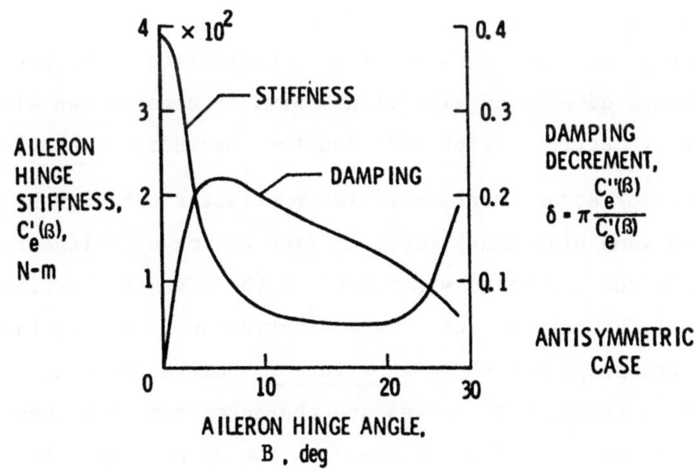

Figure 4: Hinge stiffness and damping versus hinge angle - aileron system, antisymmetric case.

The coefficients $C_e'(B)$ and $C_e''(B)$ representing stiffness and damping, respectively, can be calculated by solving the integrals

$$\left.\begin{aligned} C_e'(B) &= \frac{1}{\pi B} \int_{\varphi=0}^{2\pi} F(B\cos\varphi, -B\omega\sin\varphi)\cos\varphi \, d\varphi \\ C_e''(B) &= \frac{1}{\pi B} \int_{\varphi=0}^{2\pi} F(B\cos\varphi, -B\omega\sin\varphi)\sin\varphi \, d\varphi \end{aligned}\right\} \qquad (2)$$

with circular frequency $\omega = 2\pi f$ and integration variable $\varphi = \omega t$. Damping can also be expressed by the loss angle

$$\Theta_e(B) = \frac{C_e''(B)}{C_e'(B)} \quad . \qquad (3)$$

The functions $C_e'(B)$ and $C_e''(B)$ corresponding to the force deflection diagrams of Figures 1, 3 and 5 are plotted in Figures 2, 4 and 6, respectively. Due to the antisymmetric aileron hinge stiffness nonlinearity as shown in Figure 4, the normal frequency of the antisymmetric aileron vibration (wing assumed to be fixed) varies over a wide range, between 2.4 Hz and 7.4 Hz . At least two other antisymmetric normal modes lying in this frequency range are consequently characterized by highly amplitude-dependent portions of aileron vibrations. Similar effects can also be observed for the symmetric aileron mode and for the rudder vibration.

The effects of strong nonlinearities on flutter behavior have been demonstrated in some windtunnel tests carried out on a nonlinear wing aileron model in the low-speed windtunnel of the DFVLR in Göttingen. The nonlinear flutter boundaries for a backlash-type and for a spring-tab-type aileron hinge stiffness are shown in Figure 7. Unlike the flutter boundaries of linear systems, both curves are characterized by a considerable dependence of the critical flutter speed on the aileron amplitude. Thus, the flutter boundary of the spring-tab-type system varies between $V = 12.5$ m/s and $V = 24$ m/s . The backlash-type system shows a flutter

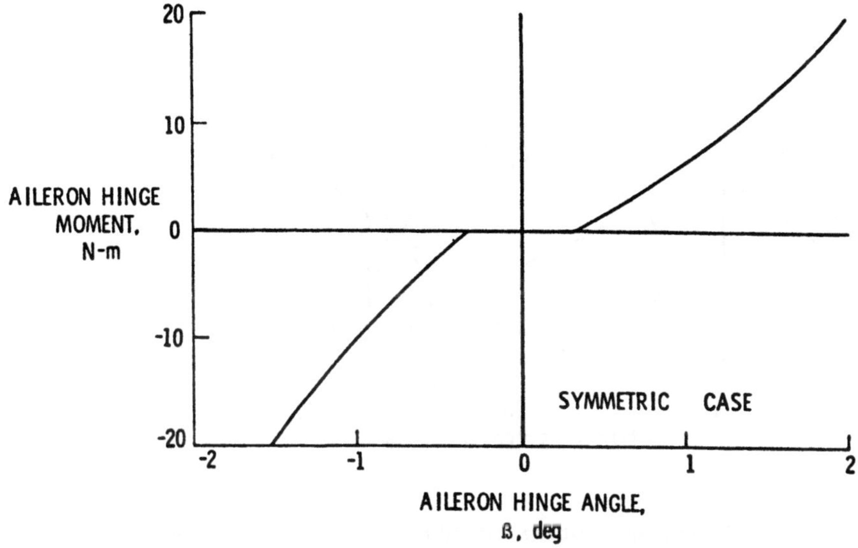

Figure 5: Force deflection diagram - aileron system, symmetric case.

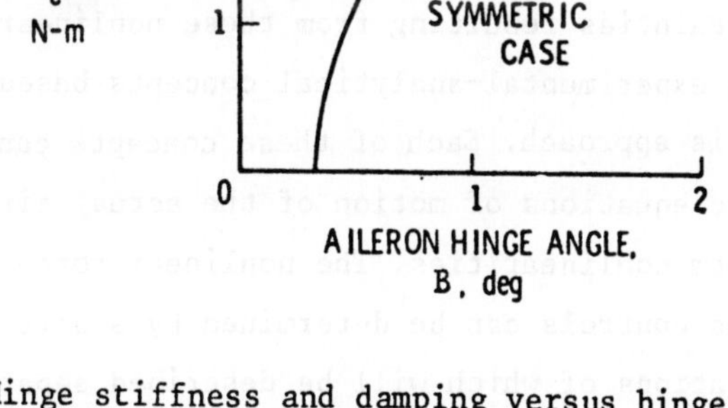

Figure 6: Hinge stiffness and damping versus hinge angle - aileron system, symmetric case.

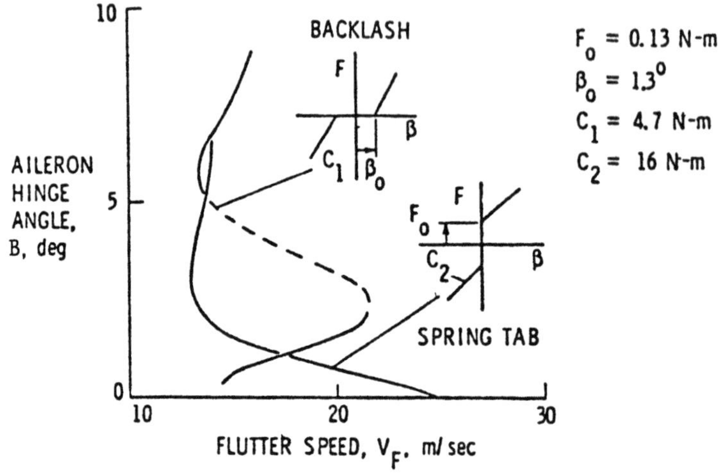

Figure 7: Measured flutter boundary of a nonlinear model.

boundary variation between $V = 13.5$ m/s and $V = 20$ m/s . For more detailed information, see Ref. [1].

3. MATHEMATICAL MODELING USING MODAL SYNTHESIS CONCEPTS

As already mentioned, determination of the elastodynamic characteristics by means of a GVT can be affected severely by localized nonlinearities in the control mechanisms. It will be shown in the following discussion that the uncertainties resulting from these nonlinear effects can be avoided by applying experimental-analytical concepts based on the well-known modal synthesis approach. Each of these concepts can be used to set up the elastodynamic equations of motion of the actual airplane including all control mechanism nonlinearities. The nonlinear force deflection diagrams of the various controls can be determined by static or dynamic tests, the fundamental relations of which will be described separately in Section 4.

Three different modal synthesis concepts will be presented, the main features of which are:

Concept I: Measurement of a set of orthogonal normal modes with the control surfaces rigidly clamped; separate determination of the control surface normal modes with the rest of the airplane rigidly fixed.

Concept II: GVT on a configuration artificially linearized by replacing the nonlinear control mechanism elements by linear and lightly damped dummy devices; thus, a set of orthogonal normal modes for the entire system can be made available.

Concept III: Measurement of a set of orthogonal normal modes with the control surfaces removed; separate determination of the normal modes of the control surfaces in uncoupled condition.

3.1 Concept I

The governing equations of motion of an aeroelastic system, formulated in terms of physical coordinates, can be written in matrix notation as follows:

$$A\ddot{u} + B\dot{u} + Cu = P \tag{4}$$

where

A	mass matrix
B	damping matrix
C	sitffness matrix
u	column matrix of the physical displacements; $\dot{u}$ and $\ddot{u}$ are first and second order derivatives with respect to time t
P	column matrix of external forces, for instance unsteady aerodynamic forces.

It is obvious that parts of the matrices B and C are nonlinear because of the localized nonlinearities of the controls.

Lifting surfaces with controls. Clamping the controls rigidly to the adjacent structure results in an artificially linearized test configuration with a stiffness matrix C_L and a damping matrix B_L whereas the mass matrix remains unchanged according to $A_L = A$. A GVT on this configuration leads to a set of n largely linear normal modes Φ_{Ar} combined in the modal matrix

$$\Phi_A = [\Phi_{A1}, \Phi_{A2}, \ldots, \Phi_{Ar}, \ldots, \Phi_{An}] \quad . \tag{5}$$

The modes satisfy the orthogonality conditions

$$\Phi_A^T A \Phi_A = M_A \quad , \quad \Phi_A^T C_L \Phi_A = \Lambda_A M_A = K_A \tag{6}$$

where

M_A	diagonal matrix of the generalized masses M_{Ar}
K_A	diagonal matrix of the generalized stiffnesses $K_{Ar} = \omega_{Ar}^2 M_{Ar}$
Λ_A	diagonal matrix of the square values of the circular normal frequencies ω_{Ar} .

The generalized damping matrix D_A (not necessarily diagonal) is defined by

$$D_A = \Phi_A^T B_L \Phi_A \quad . \tag{7}$$

Next, assuming that the control surfaces are rigid in the frequency range of interest, a number of additional control surface rotation modes with the adjacent main structure at rest can be determined and combined in the modal matrix

$$\Phi_B = [\Phi_{B1}, \Phi_{B2}, \ldots, \Phi_{B\nu}, \ldots, \Phi_{Bm}] \quad . \tag{8}$$

The physical displacements of the complete structure with freely movable controls are related to the generalized coordinates by

$$u = \Phi q \tag{9}$$

where the column matrix of the generalized coordinates q_{Ar} and $q_{B\nu}$ is

$$q = [q_A^T, q_B^T]^T \tag{10}$$

and

$$\Phi = [\Phi_A, \Phi_B] \quad . \tag{11}$$

The basic idea of this modal superposition is outlined in Figure 8.

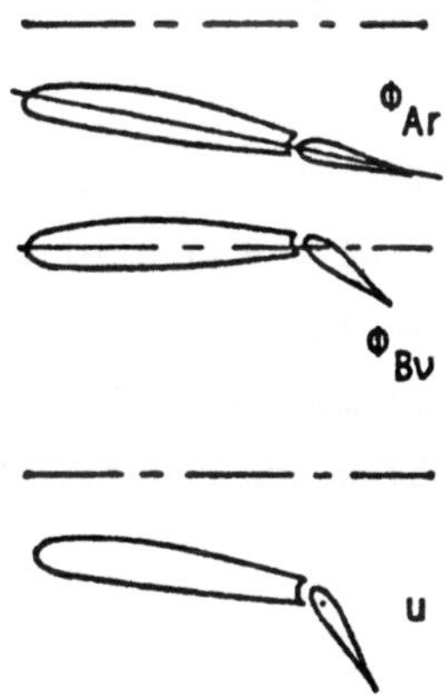

Figure 8: Modal superposition according to Concept I. Lifting surface with control.

Substituting Eq.(9) into Eq.(4) and premultiplying by Φ^T yields

$$M\ddot{q} + D\dot{q} + Kq = Q \tag{12}$$

where

$$\left.\begin{array}{ll} M = \begin{bmatrix} M_A & M_{AB} \\ M_{BA} & M_B \end{bmatrix} & Q = \Phi^T P \\ D = \begin{bmatrix} D_A & 0 \\ 0 & D_B \end{bmatrix} & K = \begin{bmatrix} K_A & 0 \\ 0 & K_B \end{bmatrix} \end{array}\right\} \quad . \tag{13}$$

The matrices M_A , K_A and D_A measured in a GVT are defined in Eqs. (6) and (7). The diagonal matrices M_B , D_B and K_B contain the generalized masses, damping values, and stiffnesses of the control surface rotation modes. In the case of nonlinear hinge stiffness and damping, the matrix elements of K_B and D_B are

$$K_{B\nu} = C'_{e\nu}(B)\, \beta^2_{\nu a} \quad , \quad D_{B\nu} = \frac{C''_{e\nu}(B)}{\omega}\, \beta^2_{\nu a} \tag{14}$$

where $C'_{e\nu}(B)$ and $C''_{e\nu}(B)$ can be determined from Eq.(2). The term $\beta_{\nu a}$ denotes the control rotation in the action line of the control actuator force. The matrix M_B can be determined by calculation or measurement taking into account not only the control surface mass but also the moving mass of such attached hardware as pushrods, cables and control stick. The elements of the coupling matrix

$$M_{AB} = M^T_{BA} = \Phi^T_A\, A\, \Phi_B \tag{15}$$

can be found by integration over surfaces $S_{B\nu}$ of the controls

$$M_{AB,r\nu} = \int_{S_{B\nu}} \left\{ e_\nu\, m_\nu\, h_r\, \beta_\nu + \left[I_\nu + e_\nu\, m_\nu (a_\nu + e_\nu) \right] \alpha_r\, \beta_\nu \right\} d\ell$$

where the following terms correspond to the ν-th control (with tab locked to the control):

m_ν	mass of the control surface per unit span
I_ν	mass moment of inertia per unit span referred to the center of gravity
e_ν	distance between center of gravity and hinge axis (see Figure 9)
a_ν	distance between hinge axis and the quarter-chord point (see Figure 9).

All these data as well as the amplitudes h_r , α_r and β_ν (see Figure 9) are functions of the span coordinate ℓ . In the case of an ideal

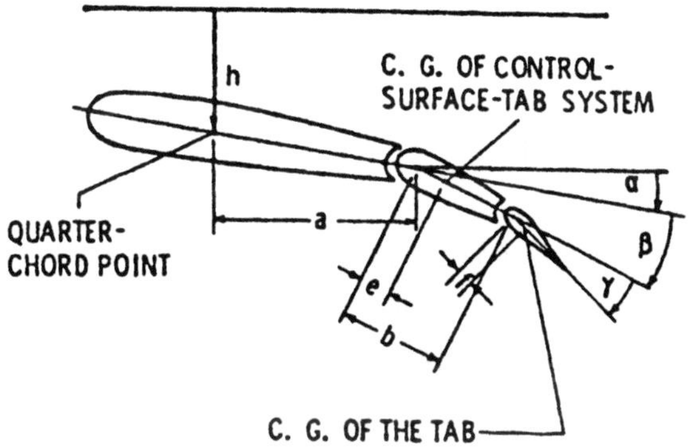

Figure 9: Lifting surface with control surface and tab.

locking of the controls, neither hinge stiffness forces nor hinge damping forces are generated in the normal modes Φ_{Ar} . Hence,

$$\left.\begin{aligned} K_{AB} &= K_{BA}^T = 0 \\ D_{AB} &= D_{BA}^T = 0 \end{aligned}\right\} \quad . \tag{17}$$

The procedure described above can easily be extended to systems with controls *and* tabs (spring tabs, trim tabs, geared tabs). Such a combination is shown schematically in Figure 9. More detailed information on this extended approach is given in Ref.[3].

3.2 Concept II

Replacement of the control nonlinearities by artificial linear stiffnesses results in a modified linearized test configuration[1,6,7] represented in matrix notation by

$$A_L\,\ddot{u} + B_L\,\dot{u} + C_L\,u = P \tag{18}$$

which is formulated in terms of physical displacements. The governing dynamic equations of the unchanged nonlinear system can be written in the same form as Eq.(4) by subdividing the matrices A , B and C as follows:

$$\left.\begin{aligned} A &= A_L - \Delta A_L + \Delta A_{NL} \\ B &= B_L - \Delta B_L + \Delta B_{NL} \\ C &= C_L - \Delta C_L + \Delta C_{NL} \end{aligned}\right\} \quad . \tag{19}$$

The term $\Delta A_{NL} - \Delta A_L$ represents the difference in the mass distribution between the artificial linear system and the real nonlinear system; ΔB_L and ΔC_L define the damping and stiffness properties of the artificial linear elements; ΔB_{NL} and ΔC_{NL} describe the damping and stiffness properties of the replaced nonlinear elements. Development of the arbitrary displacement vector u in a series expansion of the measured normal modes Φ_{Lr} of the linearized system yields

$$u = \Phi_L \, q \quad . \tag{20}$$

Inserting this modal transformation into Eq.(4), premultiplying by Φ_L^T and taking into account Eq.(19) results in generalized equations of motion in the same form as Eq.(12), but with mass, damping and stiffness matrices now defined as

$$\left.\begin{aligned} M &= M_L - \Delta M_L + \Delta M_{NL} \\ D &= D_L - \Delta D_L + \Delta D_{NL} \\ K &= K_L - \Delta K_L + \Delta K_{NL} \end{aligned}\right\} \quad . \tag{21}$$

The matrices M_L , D_L and K_L are measured in a GVT on the linearized system. Furthermore,

$$\left.\begin{aligned} \Delta M_{NL} - \Delta M_L &= \Phi_L^T (\Delta A_{NL} - \Delta A_L)\, \Phi_L \\ \Delta D_{NL} - \Delta D_L &= \Phi_L^T (\Delta B_{NL} - \Delta B_L)\, \Phi_L \\ \Delta K_{NL} - \Delta K_L &= \Phi_L^T (\Delta C_{NL} - \Delta C_L)\, \Phi_L \end{aligned}\right\} \quad . \tag{22}$$

For simplicity, consider only one control surface. For the ν-th control surface, the modal matrix Φ_L degenerates to the row matrix

$$\Phi_{\nu L} = [\beta_{\nu 1},\beta_{\nu 2},\ldots,\beta_{\nu r},\ldots,\beta_{\nu n}] \tag{23}$$

and $\Delta B_{NL} - \Delta B_L$ and $\Delta C_{NL} - \Delta C_L$ degenerate to the 1×1 matrices

$$\left.\begin{aligned} \Delta B_{NL} - \Delta B_L &= \frac{C_e''(B)}{\omega} - B_L \\ \Delta C_{NL} - \Delta C_L &= C_e'(B) - C_L \end{aligned}\right\} \tag{24}$$

where the nonlinear stiffness and damping values $C_e'(B)$ and $C_e''(B)$ can be determined again by applying Eq.(2) to the measured nonlinear force deflection diagram. The damping and stiffness matrices B_L and C_L, respectively, of the artificial linear element can be measured by means of simple tests. The matrix $\Delta M_{NL} - \Delta M_L$ can also be calculated by using the modal matrix as defined in Eq.(23), provided the two parts of the 1×1 matrix $\Delta A_{NL} - \Delta A_L$ can be defined as moments of inertia by referring the removed mass of the nonlinear system, as well as the additional mass resulting from the artificial linearization, to the hinge angle β.

3.3 Concept III

The aeroelastic equations of an airplane can also be established by means of both a set of normal modes measured in a GVT with controls removed and rigid-body and some elastic normal modes of the several controls (see Figure 10) determined experimentally or by fairly simple calculations. The equations of motion of the coupled system can be set up by means of Lagrange's equations

$$\frac{d}{dT}\left(\frac{\partial T}{\partial \dot q_r}\right) + \frac{\partial U}{\partial q_r} + \frac{\partial W}{\partial \dot q_r} = \sum_{\ell=1}^{\sigma} \lambda_\ell \frac{\partial g_\ell}{\partial q_r} \quad , \tag{25}$$

$$(r = 1,2,\ldots,n_A,n_A+1,\ldots,n_B,\ldots,n_i,\ldots)$$

where

$$\left.\begin{aligned} 2T &= \dot{q}^T M \dot{q} \\ 2U &= q^T (K + \Delta K)\, q \\ 2W &= \dot{q}^T (D + \Delta D)\, \dot{q} \end{aligned}\right\} \quad . \tag{26}$$

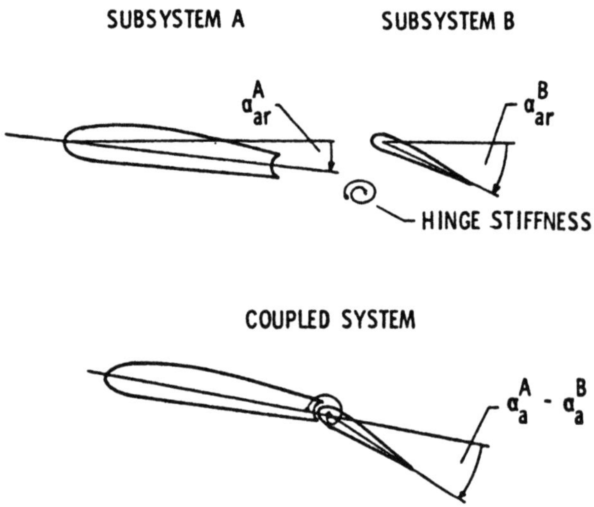

Figure 10: Modal coupling of a wing control surface system according to Concept III.

The matrices ΔK and ΔD in Eq.(26) take into account the elastic coupling between control surfaces and main structure by means of the real hinge stiffness and hinge damping elements. The term on the right-hand side of Eq.(25) is formulated in terms of Lagrange's multipliers λ_ℓ which correspond to a number of σ constraint conditions

$$g_\ell(q_1, q_2, \ldots, q_r, \ldots, q_{n_i}) = 0 \quad , \quad (\ell = 1,2,\ldots,\sigma) \quad . \tag{27}$$

Compatibility is expressed in those coupling points, where the controls can be assumed to be rigidly fixed to the main structure. Application of Eq.(25) to Eqs.(26) and (27) yields

$$M\ddot{q} + (D + \Delta D)\dot{q} + (K + \Delta K)q - \Psi^T \lambda = 0 \qquad (28)$$

where the elements of the $r \times \ell$ matrix Ψ^T are

$$\Psi^T_{\ell r} = \frac{\partial g_\ell}{\partial q_r} \quad . \qquad (29)$$

The generalized mass, damping and stiffness matrices M , D and K of the decomposed system can be considered largely linear. For the exemplary case of coupling only two subsystems A and B (wing and aileron, for instance), these matrices read:

$$\left.\begin{aligned} M &= \begin{bmatrix} M_A & 0 \\ 0 & M_B \end{bmatrix} \qquad D = \begin{bmatrix} D_A & 0 \\ 0 & D_B \end{bmatrix} \\ K &= \begin{bmatrix} K_A & 0 \\ 0 & K_B \end{bmatrix} \end{aligned}\right\} \qquad (30a)$$

where the submatrices are

$$M_i = \Phi_i^T A_i \Phi_i \quad , \quad K_i = \Phi_i^T C_i \Phi_i \quad , \quad D_i = \Phi_i^T B_i \Phi_i \quad , \qquad (30b)$$

$$(i = A,B) \quad .$$

The matrices A_i , B_i and C_i describe mass, damping and stiffness of the subsystems A and B in terms of geometrical coordinates; Φ_i is the modal matrix of subsystem i . The elements of the diagonal matrices M_i and K_i and of the damping matrices D_i , which are not necessarily diagonal, can be determined by a GVT or, as in the case of the controls, by calculation as well.

The generalized coupling matrices ΔK and ΔD can be written as follows:[7]

$$\Delta K = \Phi_{AB}^T C_{AB} \Phi_{AB} \quad , \quad \Delta D = \Phi_{AB}^T B_{AB} \Phi_{AB} \quad . \tag{31}$$

When the main structure and control surface are coupled by a nonlinear complex hinge stiffness $C_e(B)$ in the action line of the control force, we obtain

$$\Phi_{AB} = \left[\begin{array}{c|c} \alpha_{a1}^A, \alpha_{a2}^A, \ldots, \alpha_{an}^A & 0 \\ \hline 0 & \alpha_{a,n+1}^B, \alpha_{a,n+2}^B, \ldots, \alpha_{an_B}^B \end{array}\right] \tag{32}$$

$$C_{AB} = C_e'(B) \begin{bmatrix} 1 & -1 \\ -1 & 1 \end{bmatrix} \quad , \quad B_{AB} = \frac{C_e''(B)}{\omega} \begin{bmatrix} 1 & -1 \\ -1 & 1 \end{bmatrix} . \tag{33}$$

The angles of rotation α_{ar}^A and α_{ar}^B are defined in Figure 10. For the special case of coupling two systems A and B the compatibility condition for σ physical degrees of freedom can be expressed by the constraints

$$g_\ell = u_\ell^A - u_\ell^B = 0 \quad , \quad (\ell = 1,2,\ldots,\sigma) \quad . \tag{34}$$

If u_ℓ^A and u_ℓ^B are expressed in a series of the normal modes of the systems A and B , then

$$g_\ell = \sum_{r=1}^{n_A} u_{\ell r}^A q_r - \sum_{n_A+1}^{n_B} u_{\ell r}^B q_r = 0 \tag{35}$$

or in matrix notation

$$g = \Psi q = 0 \quad . \tag{36}$$

The aeroelastic equations of motion are defined now by the $n_A + n_B = m$ generalized coordinates. Due to the σ constraints there remains a number of $\varepsilon = m - \sigma$ independent generalized coordinates in terms of which the aeroelastic equations have to be formulated. To do this, the term Ψ^T in Eq.(28) has to be rearranged rowwise so that

$$\Psi^T \rightarrow \overline{\Psi}^T = \begin{bmatrix} \Psi_\varepsilon \\ \Psi_\sigma \end{bmatrix} \tag{37}$$

where Ψ_σ is a nonsingular $\sigma\times\sigma$ matrix. The matrices M , K , D , ΔK and ΔD with respect to both their columns and rows and the column matrix q have to be rearranged in the same sense. The rearranged equations can be written as

$$\overline{M}\,\ddot{\overline{q}} + (\overline{D} + \Delta\overline{D})\dot{\overline{q}} + (\overline{K} + \Phi\overline{K})\overline{q} = \overline{\Psi}^T \lambda \tag{38}$$

where

$$\overline{q} = \left(p^T, q_\sigma^T\right)^T \quad . \tag{39}$$

The new structure of the matrices $\overline{M}$, $\overline{D}$, $\overline{K}$, $\Delta\overline{D}$ and $\Delta\overline{K}$ is shown in the following equation using $\overline{M}$ as an example:

$$\overline{M} = \begin{bmatrix} M_{\varepsilon\varepsilon} & M_{\varepsilon\sigma} \\ M_{\sigma\varepsilon} & M_{\sigma\sigma} \end{bmatrix} \quad . \tag{40}$$

Thus, λ can be determined as follows:

$$\lambda = (\overline{\Psi}^T)^{-1} \Big\{ [M_{\sigma\varepsilon}, M_{\sigma\sigma}]\ddot{\overline{q}} + ([D_{\sigma\varepsilon}, D_{\sigma\sigma}] + [\Delta D_{\sigma\varepsilon}, \Delta D_{\sigma\sigma}])\dot{\overline{q}} + \ldots$$
$$\ldots + ([K_{\sigma\varepsilon}, K_{\sigma\sigma}] + [\Delta K_{\sigma\varepsilon}, \Delta K_{\sigma\sigma}])\overline{q} \Big\} \quad . \tag{41}$$

From Eqs.(36) and (37) it follows that

$$q_\sigma = -\Psi_\sigma^{-1} \Psi_\varepsilon \, p \quad . \tag{42}$$

Inserting Eq.(41) into the first ε rows of Eq.(38) and taking into account Eq.(42) results in the following equation:

$$(M_{\varepsilon\varepsilon} - M_{\varepsilon\sigma} X - X^T M_{\sigma\varepsilon} + X^T M_{\sigma\sigma} X) \ddot{p} + \Big\{ D_{\varepsilon\varepsilon} + \Delta D_{\varepsilon\varepsilon} - (D_{\varepsilon\sigma} + \Delta D_{\varepsilon\sigma}) X -$$
$$- X^T(D_{\sigma\varepsilon} + \Delta D_{\sigma\varepsilon}) + X^T(D_{\sigma\sigma} + \Delta D_{\sigma\sigma}) X \Big\} \dot{p} + \Big\{ K_{\varepsilon\varepsilon} + \Delta K_{\varepsilon\varepsilon} - (K_{\varepsilon\sigma} + \Delta K_{\varepsilon\sigma}) X -$$
$$- X^T(K_{\sigma\varepsilon} + \Delta K_{\sigma\varepsilon}) + X^T(K_{\sigma\sigma} + \Delta K_{\sigma\sigma}) X \Big\} p = 0 \tag{43}$$

where

$$X = \Psi_\sigma^{-1} \Psi_\varepsilon \quad . \tag{44}$$

It can easily be shown that Eq.(43) can be transformed into the more convenient equation

$$Y^T \bar{M} Y \ddot{p} + Y^T(\bar{D} + \Delta\bar{D}) Y \dot{p} + Y^T(\bar{K} + \Delta\bar{K}) Y p = 0 \tag{45}$$

where

$$Y = \begin{bmatrix} I \\ \\ -X \end{bmatrix} \tag{46}$$

with the unity matrix I . It should be mentioned that a nonsingular matrix Ψ_σ can be determined optimally by applying common mathematical tools for determination of the linear independence of a given number of vectors, as described, for example, in Ref.[8]. These methods are also applicable to cases with the number of constraints higher than the rank of matrix Ψ_σ . Practical applications to structural dynamics problems are presented in Ref.[9]. A detailed examination of the possibilities offered by the three concepts makes it obvious that sometimes their combined application may be very beneficial. Four possible variations are outlined in detail in Ref.[3].

4. IDENTIFICATION OF THE DYNAMIC BEHAVIOR OF NONLINEAR JOINTS AND LINKAGE ELEMENTS

The principle of energetic equivalence, the basic relations of which are given in Section 2.2 without derivation, can be used successfully to describe the elastomechanic behavior of nonlinear elements with good approximation in terms of equivalent amplitude-dependent stiffness and damping values. The latter can be calculated by means of Eqs.(2) and (3) if the nonlinear force deflection diagram $F(\beta,\dot{\beta})$ is available in measured or analytical form. The force function F describes the periodic time history due to an harmonic deflection β. Fortunately, F is for a preponderant number of technical problems merely a function of deflection β, i.e. $F = F(\beta)$. This property features not only a simpler solution of the integrals in Eq.(2) but also less test effort.

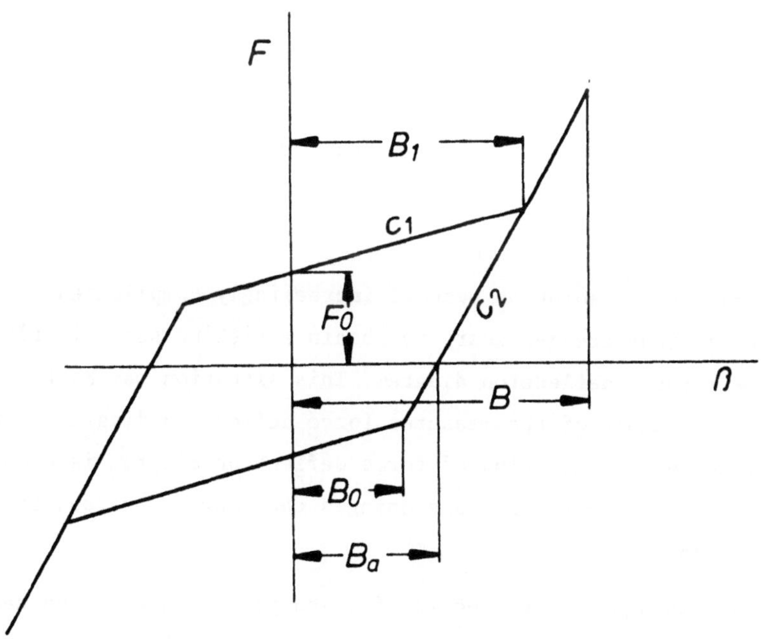

Figure 11: Bilinear force deflection diagram.

An analytical solution of the integrals in Eq.(2) is possible if function F is at least sectionwise at hand in analytical formulation. This may be demonstrated by the example of a simple bilinear force deflection diagram such as shown in Figure 11. For this case which describes in a simplified manner the most frequently occuring nonlinearity with backlash and dry friction, the following solutions can be found for the equivalent stiffness and damping values:

$$C'_e(B) = \frac{1}{\pi B}\left[2F_0(\sin\varphi_0 - \cos\varphi_1) - 2B_a c_2(\sin\varphi_0 + \cos\varphi_1) + \right.$$

$$+ \frac{B}{2}(c_2 - c_1)(\sin 2\varphi_0 + \sin 2\varphi_1 + 2\varphi_0 + 2\varphi_1) +$$

$$\left. + \frac{B\pi}{2}(c_2 + c_1)\right] \tag{46a}$$

$$\gamma_e(B) = \frac{1}{\pi B\, C'_e(B)}\left[2F_0(\sin\varphi_1 + \cos\varphi_0) - 2B_a c_2(\cos\varphi_0 - \sin\varphi_1) + \right.$$

$$\left. + \frac{B}{2}(c_2 - c_1)(\cos 2\varphi_0 + \cos 2\varphi_1)\right] \tag{46b}$$

where

$$\varphi_0 = \text{arc}\cos\frac{B_0}{B} \quad , \quad \varphi_1 = \text{arc}\sin\frac{B_1}{B} \quad . \tag{47}$$

Such an analytical solution becomes increasingly complicated as additional linear sections are necessary to obtain a fitting sufficiently close to the nonlinear force deflection diagram. This situation may best be illustrated by the example of the measured force deflection diagram shown in Figure 12. Analysing this kind of force deflection diagram is even more aggravated by the fact that the shape changes considerably versus the deflection amplitude.

These disadvantages of the sectionwise analytical integration can be avoided by means of an experimental-numerical approach which is characterized by greater accuracy, broader general applicability and increased simplicity.

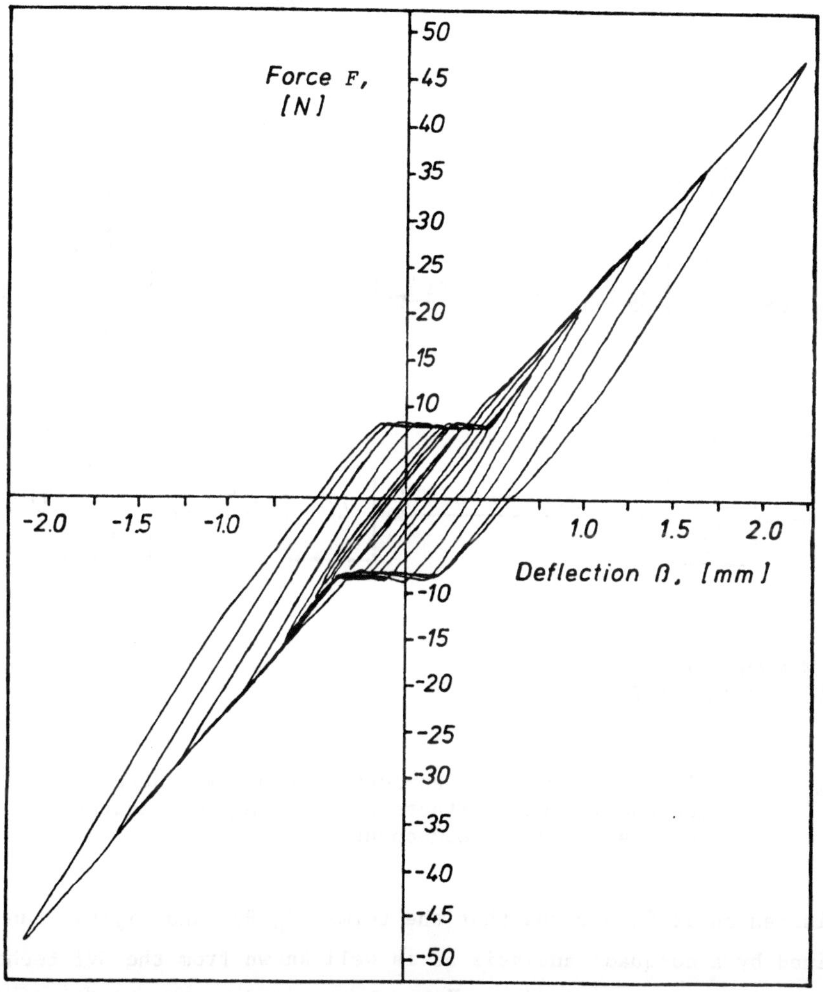

Figure 12: Typical nonlinear force deflection diagram measured on the sweep mechanism of a half-wing flutter model of a sweepable wing aircraft.

4.1 Determination of the equivalent stiffness and damping values by means of an experimental-numerical approach

A closer look at the integrals in Eq.(2) discloses the surprising fact that $C_e'(B)$ is merely the real part of $F_1'(B)$ of the first harmonic of the periodic force F divided by B. In the same way $\gamma_e(B)$ can be interpreted as the imaginary part $F_1''(B)$ of F_1 divided by $C_e'(B)$:

$$C'_e(B) = \frac{F'_1(B)}{B} \quad , \quad \gamma_e(B) = \frac{F''_1(B)}{C'_e(B)B} \quad . \tag{48}$$

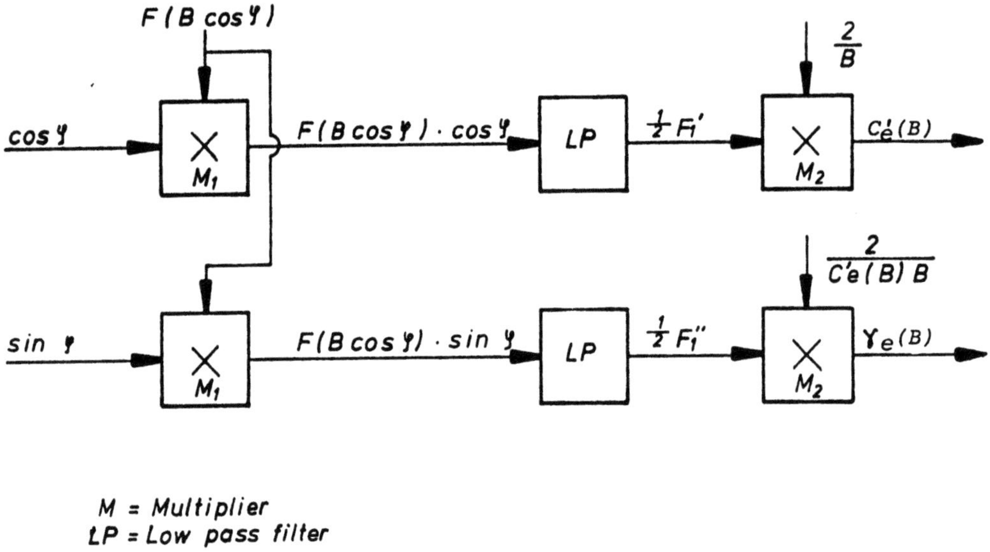

Figure 13: Block diagram of an analogue circuit for determining the equivalent stiffness and damping coefficients of nonlinear linkage components.

For this reason it is obvious that the terms $C'_e(B)$ and $\gamma_e(B)$ can be determined by a co-quad[1] analysis as is well known from the GVT technique. This analysis can be carried out either by the modern means of a digitized time series analysis or by applying a more conventional analogue circuit such as used occasionally in analogue co-quad analysers. The block diagram of such a circuit is shown in Figure 13. It consists of an upper and a lower part which are identically composed of a multiplier M_1 , an extreme low-pass filter LP and a second multiplier M_2 . The following operations are consecutively executed:

---

[1] Co(incidence)-quad(rature) analysis is identical to a vector component analysis delivering real and imaginary parts of the first harmonic of a periodic signal in relation to a given phase reference.

- *Multiplication of the periodic force function* $F = F(B \cos \varphi)$ *with the reference voltage* $\cos \varphi$ *in phase with the harmonic deflection function* $\beta = B \cos \varphi$, *i.e.*

$$F(\varphi) \cos \varphi = \cos \varphi \sum_{n=1}^{\infty} (F_n' \cos n\varphi + F_n'' \sin n\varphi) \tag{49}$$

- *Filtering out all the time-variant components (multiplier* M_1 *) leads to the time-invariant term* $\frac{1}{2} F_1'$ *which after multiplication with* $2/B$ *results in the equivalent stiffness value*

$$C_e'(B) = \frac{F_1'}{B} \quad . \tag{50}$$

Accordingly, the lower part of the circuit delivers the term $F_1''/2$ which, after low-pass filtering and multiplication with $2/C_e'(B)B$, results in the equivalent damping loss angle

$$\gamma_e(B) = \frac{F_1''}{C_e'(B)B} = \frac{F_1''}{F_1'} \quad . \tag{51}$$

This procedure requires a special (displacement controlled) excitation forcing the nonlinear test specimen to carry out an harmonic deflection $B \cos \varphi$. In reacting to this constraint the exciting force becomes periodic and can be measured by commercial force gauges. Furthermore, the excitation frequency must be extremely low (quasi-static) to avoid disturbing effects due to inertia forces.

The way in which this concept works may best be demonstrated on a practical problem. Figure 14 shows the force deflection diagram of a typical hysteresis-type nonlinearity measured at 5 different amplitude levels. The related time histories of both the harmonic deflection and the pertaining periodic force function are plotted in Figure 15. The resulting equivalent stiffness and damping coefficients are given in Figure 16 as functions of deflection amplitude B .

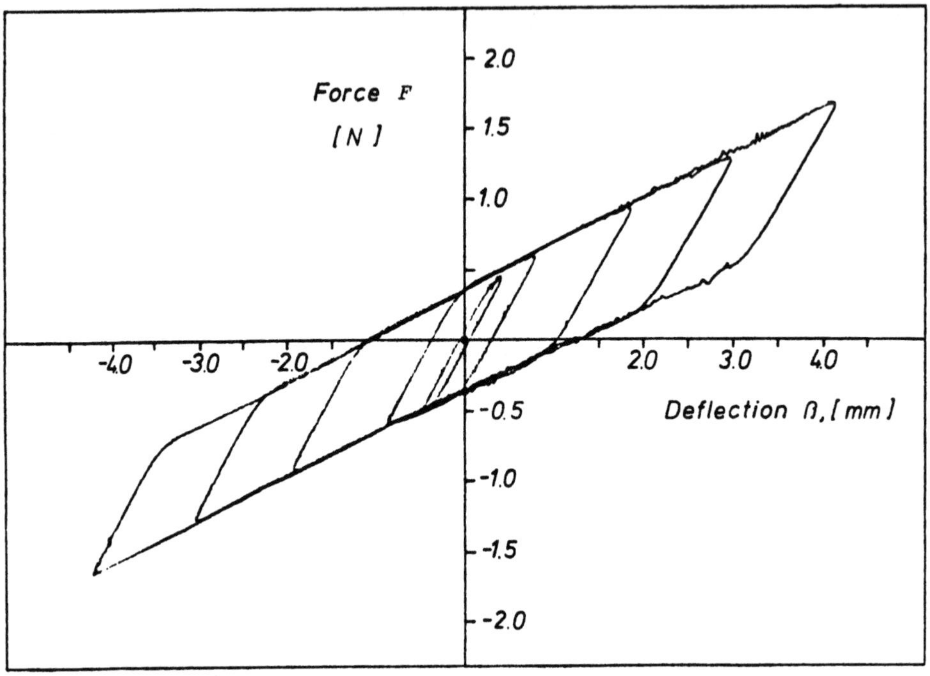

Figure 14: Force deflection diagram of an hysteresis-type nonlinearity measured at 5 various force levels.

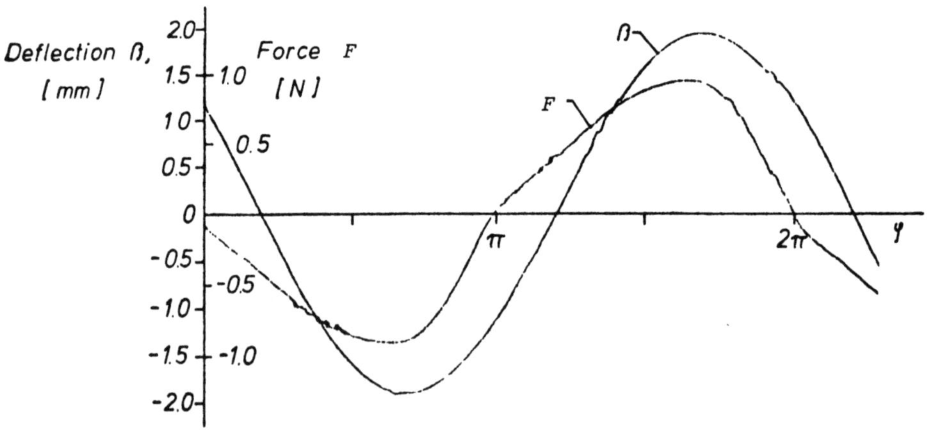

Figure 15: Time histories of force and deflection related to an hysteresis-type force deflection diagram.

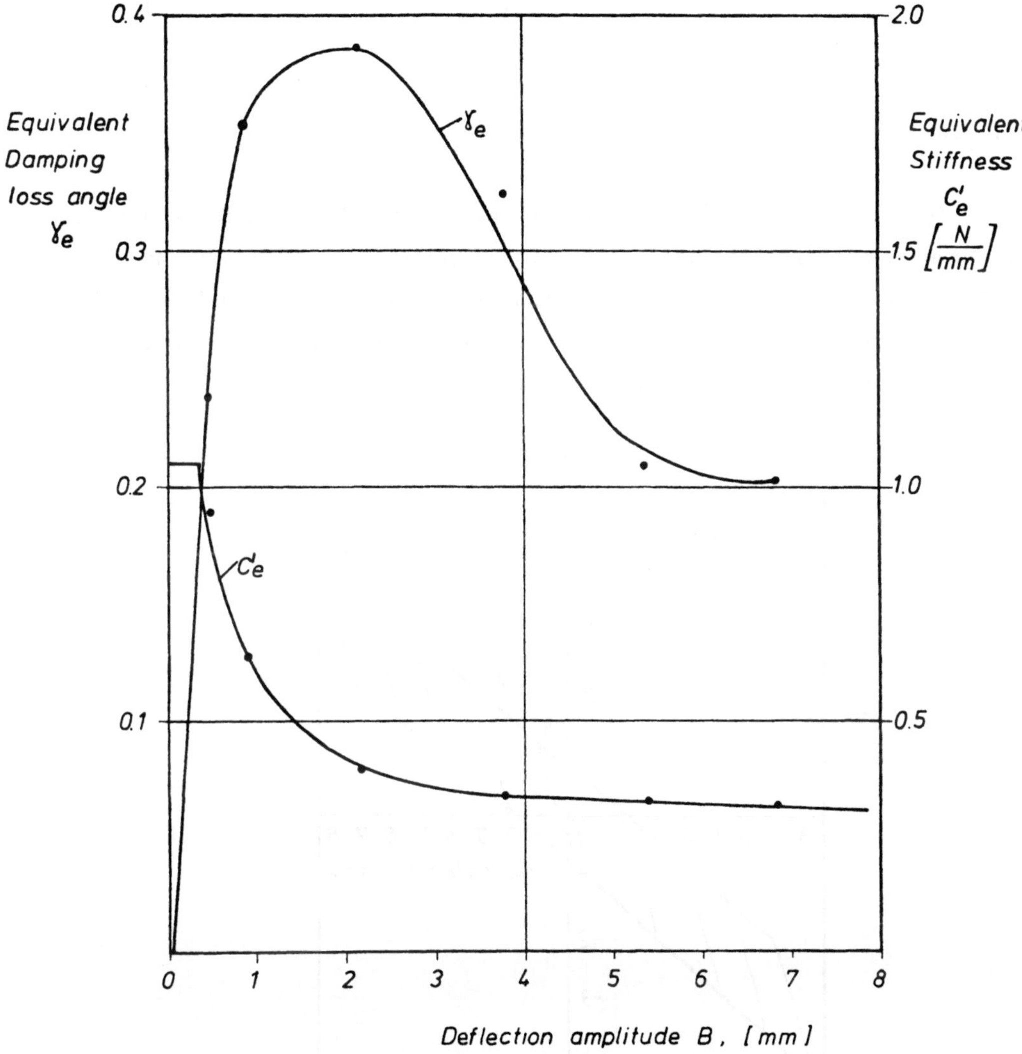

Figure 16: Equivalent stiffness and damping loss angle as functions of deflection amplitude B .

4.2 Experimental verification

To get some idea of the accuracy and limitations of the approach just described, tests were performed on a simple one-degree-of-freedom oscillator with a strong nonlinear element. The system was harmonically excited at several amplitude levels for each of which both the resonance frequency and the corresponding input energy were measured by means of the phase resonance criterion. With these data and the correct mass inertia properties in hand the equivalent stiffness and damping coefficients can be calculated.

The test specimen in question is characterized by the force deflection diagram plotted in Figure 17. The related equivalent stiffness and damping values are given in Figure 18 as a function of deflection amplitude B showing good agreement between the results of the equivalent linearization approach and the dynamic verification.

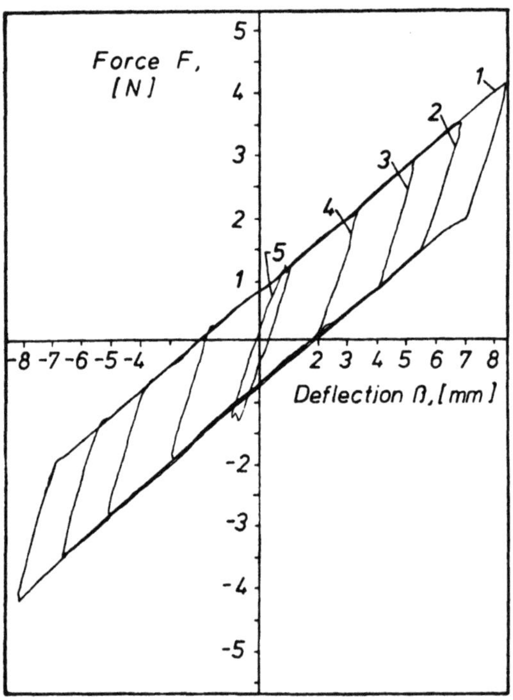

Figure 17: Nonlinear force deflection diagram.

Consequent to this investigation one could draw the conclusion to maintain the above-described dynamic approach as an useful means to measure the equivalent stiffness and damping characteristics of nonlinear elements.

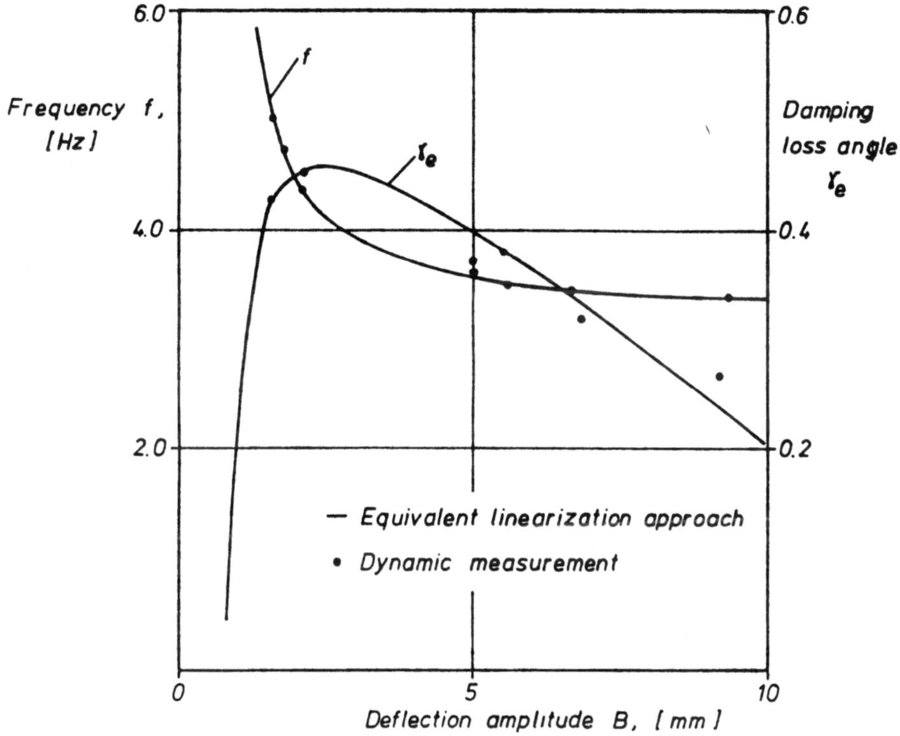

Figure 18: Equivalent stiffness (in terms of resonance frequency) and damping loss angle of a typical hysteresis-type nonlinearity. Comparison between the equivalent linearization approach and dynamic measurements.

However, some contrary arguments are:

- *The dynamic approach cannot be applied to systems with instabilities, i.e. jump phenomenon in the transfer function;*
- *An important prerequisite for using the dynamic approach is the full knowledge of all components of the vibration system of which the unknown nonlinear element is part. This requirement cannot be fulfilled especially in such practical cases like wing/external stores connections or control surface linkage mechanisms;*
- *If applied to real systems, the dynamic approach implies also an excitement of the adjacent structural parts of the nonlinear test specimen.*

All these disadvantages of the dynamic approach are arguments for the quasi-static method described above. However it should be mentioned that even this method has a limited applicability in cases of force deflection diagrams which are functions not only of β but also of $\dot{\beta}$. This type of force deflection diagram becomes important if the dynamic behavior of hydraulic actuators is to be identified. To solve these problems new approaches must be developed.

5. CONCLUDING REMARKS

It has been known for many years that structural dynamics investigations can be greatly afflicted with uncertainties stemming from strong localized nonlinearities in joints and linkage elements. It is shown that the establishment of more reliable and accurate mathematical models for structural analysis requires modified ground vibration test procedures combined with suitably adapted modal synthesis approaches. Three basic concepts have been described in detail. They offer a diverse choice of tools for carrying out both approximately linearized and nonlinear investigations.

It is obvious that a final evaluation of the applicability and accuracy of the different concepts is rather difficult because, up to the present time, only Concept I has been applied to some extent to real airplane structures. Relatively little experience with the other concepts is at hand. Only Concept II has recently been employed in the course of nonlinear investigations of a sailplane.[11] In this context it is worth mentioning that the ground vibration tests carried out according to this concept took far less test time than a normal test on the unchanged structure (reduction of about 80%).

The first comparative investigation of the Concepts I, II and III has been the special concern of Ref.[10], where results are reported for a simple plate-type wing aileron model with linear elastodynamic properties. Although this model cannot be considered representative in all respects of the elastodynamic behavior of real airplanes, it seemed opportune to use the results of this investigation together with the present experience with

Concepts I and II as a basis for a preliminary assessment concerning the advantages and the weak points of different methods.[3]

Each concept requires knowledge of the dynamic behavior of the non-linear joints and linkage elements in such a manner that they can be incorporated into the mathematical model of a nonlinear system. It was shown how this problem can be solved by means of an experimental-numerical approach based on the principle of energetic equivalence, the validity of which was proved successfully by dynamic tests.

6. REFERENCES

1. Breitbach, E., Effects of Structural Non-Linearities on Aircraft Vibration and Flutter. *AGARD Report No. 665*, Sept. 1977.

2. Bogoljubow, N.N.; Mitropolski, J.A., Asymtotische Methoden in der Theorie der nichtlinearen Schwingungen. *Akademie-Verlag*, Berlin, 1965.

3. Breitbach, E.J., Treatment of the Control Mechanisms of Light Airplanes in the Flutter Clearance Process. *NASA CP - 2085*, Pt. II, 1979, pp. 437 - 466.

4. Scanlan, R.H., Rosenbaum, R., Introduction to the Study of Aircraft Vibration and Flutter. *Mac Millan Co.*, 1951.

5. Küssner, H.G., Göllnitz, H., Theorie und Methode der Flatterrechnung von Flugzeugen unter Benutzung des Standschwingungsversuchs. *AVA - Forschungsbericht 64-01*, 1964.

6. Küssner, H.G., Breitbach, E., Bestimmung der Korrekturglieder der Bewegungsgleichungen bei Änderungen eines elastomechanischen Systems. *AVA-Report 69 J 01*, 1969.

7. Breitbach, E., Investigation of Spacecraft Vibrations by Means of the Modal Synthesis Approach. *ESA-SP-121*, Oct. 1976, pp. 1-7.

8. Courant, R., Hilbert, I., Methoden der mathematischen Physik I. *Heidelberger Taschenbücher*, Springer-Verlag, 1968.

9. Walton, W.C., Steeves, E.C., A New Matrix Theorem and its Application for Establishing Independent Coordinates for Complex Dynamical Systems with Constraints. *NASA TR R-326*, 1969.

10. Hüners, H., Berücksichtigung der Ruderfreiheitsgrade im Flatternachweis von Flugzeugen. *DFVLR-Report IB 253-78 J 07*, 1978.

11. Breitbach, E.J., Flutter Analysis of an Airplane with Multiple Structural Nonlinearities in the Control System. *NASA TP - 1620, May 1980.*

SOME MODERN DEVELOPMENTS IN SYSTEM IDENTIFICATION USING PARAMETER ESTIMATION METHODS

H. Unbehauen
Ruhr-Universität
Bochum

Contribution to the book
"Identification of Vibrating Structures"
(Ed.: H.G. Natke) Springer Verlag, 1982.

CONTENTS

1. Introduction

2. Identification of time-variant systems

3. Identification of nonlinear systems

4. Closed-loop systems identification
   4.1. Indirect identification of closed-loop systems
   4.2. Direct identification of closed-loop systems
   4.3. Joint process identification of closed-loop systems

5. Identification of multivariable systems
   5.1. Identification using state space models
   5.2. Identification using input-output models
      5.2.1. Submodel technique
      5.2.2. Subsubmodel technique
      5.2.3. Identification algorithm
      5.2.4. Some results of simulation
   5.3. Practical application of MIMO identification

6. Conclusions

References

1. INTRODUCTION

Process identification by parameter estimation methods has been used successfully in practical application during recent years. Variations and modifications of the models and the estimation algorithm often had to be developed to overcome the special needs of the investigated processes, because 'classical' parameter estimation methods are limited usually to linear, time-invariant single-input/single-output systems. Most practical processes, however, belong to one of the following system classes:

- time-variant systems,
- nonlinear systems,
- closed loop systems,
- multi-input/multi-output systems, or
- distributed parameter systems.

Since 1975 much attention has been paid to the identification of systems with the above-mentioned characteristics. In this chapter some distinctive ideas of this modern development will be presented, but the number of problems must be limited. For time-variant systems it is very important that the variations of the parameter is slower than the convergence to the new estimated parameters. If this is the case, it is possible to use models and methods of time-invariant systems with so-called weighting factors to improve the convergence. This procedure is presented in section two.

Non-linear processes have the disadvantage that a great number of models are necessary to describe the different modes of those processes. In section three different kinds of non-linear models are presented and the possibility of identifying their parameters is discussed.

For systems operating in closed loop examination must be made of which signals are appropriate for identification for reasons of independence and permanent excitation. Often the known regulator of the closed loop system is part of the estimated transfer function, which leads to an increase of the numerical treatment. Some methods for identification of closed loop systems are presented in section four.

The identification of multi-input/multi-output (MIMO) systems is not a simple extension of the models and methods of single-input/single-output (SISO) systems. To reduce the number of parameters, two systematical forms of model design have been developed during recent years and are presented in chapter five.

For the identification of systems with distributed parameters, models are used which are not based on models derived in the previous chapter "Introduction to System Identification". Therefore, the interested reader is referred to the following literature:

Survey papers: Goodson (1976), Kubrusly (1977)

Methods: Chavent (1979)

Conference: Ruberti (1978) (Editor): IFIP Working Conference on "Distributed Parameter Systems", Rome (1976).

2. IDENTIFICATION OF TIME-VARIANT SYSTEMS

In many cases, especially for adaptive systems, identification procedures are used in on-line operation to analyse the time-variant behaviour of an industrial plant, which may be caused, for example, by changes in the operating point, material attrition, dirt etc. In order to describe such systems by a mathematical model, the parameters of the model must be adapted on-line to the corresponding system operation. Thus a time-variant model is obtained. The case when the model structure itself is time-variant, will not be considered in the following, because this would mean that without exception the estimation algorithm must be started again, because all matrices and vectors will be changed. Thus the time-variant behaviour is limited to parameter changes in a fixed model structure.

Before a system is defined as time-variant, investigations have to be made into whether or not other system properties could be responsible for changes in the system behaviour. Such changes could also be caused by

- nonlinearities,
- additional input signals, or
- disturbances.

The aim of identification of time-variant systems is to estimate the varying parameters as fast as possible. The observation of changes in the parameters is usually not difficult, as the error between measured system output and estimated model output is directly used in the estimation algorithm. However, these changes will only cause an adaptation of the model if the correction vector is also sufficiently large. This is the crucial reason why the usual estimation methods cannot be applied directly for the estimation of systems with time-varying parameters. It was already mentioned in section 3.6 of the chapter on "Introduction to System Identification", that the matrix $\underline{P}$, which considerably influences the correction vector, decreases too rapidly. In order to regard parameter changes at a later time as well, the coefficients of this matrix must be increased. This can be obtained, for example, by introducing the weighted version of the recursive IV-estimation algorithm according to appendix B. By introducing a *weighting term* ρ, the elements of $\underline{P}$ can be increased so that the estimation results can be influenced better by new measuring values. This is advantageous for the estimation of time-varying parameters. However, if additional disturbances occur, these will also disturb the accuracy of the parameter estimation. Therefore an optimum compromise must be made between a good parameter adaptation in the model and the elimination of disturbances. Different methods for choosing the weighting term ρ have already been discussed in section 3.6 of the above-mentioned chapter. For time-variant systems optimum weighting factors have been proposed by Bauer (1977), and are listed in Table 2.1.

noise level	parameter changes	modified methods				
		R	G1	G2	G3	M3
		k_R	ρ_o	$\rho(0)/\Delta\rho$	ρ	g_{ii}
λ_s=0	slow	200	0,95	0,5/0,005 0,6/0,004	0,06-0,08	0,0005-0,001
	medium	200	0,9	0,5/0,003 0,6/0,002	0,1	0,002
	fast	50	0,83	0,8/0,001	0,16-0,18	0,01
λ_s=0,1	slow	-	0,975	-	0,03	0,0005
	medium	-	0,96	-	0,05	0,0012
	fast	-	0,92	-	0,09	0,005

Table 2.1. Optimum weighting factors for the investigated methods

If the speed of parameter changes is higher than the convergence rate, an adaptation of the model parameters cannot also be quaranteed when using optimal weighting factors. As the convergence rate decreases considerably with increasing model order, the identification of systems of 4th order already causes problems. Investigations conducted earlier by Bauer (1977) and Schenk (1976) were devoted only to systems of 2nd order. But even with this order the speed of parameter variation may not become too high, i. e. after a step change of one parameter at least 100 samples should be processed before the next step change takes place. Fig. 2.1 shows as an example the adaptation of the estimated gain factor $\hat{K}_S$ of a 2nd order model after applying rectangular changes of the gain factor K_S of the real system.

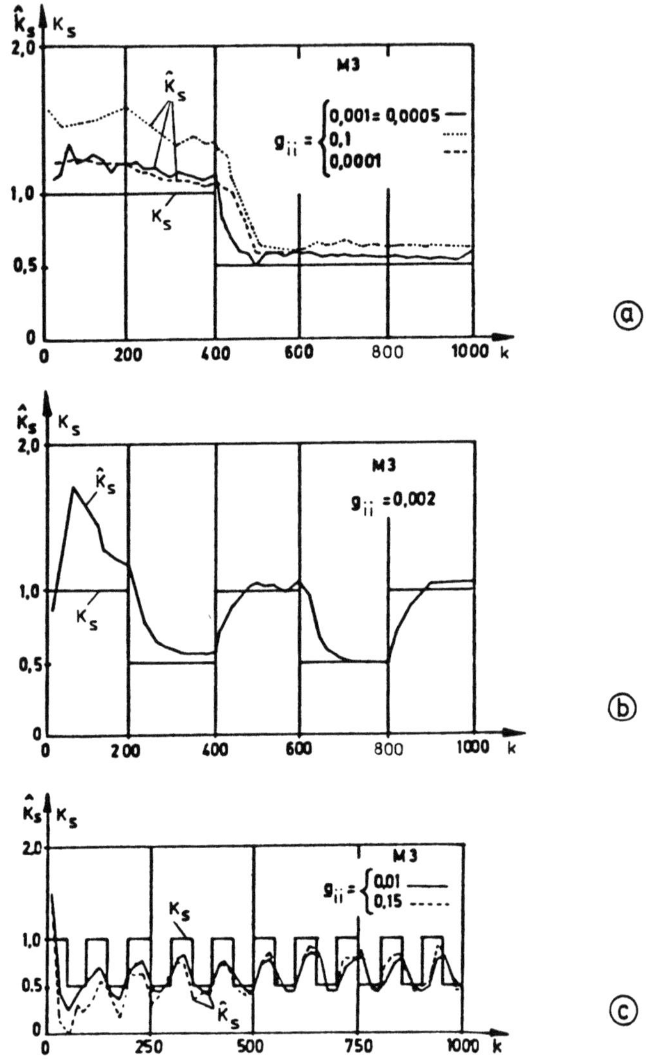

Fig. 2.1. Results for the identification of the gain factor K_S of a 2nd order system for (a) slow, (b) medium and (c) fast changes using the IV-method in combination with the M3-weighting method

From this example and the previous discussion it follows that very fast parameter changes can no longer be identified precisely by the parameter estimation methods presented here. Therefore Kreuzer (1977) proposed distinguishing between *variant* and *quasi-variant* systems. According to his proposal, quasi-variant systems are defined to be estimable by the parameter estimation methods discussed above, whereas for the considerably faster variant systems a new identification method is proposed, which is similar to the linear filtering method introduced by Schaufelberger (1969). However, Kreuzer does not discuss the permissible speed of parameter changes using this method.

It is very important to know that the identification of time-variant systems can be performed much more easily if the parameter changes occur at a time when the previous estimates are already good. Bauer (1977) has shown that the speed of parameter changes can then be increased for the same estimation accuracy.

Finally it should be mentioned that most industrial processes with time-variant parameters show only slowly varying parameters, which are very often attributed to material attrition, contamination, material or energy consumption etc. Environmental influences on a system cannot generally be the cause of time-varying system behaviour, but even these changes are usually not too fast.

3. IDENTIFICATION OF NONLINEAR SYSTEMS

The term "nonlinear" comprises a very large variety of diverse system characteristics. Many different forms for describing nonlinear systems had been developed. For the identification of nonlinear processes it is necessary to find models which describe the real system behaviour as simply and precisely as possible. In recent years a series of papers has been published including proposals for modelling various kinds of nonlinearities. A good survey was published by Haber (1976). He classifies nonlinear systems according to the following aspects:

- state space description,
- input/output canonical forms,
- block-oriented models,
- models linear in the parameters,
- models with signal-dependent parameters,
- models with static nonlinearities and
- models with stepwise characteristics.

Because in practice it is often more useful to assume simple models instead of truncating the series of the first two describing forms, e.g. Taylor series, Hammerstein series, Volterra series, etc., the block-oriented models have been introduced. Models are linear in the parameters when the nonlinearity is caused by signal connexions and the parameters are independent of these signals.

The most important group of nonlinear systems is the one with static nonlinearities. In Fig. 3.1 several possibilities for static nonlinearities are given in graphical and numerical description. In order to analyse

	symbol	formula
1	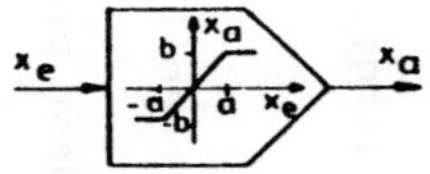 saturation	$x_a = \begin{cases} -b & \text{for } x_e \leq -a \\ \frac{b}{a} x_e & \text{for } -a \leq x_e \leq a \\ b & \text{for } x_e \geq a \end{cases}$
2	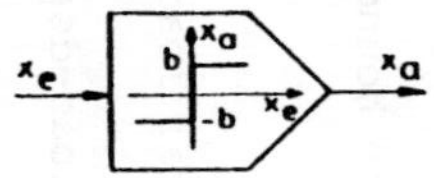 ideal relay	$x_a = b \operatorname{sgn} x_e = \begin{cases} -b & \text{for } x_e < 0 \\ b & \text{for } x_e > 0 \end{cases}$
3	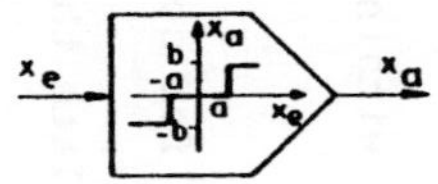 relay with dead zone	$x_a = \begin{cases} -b & \text{for } x_e < -a \\ 0 & \text{for } -a < x_e < a \\ b & \text{for } x_e > a \end{cases}$
4	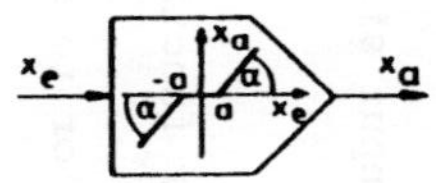 dead zone	$x_a = \begin{cases} (x_e + a)\tan\alpha & \text{for } x_e < -a \\ 0 & \text{for } -a \leq x_e \leq a \\ (x_e - a)\tan\alpha & \text{for } x_e > a \end{cases}$
5	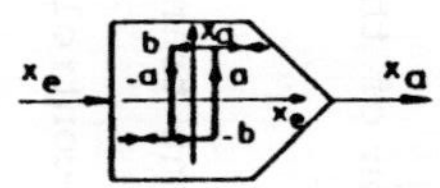 relay with hysteresis	$x_a = \begin{cases} -b & \text{for } x_e < -a \\ b \operatorname{sgn}(x_e - a \operatorname{sgn} \dot{x}_e) & \text{for } -a < x_e < a \\ b & \text{for } x_e > a \end{cases}$
6	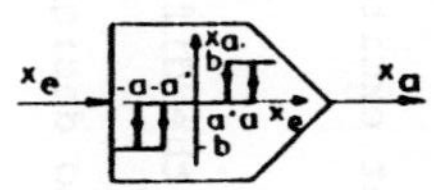 relay with dead zone and hysteresis	expensive mathematical description
7	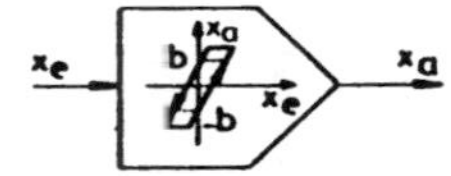 backlash	expensive mathematical description
8	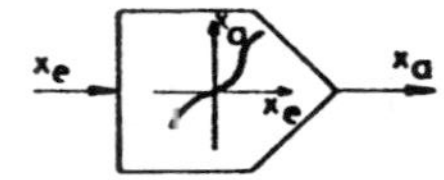 arbitrary nonlinearities	$x_a = f(x_e)$
9	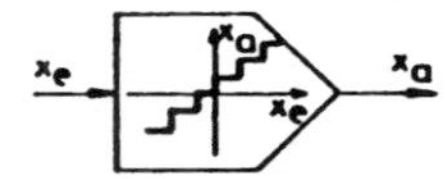 quantization	x_e has only quanticized value
10	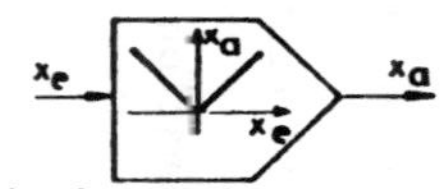 absolute value quantization	$x_a = \lvert x_e \rvert$
11	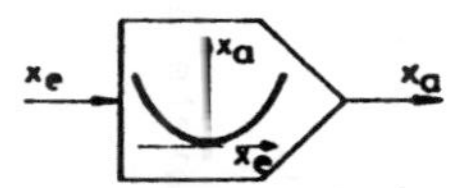 square characteristic	$x_a = x_e^2$
12	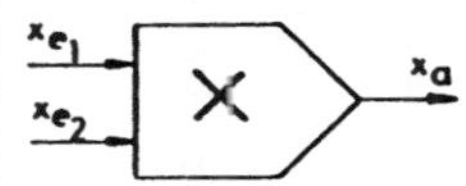 multiplication	$x_a = x_{e1} x_{e2}$
13	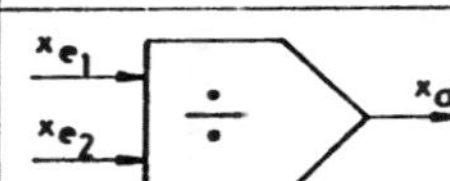 division	$x_a = \frac{x_{e1}}{x_{e2}}$

Fig. 3.1. Some statistic nonlinearities

these systems the input signal must excite the whole range of nonlinearity. Then the connection between the input and output signals can be analysed by simple graphical identification. In some practical applications a static nonlinearity is superposed to a linear dynamic model, especially in closed loop systems. In these cases there is no unique identification algorithm to estimate the nonlinear static and the linear dynamic behaviour. One possibility for solving this problem is a piecewise approximation of the system. By comparison of the approximated models it is perhaps possible to identify the nonlinear part of the system and afterwards the linear part by separate estimation.

The piecewise approximation is also one possibility for estimating nonlinear dynamic behaviour. In many industrial control processes, for example load control of nuclear plants, it can be assumed that the global system behaviour can be described for several load rates by corresponding linear models. So every load rate has to have its own discrete transfer function estimated for it. A good dynamic description can be obtained by this procedure, when the following two conditions are fulfilled:

- no load change during the identification and
- permanent excitation of the input signal within the bounded range.

With the different transfer functions for several load ranges it is possible to design a controller or to simulate the load-dependent process. This procedure is usual.

Model verification is very important for nonlinear systems in particular. Besides structure determination, the verification also includes the examination of the region of validity of the estimated model. Here it is necessary to test the model for all possible operating points and input signals. For nonlinear systems it is always critical whether or not the available measuring data contain the full information which is necessary for system identification.

4. CLOSED-LOOP LINEAR SYSTEMS IDENTIFICATION

During the discussion in the previous sections it was assumed that the systems to be identified are of open-loop type, which means that no feedback exists between input and output signals. However, most industrial processes are parts of feedback systems, which, for of safety and economic reasons should not be disconected. It is therefore also very important to have methods available which offer process identification of closed-loop systems. In addition to a method proposed by Caines (1975, 1976) for economic systems, three parameter estimation procedures in particular have been proposed during recent years for the identification of closed-loop systems:

- indirect identification (ID-method),
- direct identification (DI-method),
- joint process identification (JPI-method).

The parameters of the complete closed-loop system are estimated with *indirect identification*. The plant transfer function can be calculated from the global transfer function if the transfer function of the feedback path, i.e. usually the controller transfer function, is known, e.g. see Krebs (1974), Graupe (1975), Söderström (1975), Kurz (1975), Ljung (1974) and Gustavsson (1976).

In *direct identification* the selected parameter estimation method will be used as if the measurements of the input and output signals of the process came from an open loop system, see e.g. Söderström (1975), Kurz (1975), Ljung (1974), Gustavsson (1976), Schulze (1971), Thöm (1975) and Wellstead (1976). The investigated process can be identified unequivocally by the DI-method, if either

a) an additional test signal is used, and
b) the controller settings are varied or
c) special dependents between the orders of model and controller are considered, see e.g. Söderström (1975, 1978), Ljung (1976), Gustavsson (1976) and Wellstead (1975).

In *joint process identification* it is assumed that the disturbances superposed to the closed-loop system are the "true" inputs, and the process input and output signals u and y respectively would be "true" outputs. The transfer matrix between these new input and output signals (jointed signals) will then be identified, and again there is a conclusion to the transfer function of the feedforward path by a reverse calculation. This method was proposed by Phadke (1974) and Caines (1976).

4.1. Indirect identification of closed-loop systems

Assuming that in the structure of the closed-loop system of Fig. 4.1a, the measurable test signal $r_2(k)$ added to the output of $G_c(z)$ is stationary and not correlated with the disturbance signal $r_1(k)$, then $r_2(k)$ can be interpreted as an independent input variable of the global system. The input signal u(k), correlated in the closed-loop with $r_1(k)$ and with the error signal e(k) respectively, can now be eliminated, and according to Fig. 4.1a we obtain the equation

$$Y(z) = \frac{G_m(z)}{1+G_c(z)G_m(z)} R_2(z) + \frac{1}{1+G_c(z)G_m(z)} R_1(z) , \qquad (4.1)$$

which describes the open-loop system according to Fig. 4.1b. This transformation provides a simple determination of the estimation value $\hat{G}$ of

$$G(z) = \frac{G_m(z)}{1+G_c(z)G_m(z)} , \qquad (4.2)$$

which represents the transfer function of the closed-loop system $Y(z)/R_2(z)$. With the known transfer function of $G_c(z)$ (e.g. usually the controller transfer function), the coefficients of the desired transfer function $G_m(z)$ of the feed-forward-path can be finally obtained *indirectly*

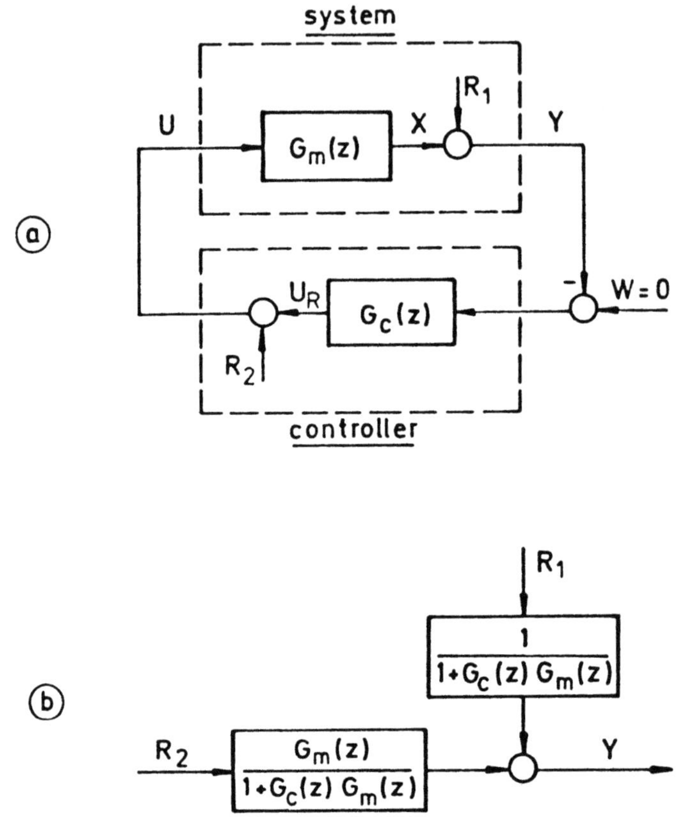

Fig. 4.1. Closed-loop system (a) and transformed representation (b)

by reverse computation from $\hat{G}(z)$ according to Eq.(4.2):

$$\hat{G}_m(z) = \frac{\hat{G}(z)}{1-G_c(z)\,\hat{G}(z)} . \tag{4.3}$$

Using the recursive version of the instrumental variable method as the estimation method, the following steps are taken to determine the parameters of $\hat{G}_m(z)$:

- Determination of $\hat{G}(z)$ by estimation in the open-loop system according to Fig. 4.1b and Eq.(4.1) using the IV-algorithm.
- Indirect calculation of the unknown parameters from Eq.(4.3), where knowledge of $G_c(z)$ is assumed.

While in the case of indirect identification, on the one hand, all difficulties of identification in a closed-loop system are avoided by the reduction to an estimation in the open-loop, this causes, on the other hand, considerable numerical expense which can result in worse convergence. The unknown plant parameters of the forward path have to be estimated by a model which is of higher order than the investigated process. The required transformation of the parameters can partially cause numerical difficulties; in any case, this transformation results in a considerable increase of computing time.

4.2. Direct identification of closed-loop systems

To determine the model parameters by the application of the IV-method it is now suggested that not only the instrumental variables $y_H(k)$ should be used for the output signal y(k), but also the input signal u(k) which is correlated to $r_1(k)$ and e(k) because of the feedback. This new modification of the IV-method was stimulated by the similar statement of a special case of the "least-square-method" according to Wellstead (1975).

If a test signal $r_2(k)$, statistically independent of $r_1(k)$ and $u(k)$, is added at the output of $G_c(z)$ as in the direct method according to Fig. 4.1a, then $r_2(k)$ can be used as the instrumental variable for $u(k)$, because it is correlated to the signal itself, but not to the disturbance part of $u(k)$. By determining the corresponding instrumental variables $y_H(k)$ for the output signal $y(k)$ according to Fig. 4.2, the conditions of Eqs. (2.55) and (2.56) in the chapter 'Introduction to System Identification' can be fulfilled if

$$\underline{w}^T(k) = \left[-y_H(k-1)\ldots -y_H(k-n)\,|\,r_2(k-1)\ldots r_2(k-n)\right]$$

$$\text{with } k=n+1,\ldots,n+N \tag{4.4}$$

is selected as rows of the instrumental matrix $\underline{W}$. This results in a consistent estimation of the parameters of the forward path, see Bauer (1977).

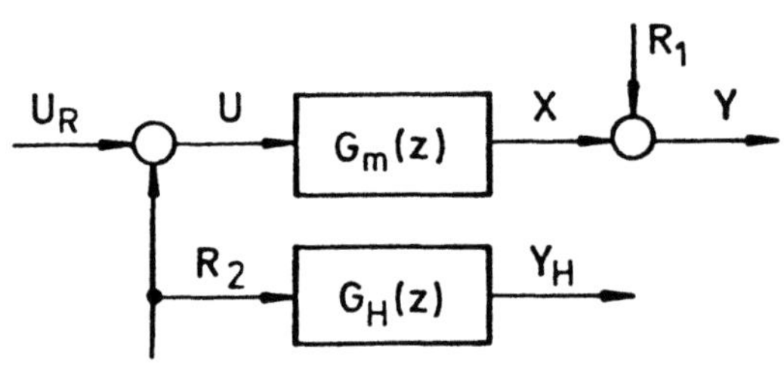

Fig. 4.2. Special selection of the auxiliary model in the closed-loop

For identification, the recursive IV-algorithm is used as in the indirect method. The identification procedure is thus completely identical with that in the open-loop, but only in $\underline{w}^T(k)$ the values $u(k-i)$ of the input signal have to be replaced by the corresponding values $r_2(k-\nu)$, where $\nu = 1,2,\dots,n$.

Compared with the indirect identification method, the direct identification method has not only numerical advantages, especially because the parameters do not have to be converted, but it also offers a considerably lower consumption of computing time, which does not increase when compared with estimation in the open-loop system. But there might be a deterioration of the numerical stability by now using two instrumental variables.

4.3. Joint process identification of closed-loop systems

Joint process identification can be applied to linear closed-loop systems which do not have an additional input signal, i.e. $w = 0$, similar to indirect identification. Compared with indirect identification it is, however, assumed that the disturbances r_1 and r_2 pass through filters with the transfer functions F and H respectively, as shown in Fig. 4.3. The system can then be described completely by

$$Y(z) = G_m(z)\ U(z) + F(z)\ R_1(z) \tag{4.5}$$

and

$$U(z) = -G_c(z)\ Y(z) + H(z)\ R_2(z)\ . \tag{4.6}$$

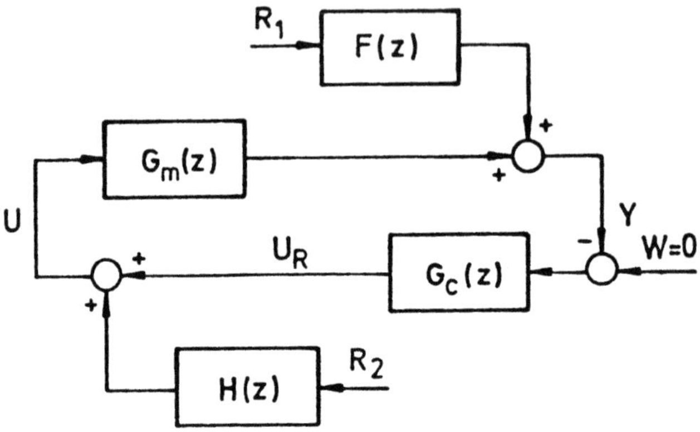

Fig. 4.3. Closed-loop structure for joint process identification

Here r_1 and r_2 are stochastic signals with zero mean and the covariances ε_1 and ε_2.

The basic idea of the joint process identification method consists, as already mentioned, in defining the signals r_1 and r_2 as being the "true" input signals, and the measurable input and output signals u and y as being the "true" output signals. Thus the z-transformed new output vector is

$$\underline{Z}(z) = \begin{bmatrix} Y(z) \\ U(z) \end{bmatrix} , \tag{4.7}$$

whereas the new z-transformed input vector is

$$\underline{R}(z) = \begin{bmatrix} R_1(z) \\ R_2(z) \end{bmatrix} . \tag{4.8}$$

The input/output description of the closed-loop system is given by

$$\underline{Z}(z) = \underline{\underline{G}}(z)\ \underline{R}(z)\ , \tag{4.9}$$

where $\underline{\underline{G}}(z)$ represents the transfer matrix, which can be obtained by inserting Eqs.(4.5) to (4.8) into Eq.(4.9). The task of identification is to determine the transfer function matrix $\underline{\underline{G}}(z)$. The estimation procedure is presented in detail by Gevers (1976).

5. IDENTIFICATION OF MULTIVARIABLE SYSTEMS

The development of practicable algorithms for the identification of multivariable systems actually started very late, in about 1976. One essential reason for this relatively late development when compared to linear single-input/single-output systems (SISO-systems), is based on the incorrect assumption, which partly still exists today, that the estimation algorithms for SISO-systems can simply be applied by the extension of the corresponding vectors and matrices. Thus there seemed to be no difficulties within the theoretical treatment. However, this is not true. The real difficulties in the identification of multi-input/multi-output systems (MIMO-systems) are connected with the following two tasks:

- model structure determination and
- estimation algorithms.

Within the model structure determination the causal connexions in the system are examined by analysing cause and action. A connexion between two or more variables (signals or states) is described by a model structure, the coefficients of which are determined by estimation algorithms.

The identification of MIMO-systems can be divided into two principal blocks which orient themselves according to the model structure:

- the state space formulation and
- the input-output description in terms of polynomial matrices.

5.1. Identification using state space models

Since multi-variable processes are presented mainly in state space in modern control theory, an initial attempt was made to use this description for identification, too. The pioneering work in this field was done by Gopinath (1969) and Budin (1971, 1972), and later by Gupta (1974) and Passeri (1972). These state space methods have two common disadvantages:

- they are not suitable for on-line application and
- they are limited due to the assumption that the measurements were free of noise.

Another possibility for estimating a state space model consists of selecting a canonical state space representation and of using the estimation methods described in the 'Introduction to System Identification'. This procedure has been mainly applied in flight control, where the maximum-likelihood method was chosen by Mehra (1970) and Schulz (1975). Extended surveys on possible state space representations are given by Ackermann (1972) and Hajdasinski (1979). However, the application of state space models for multivariable systems very early proved to consume a high amount of computing memory as well as computing time even for the identification of low order systems.

5.2. Identification using input-output models

To circumvent the above mentioned disadvantages, in the last few years the input-output description in terms of polynomial matrices has been successfully explored. To realize a uniform description of the different systems, two models have been introduced in recent years (Guidorzi (1975), Diekmann (1979))

- description with submodels and
- description with subsubmodels.

The question of which of the two descriptions should be selected depends on the objectives of the identification. To run *system identification*, i.e. if just the simulation or the prediction of the output signals is of importance without knowledge the inner structure, then it is sufficient to use a submodel for identification. For some reason, however, the estimation of single or of all system orders or the estimation of the parameters might be required, and then *parameter identification* has to be performed, which requires a description with subsubmodels. Both models are now introduced.

5.2.1. Submodel technique

Consider a linear, time-invariant, discrete-time MIMO-system with r inputs and m outputs, as shown in Fig. 5.1, where G_{ij} is the transfer function from input j to output i. For identification of this system a

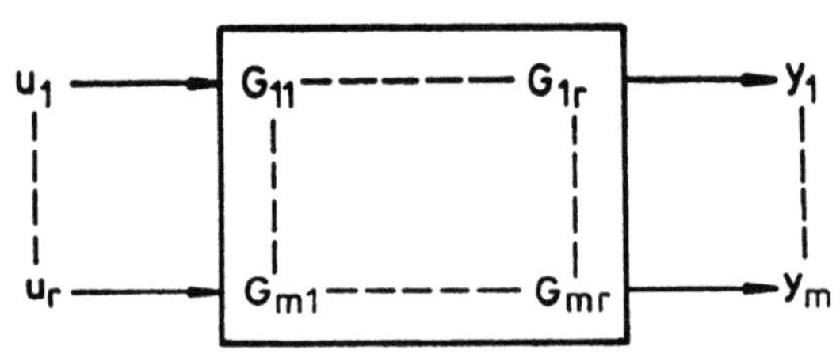

Fig. 5.1. A MIMO-system

corresponding model is decomposed into m submodels, as shown in Fig. 5.2. The structure of the submodel i is presented in Fig. 5.3, where the inputs are defined as $\bar{u}_{ij}$ and the outputs as $\bar{y}_i$. By derivation from observable canonical state space forms, the submodel outputs $\bar{y}_i$ are the system outputs y_i, which are further used in order to increase the readability. The submodel inputs $\bar{u}_{ij}$ may be the system inputs u_i or the outputs of the other submodels y_i. To provide redundancy, the number r_i of the submodel inputs is $r_i \leq r$. This definition of a submodel includes P-, V- and non-

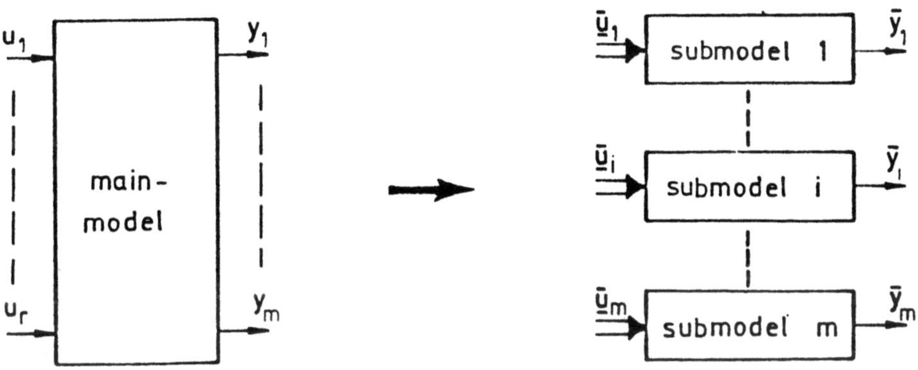

Fig. 5.2. Decomposition of the main model into m submodels

canonical forms as defined by Mesarović (1960). It should be noted that the transfer functions $\bar{G}_{ij}$ of the submodels correspond only to the transfer functions G_{ij} of the system in P-canonical form.

The output feedback in V- and non-canonical forms must be handled carefully, so that there is no rise of a closed-loop or a dependence on future values. However, a simpler system meaning can often be obtained by output feedback.

The i-th submodel of the main model is described by

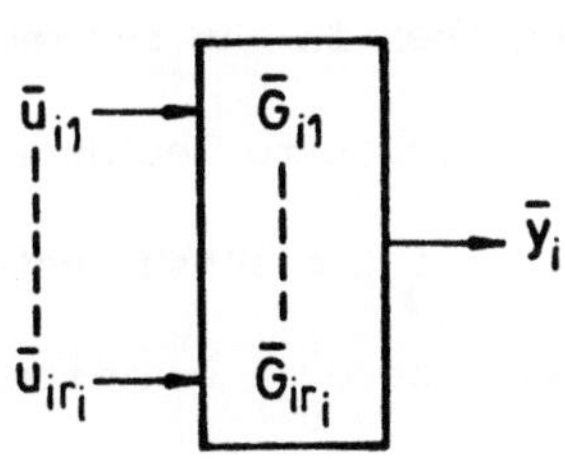

Fig. 5.3. A decomposed submodel

$$Y_i(z) = \left[\bar{G}_{i1}(z^{-1}),\ldots,\bar{G}_{ir_i}(z^{-1})\right] \begin{bmatrix} \bar{U}_{i1}(z) \\ \vdots \\ \bar{U}_{ir_i}(z) \end{bmatrix} \tag{5.1}$$

$$Y_i(z) = \left[\frac{B_{i1}(z^{-1})}{A_{i1}(z^{-1})},\ldots,\frac{B_{ir_i}(z^{-1})}{A_{ir_i}(z^{-1})}\right] \begin{bmatrix} \bar{U}_{i1}(z) \\ \vdots \\ \bar{U}_{ir_i}(z) \end{bmatrix} . \tag{5.2}$$

In order to define the general equation error E(z), according to section 2 of the chapter on 'Introduction to System Identification', described by

$$E(z) = A_i^*(z^{-1})\, Y_i(z) - B_{i1}^*(z^{-1})\, \bar{U}_{i1}(z) - \ldots - B_{ir_i}^*(z^{-1})\, \bar{U}_{ir_i}(z) , \tag{5.3}$$

the characteristic polynomial of the submodel $A_i^*(z^{-1})$, defined as the least common denominator of all minors, must be computed. Besides greatly increasing the numbers of parameters to be estimated, this description has another important drawback in identification: when the poles of two or more subsubsystems lie close together in the z-domain, for the identification algorithms there is no criterion to decide when a pole is common and when not. So it is possible that two subsubsystems with very different poles in the frequency-domain will have an estimated model with the same poles in the z-domain. The pole estimation error will be covered by an appropriate pole-zero compensation. This phenomenon will be explained by an example.

Example: We consider a three-input/one-output subsystem, as shown in Fig. 5.4. Least-Squares (LS) identification estimates the output signal $\bar{y}_1$ after 100 samples with an accuracy of 1 %. The estimation of the parameters

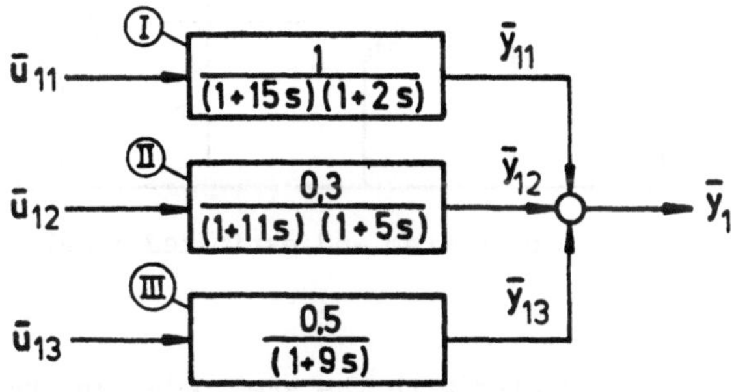

Fig. 5.4. Three input- one output subsystem

will be compared with the true parameters in pole-zero pattern in Table 5.1. Instead of 5 real poles only 3 real and one complex pair of poles are estimated. The three estimated real poles lie exactly between the 5 true poles. This means that two real poles are used as common poles for two subsubsystems although the real system has no common pole.

During identification, the output signal $\bar{y}_i$ is simulated with the estimated transfer functions. The simulation of the three systems is cascaded. Therefore only two poles are used simultaneously. This is the reason why all identification algorithms cannot recognize that all three

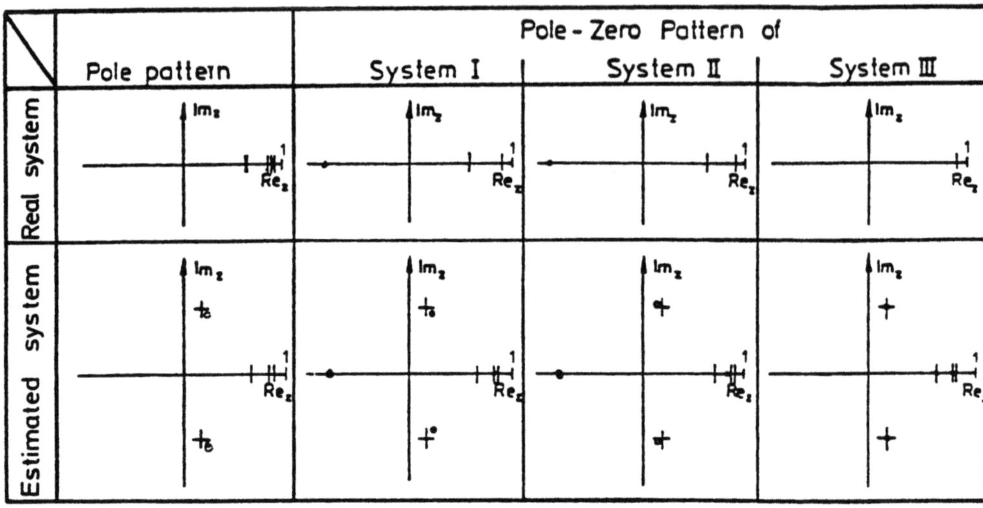

Table 5.1. Pole-zero pattern of system and estimated model

transfer functions were simulated with the same poles and that two real poles are missing. The pole approximation error that arose was covered by an appropriate pole-zero compensation.

We will generalize the results of this example: recursive and non-recursive identification algorithms which use a model with a normalized input-output description, as defined in Eq.(5.3), may estimate neither a better nor a worse output signal y_i with *false* parameters. When the true parameters are required, a simple examination of the pole-zero pattern will point out whether the phenomenon explained above occurs. If the phenomenon occurs it is impossible to modify the identification algorithm to pick up the true parameters.

5.2.2. Subsubmodel technique

To circumvent the phenomenon explained above, a new identification model must be derived when the true parameters are required. Therefore the following simple model design was suggested by Diekmann (1979):

1) Decompose the main model in m submodels as described in the last section.
2) Decompose the submodel i in r_i single-input/single-output (SISO) subsubmodels, as shown in Fig. 5.5.

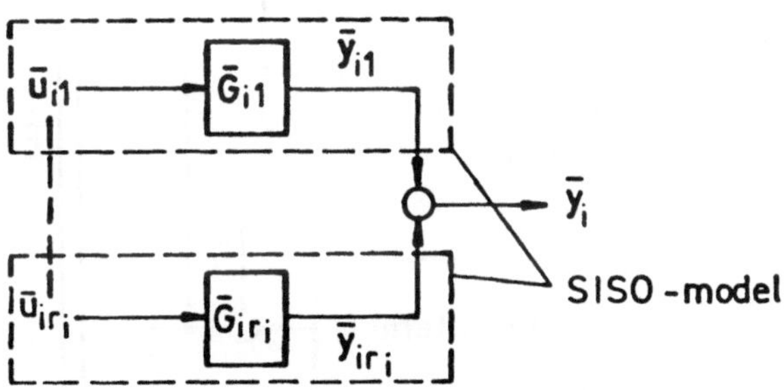

Fig. 5.5. Decomposition of a submodel

MIMO-identification is reduced by this decomposition to SISO-identification. There is only one problem here: the subsubmodel outputs $\bar{y}_{ij}$ cannot be measured. However, based on the idea of the IV-method, they can be estimated recursively. The difficulties of such an recursive estimation algorithm are

- good initial values and
- an error criteria to control the estimation.

These problems have been solved by Diekmann (1979).

5.2.3. Identification Algorithm

MIMO-systems are decomposed in multi-input/single-output (MISO) subsystems as shown in Fig. 5.2. With regard to a MISO-subsystem, using the model mentioned above, identification can be divided into three algorithms (according to Fig. 5.6):

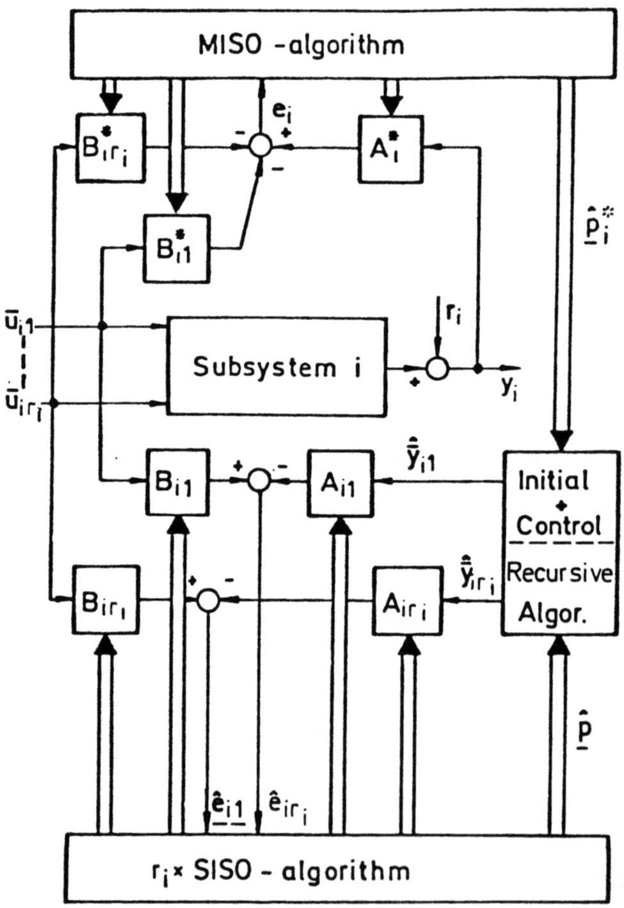

Fig. 5.6. Block diagram of MISO-algorithm

1) MISO-algorithm: to estimate the initial values $\hat{\bar{y}}_{ij}$ and to control the recursive algorithm.
2) Recursive algorithm: to estimate $\hat{\bar{y}}_{ij}$ iteratively.
3) SISO-algorithms: to estimate the parameters $\hat{a}_i$, $\hat{b}_i$.

5.2.4. Some results of simulation

The three-input/one-output submodel, as shown in Fig. 5.4, has been considered. The inputs $\bar{u}_{11}$, $\bar{u}_{12}$ and $\bar{u}_{13}$ are pseudo random binary sequences (PRBS), which are generated by shifting one sequence by 1/3 sequence length. In this example we used undisturbed measurements. Using the Least-Squares method for the MISO and the SISO-models, the results after 256 samples are compared. The pole-zero pattern in Table 5.2 shows the phenomenon for the submodels explained above, while true values are estimated using the SISO-subsubmodels.

5.3. Practical application of MIMO-identification

The possibilities of MIMO-identification will be briefly shown with an example from the steel industry, the control of a blast furnace. A blast furnace is a complicated multivariable system which can be described as a comprehensive quantity/quality model according to Fig. 5.7a. For control and identification purposes this complex model can be reduced to a simpler one, according to Fig. 5.7b, which contains as input

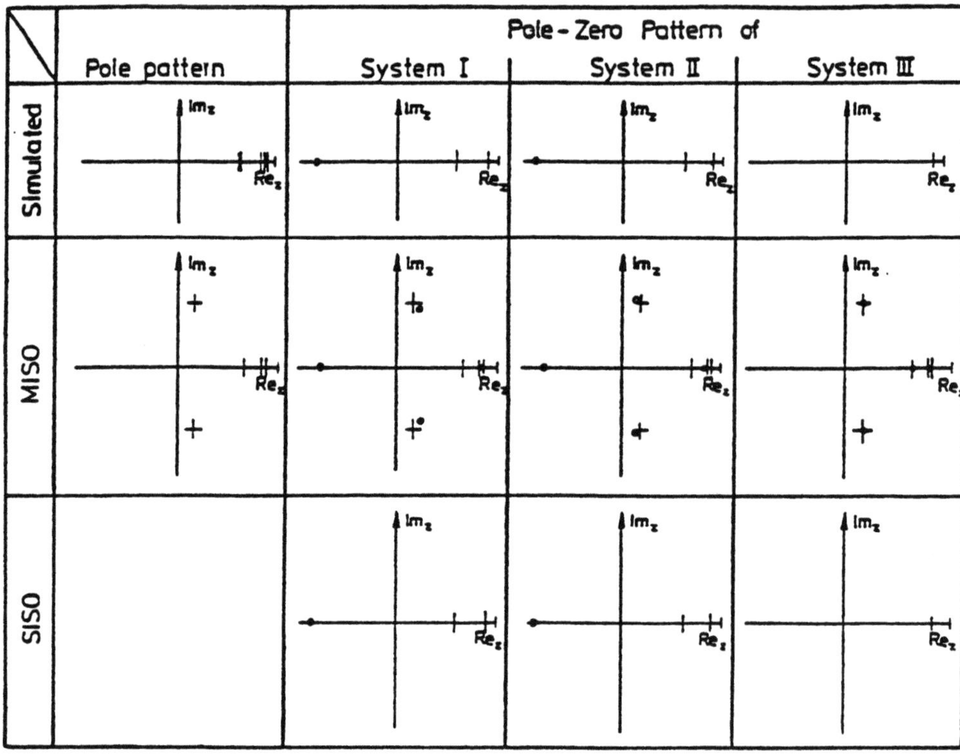

Table 5.2. Pole-zero-pattern comparison

signals only those which influence the quality of the crude iron in the sense of an automatic control. A further reduction (see Fig. 5.7c) can be obtained when only the crude iron temperature and the Si-content are used as the most important system output signals.

This reduction shows the importance of model structure determination for MIMO identification. Only those signals are used as input-output signals which have an observable influence on the control.

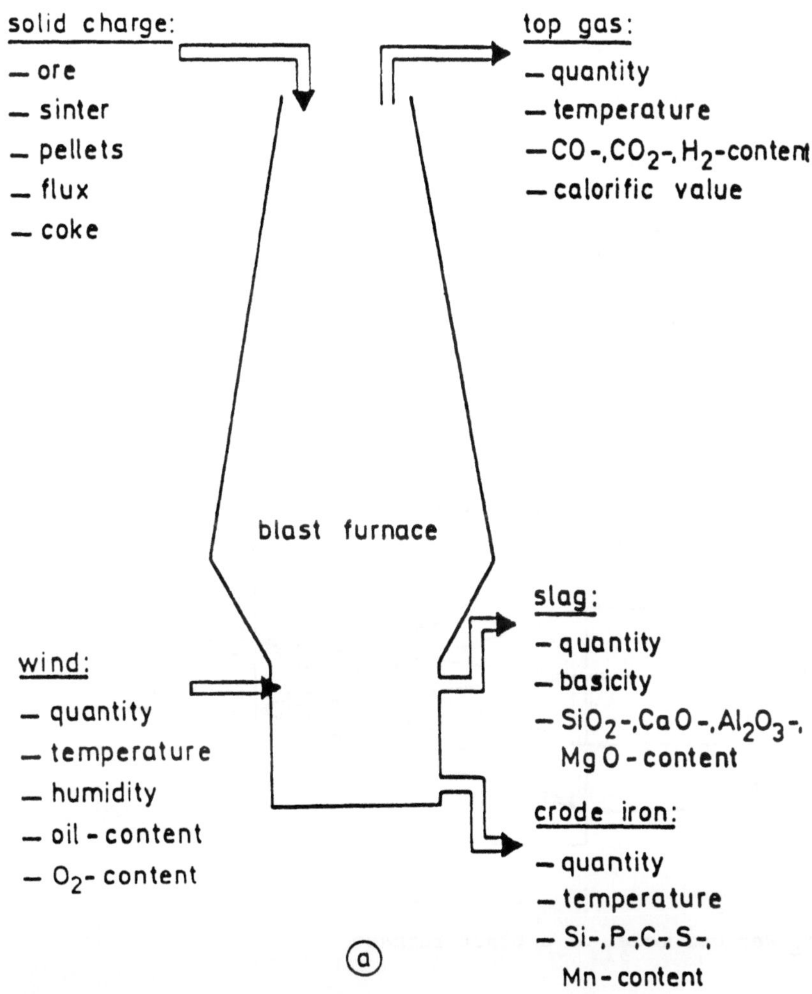

Fig. 5.7a. Input and output variables of a blast furnace

The remaining model is divided into two submodels according to Fig. 5.7d:

- submodel 1 has five input signals, the wind variables, and the Si-content as the output signal;
- submodel 2 has the same input singals and the Si-crude iron temperature

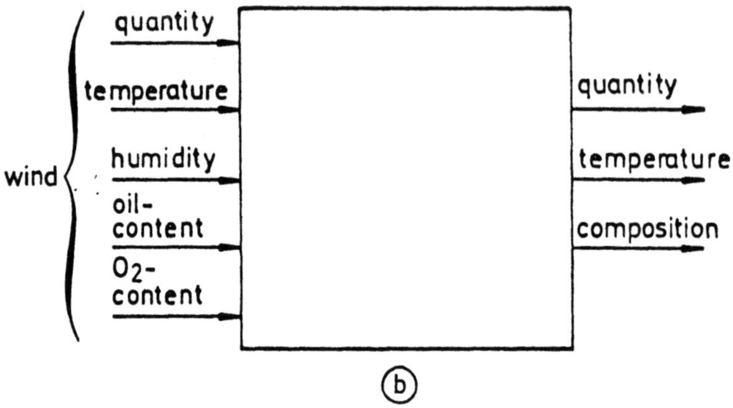

Fig. 5.7b. Reduced quantity/quality model of a blast furnace

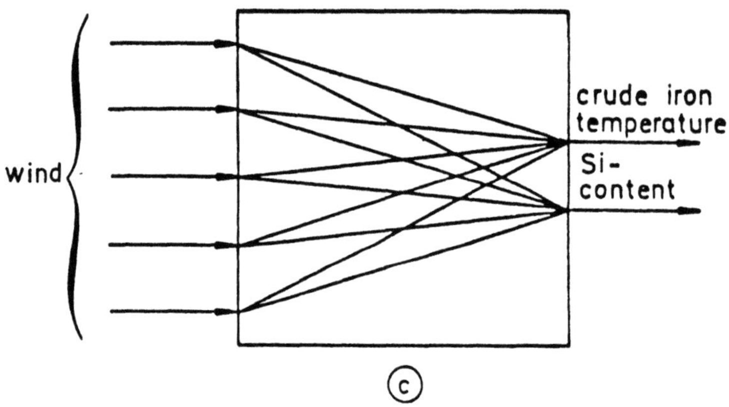

Fig. 5.7c. Reduced model of a blast furnace

as the output signal.

Both submodels have P-canonical structure and can be identified by the submodel - or the subsubmodel technique according to the purpose of identification. Detailed results of both techniques have been presented by Diekmann (1981). To show, as an example, the quality of the estimated models, the measured and the estimated output signals are compared in

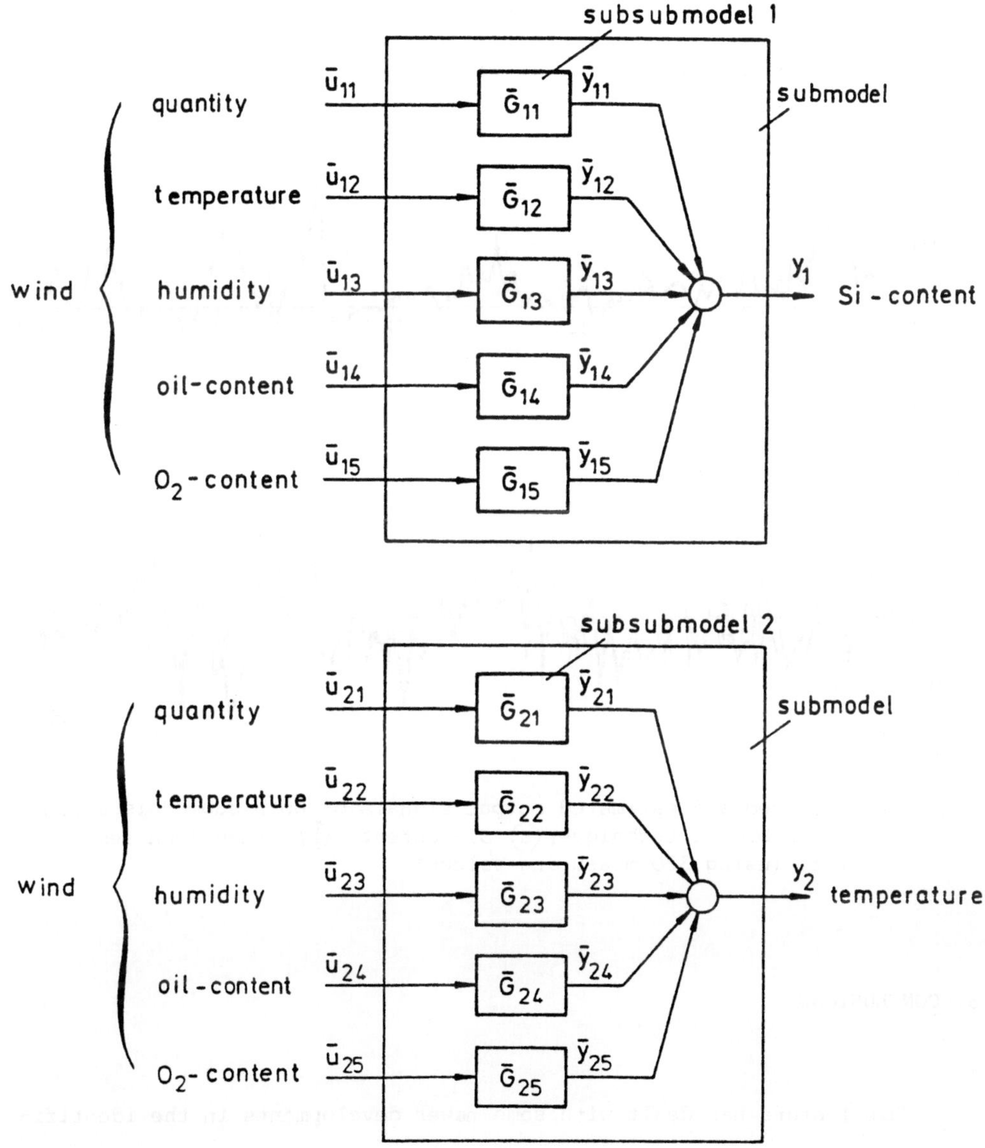

Fig. 5.7d. Design of submodels and subsubmodels for a blast furnace

Fig. 5.8 using the submodel technique. It is obvious that the output signals can be simulated by the models with very high accuracy.

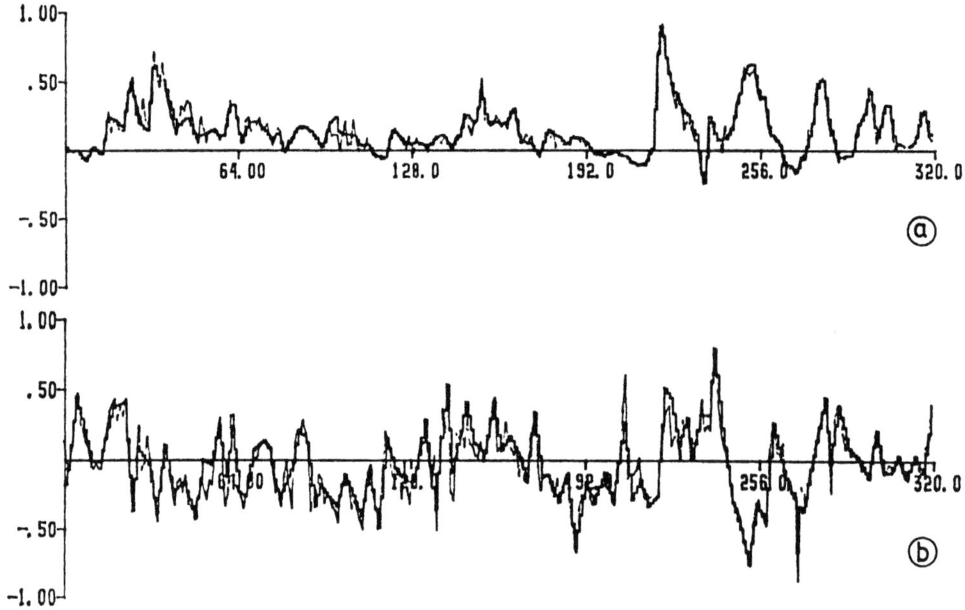

Fig. 5.8. Measured and estimated output signals of both submodels using the submodel technique; ⓐ Si-content; ⓑ crude iron temperature (using 320 measuring values)

6. CONCLUSIONS

This lecture has dealt with some newer developments in the identification of dynamical systems, especially from the point of view of a control engineer. The problems connected with system identification using parameter estimation methods for the evaluation of stochastic measuring data from time variant systems, nonlinear systems, closed-loop systems, and multivariable systems have been discussed. As a result of this chapter it can be concluded that some powerful methods are already available. However, much effort is necessary for the solution of problems which are still open or for the improvement of already existing solutions. In the field of the identification of nonlinear systems and closed-loop MIMO-systems in particular there are still many unsolved problems.

REFERENCES

Ackermann, J. (1972), Abtastregelungen, Springer Verlag.

Bauer, B. (1977), Parameterschätzverfahren zur on-line Indent. dynamischer Systeme im offenen und geschlossenen Regelkreis. Diss.Ruhr-Universität Bochum.

Budin, M.A. (1971), Minimal realization of discrete linear systems from input-output observation. IEEE Tr. AC, 16, pp. 395-401.

Budin, M.A. (1972), A new approach to system ident., IEEE Tr. SMC, 2, pp. 396-402.

Caines, P.E., Chan, C.W. (1975), Feedback between stationary stochastic processes. IEEE Tr. AC, 20, pp. 498-508.

Chavent, G. (1979), On the identification of distributed parameter systems, IFAC Symp. Darmstadt, FRG.

Diekmann, K., Unbehauen, H. (1979), Recursive identification of multi-input, multi-output systems, IFAC Symp. Darmstadt, FRG.

Diekmann, K., Unbehauen, H. (1981) Identif. of submodels of multi-input, multi-output ystems. Int. Conf. on 'Control and its Applications', Warwick.

Gevers, M.R. (1976), On the identification of feedback systens. IFAC Symp. Tiflis, USSR.

Goodson, R.W., Polis, M.P. (1976), Parameter identification in distributed systems, Proc. IEEE, V. 64.

Gopinath, B. (1969), On the identi. of linear time-invariant systems from input-output data. Bell System Techn. J., 48, pp. 1101-1113.

Graupe, D. (1972), Identification of systems. Van Nostrand Reinhold Comp. New York.

Graupe, D. (1975), On identifying stochastic closed-loop systems. IEEE Trans. AC-20, Np. 4, pp. 553-555.

Guidorzi, R. (1975), Canonical structure in the ident. of multi-variable systems. Automatica, 11, pp. 361-398.

Gupta, R.D., Fairman, F.W. (1974), Param. est. for multivariable systems. IEEE Tr. AC, 19, pp. 546-549.

Gustavsson, I., Ljung, L., Söderström, T. (1976), Identification of processes in closed loop-identifiability and accuracy aspects. IFAC-Symposium, Tbilisi, paper S-3.

Haber, R., Keviczky, L. (1976), Ident. of nonlinear syst. IFAC Symp. Tbilisi, pap. S-4.

Hajdasinski, A.K., Niederlinski, A. (1979), Multivariable system identification, IFAC Symp. Darmstadt, FRG.

Krebs, V., Thöm, H. (1974), Parameteridentifizierung nach der Methode der kleinsten Quadrate - ein Überblick. Regelungstechnik und Prozeßdatenverarb. 22, No. 1, pp. 1-32.

Kreuzer, W. (1977), Parameteridentifikation bei linearen zeitinvarianten Systemen. Diss. Universität Karlsruhe.

Kubrusly, L.S. (1977), Distributed parameter system identification, a survey. Int. J. Control, V. 26, pp. 509-535.

Ljung, L., Gustavsson, I., Söderström, T. (1974), Identification of linear multivariable systems operating under linear feedback control. IEEE Trans. AC-19, pp. 836-840.

Ljung, L. (1976), Consistency of the LS ident. method. IEEE Tr. AC, 21, pp. 779-781.

Mehra, R.K. (1970), ML ident. of aircraft param., JACC Atlanta, pp.442-444.

Mesarović, M.D. (1960), The control of multivariable systems.Wiley, N.Y.

Passeri. D.P., Herget, C.J. (1972), Param. ident. of a class of multiple input/output linear discrete-time systems. JACC Stanford, pp. 786-793.

Phadke, M.S., Wu, S.M. (1974), Identification of multi-input, multi-output transfer function and noise model of a blast furnace from closed loop data, IEEE Trans. AC-19, pp. 944-951.

Ruberti, R. (1976), Distributed parameter systems: modelling and identification, IFIP Working Conference, Rome, Italy.

Schaufelberger, W. (1969), Modelladaptive Systeme, Diss. ETH Zürich.

Schenk, R., Scheurer, H.G. (1976), Vergleich einiger moderner Parameterschätzverfahren. Regelungstechnik 24, No. 1, pp. 11-19.

Schulz, G. (1975), ML Ident. mittels Kalman Filterung. DFVLR, Rep.75-54.

Schulze, H. (1971), Anwendung von Schätzverfahren für die Kenngrößen von Regelstrecken aufgrund von Messungen im geschlossenen Regelkreis. Regelungstechnik und Prozeßdatenverarbeitung, 19, No. 3, pp.113-119.

Söderström, T., Gustavsson, I., Ljung, L. (1975), Identifiability conditions for linear systems operating in closed loop. Int. J. Control, 21, No. 2, pp. 243-255.

Söderström, T. (1978), A theoretical analysis of recursive ident. methods. Autom., 14, pp. 231-244.

Thöm, H., Krebs, V. (1975), Ident. im geschlossenen Regelkreis - Korrelationsanalyse oder Parameterschätzung. Regelungstechnik, 23, pp.17-19.

Wellstead, P.E., Edmunds, J.M. (1975), Least squares identification of closed-loop systems. Int. J. Control 21, pp. 689-699.

Wellstead, P.E. (1976), The identification and parameter estimation of feedback systems. IFAC-Symposium, Tbilisi, paper 16.1.

MODERN METHODS OF COVARIANCE ANALYSIS AND APPLICATIONS TO THE ESTIMATION THEORY

WALTER WEDIG
University of Karlsruhe

1. INTRODUCTION

The classical spectral method of the stochastic analysis is connected with transfer functions and power spectra describing the excitation and response processes of dynamic systems in the frequency domain. Naturally, such a description is most adequate to vibrating systems. However, it possesses the significant disadvantages that it is restricted to stationary processes and to linear time-invariant systems.

Therefore, we introduce the wellknown method of Ljapunow equations by which we have the advantage to study the stationary and the nonstationary covariances of the solution processes in a direct manner and without any integration of the associated power spectra. This approach, too, is restricted to linear systems under white noise or more realistic stochastic excitations generated by means of shaping filters.

A generalization of the Ljapunow equations is achieved by the theory of stochastic differential equations which is based on the Wiener process and Itô's calculus. The most important advantage of this modern method consists in its applicability to nonlinear systems and to dynamic systems with distributed parameters. Finally, we give some instructive examples and applications in the parameter estimation theory in order to show the simple handling with Itô's calculus.

2. STOCHASTIC SYSTEMS UNDER WHITE NOISE

The classical way to describe white noise is starting with an exponentially correlated stationary process having zero mean and a Gaussian density distribution. Its correlation function $R_{zz}(\tau)$ and the corresponding power spectrum $S_{zz}(\omega)$ are given by

$$R_{zz}(\tau) = \frac{\rho}{2} \exp(-\rho|\tau|), \qquad S_{zz}(\omega) = \frac{\rho^2}{\rho^2+\omega^2} \tag{1}$$

As shown in Fig. 1, ρ is a limiting frequency which determines the bandwidth of the process in the spectral range ω as well as the correlation decay at any time difference $\tau = |t_2-t_1|$.

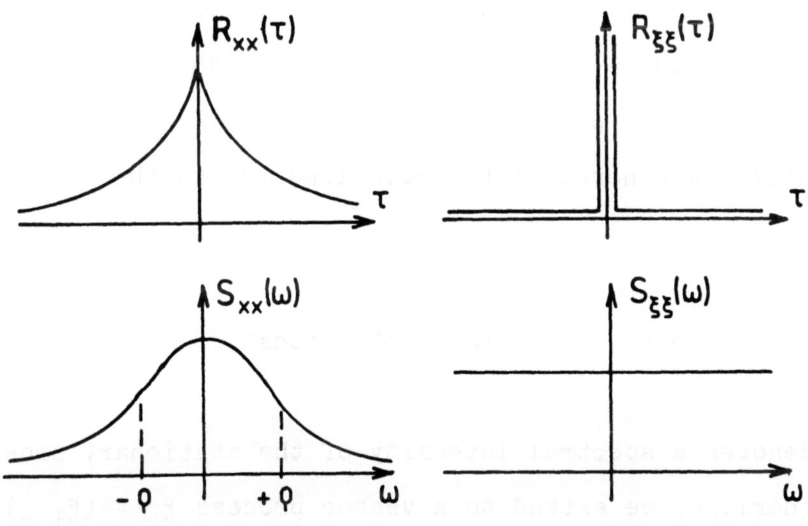

Fig.1 Characteristics of coloured and white noise

In both statistical characteristics, it is possible to pass to $\rho \to \infty$ in obtaining the following limiting case:

$$R_{\xi\xi}(\tau) = \delta(\tau), \qquad S_{\xi\xi}(\omega) = 1 = \text{const.} \tag{2}$$

It characterizes a stationary process called Gaussian white noise which is completely uncorrelated for $|\tau| > 0$ and constantly distributed in the frequency domain $|\omega| < \infty$, respectively. Because of the Dirac's delta function $\delta(\tau)$, white noise has an unbounded variance $\sigma_\xi^2 = R_{\xi\xi}(0)$ so that the associated density distribution is not defined in the classical sense of the probability theory. To overcome this difficulty, the statistical characteristics (2) are usually considered as an approximation of the more realistic excitation (1) and the limiting procedure $\rho \to \infty$ is performed in the characteristic of the response processes.

Finally, we generalize the relations (2) to the non-normed case

$$R_{\xi\xi}(\tau) = \sigma^2\delta(\tau), \qquad S_{\xi\xi}(\omega) = \sigma^2 = \text{const.} \tag{3}$$

where σ^2 denotes a spectral intensity of the stationary process. Furthermore, we extend to a vector process $\underset{\sim}{\xi}_t = (\xi_{i,t})$ with $i = 1,2,\ldots n$ components $\xi_{i,t}$, with vanishing mean values and a covariance matrix of the form

$$E(\underset{\sim}{\xi}_t\underset{\sim}{\xi}_s^T) = \Sigma\,\delta(t-s), \qquad \Sigma = (\sigma_{ij}), \quad i,j=1,2,\ldots n. \tag{4}$$

Herein, E is the expectation operator, the indices t and s denote the time dependences and $\underset{\sim}{\xi}_t^T$ is the transposed form of

$\underset{\sim}{\xi}_t$. The matrix Σ contains the spectral intensities σ_{ii} of each component $\xi_{i,t}$, as well as the cross spectra σ_{ij} of two different processes $\xi_{i,t}$ and $\xi_{j,t}$. If all components of $\underset{\sim}{\xi}_t$ are statistically independent and normed, the covariance matrix Σ degenerates to the unit matrix I in correspondence to (2).

The vector process $\underset{\sim}{\xi}_t$ is now applied to a dynamic system described by the following matrix form of ordinary differential state equations:

$$\underset{\sim}{X}_t = A\underset{\sim}{X}_t + B(t)\underset{\sim}{\xi}_t, \qquad \underset{\sim}{X}_t = (X_{1,t} \; \ldots \; X_{n,t})^T \tag{5}$$

Herein, $\underset{\sim}{X}_t$ is a n-dimensional vector containing the state processes of the system. As usual in the stochastic analysis[1], they are noted by a capital letter or more sophisticated by $\underset{\sim}{X}_t(\omega)$ where ω is the realization parameter. Furthermore, A and B(t) are n×n-matrices with deterministic elements. The matrix A contains the constant parameters of the dynamic system. B(t) may be time-dependent or constant. The differential equation (5) has a matrix solution of the form[2]

$$\underset{\sim}{X}_t = e^{A(t-t_o)}\underset{\sim}{X}_o + \int_{t_o}^{t} e^{A(t-\tau)}B(\tau)\underset{\sim}{\xi}_\tau d\tau, \qquad t \geq t_o, \tag{6}$$

where $\underset{\sim}{X}_o$ is an initial state vector which may be deterministic or stochastic. For sake of brevity, we assume that $\underset{\sim}{X}_o = 0$ at the initial time t_o.

From the explicit solution (6), we are able to derive the mean value, the covariances or higher order moments of the system (5). For the calculation of the covariances e.g., the solution $\underset{\sim}{X}_t$ is combined with its transposed form at time s.

$$\underset{\sim}{X}_t = \int_{t_o}^{t} e^{A(t-\tau)} B(\tau) \underset{\sim}{\xi}_\tau \, d\tau, \quad \underset{\sim}{X}_s^T = \int_{t_o}^{s} \underset{\sim}{\xi}_\sigma^T B^T(\sigma) e^{A^T(s-\sigma)} d\sigma. \tag{7}$$

Multiplying both forms and taking the expectation leads to

$$E(\underset{\sim}{X}_t \underset{\sim}{X}_s^T) = \int_{t_o}^{t}\int_{t_o}^{s} e^{A(t-\tau)} B(\tau) E(\underset{\sim}{\xi}_\tau \underset{\sim}{\xi}_\sigma^T) B^T(\sigma) e^{A^T(s-\sigma)} d\sigma d\tau. \tag{8}$$

Subsequently, we introduce the given covariance matrix (4) and perform a first integration by utilizing the property of the Dirac's delta function.

$$\begin{aligned} E(\underset{\sim}{X}_t \underset{\sim}{X}_s^T) &= \int_{t_o}^{t} e^{A(t-\tau)} B(\tau) \Sigma B^T(\tau) e^{A^T(s-\tau)} d\tau, \quad (t \leq s) \\ E(\underset{\sim}{X}_t \underset{\sim}{X}_s^T) &= \int_{t_o}^{s} e^{A(t-\sigma)} B(\sigma) \Sigma B^T(\sigma) e^{A^T(s-\sigma)} d\sigma, \quad (t \geq s) \end{aligned} \tag{9}$$

Finally, we pass to the special case t = s and obtain the associated covariance matrix $E(\underset{\sim}{X}_t \underset{\sim}{X}_t^T)$. Differentiated with respect to the time t, it results to an ordinary matrix equation of the form[3]

$$\dot{E}(\underset{\sim}{X}_t \underset{\sim}{X}_t^T) = AE(\underset{\sim}{X}_t \underset{\sim}{X}_t^T) + E(\underset{\sim}{X}_t \underset{\sim}{X}_t^T) A^T + B(t) \Sigma B^T(t). \tag{10}$$

It is the so-called Ljapunow equation and describes the time behaviour of all variances $E(X_{i,t}^2)$ and covariances $E(X_{i,t}X_{j,t})$ of the solution processes $X_{i,t}$ for an arbitrary initial condition $\underset{\sim}{X}_o$.

In the special case of a constant matrix B, we are interested in the stationary covariances. Provided their existence, they are time-independent and therefore given by

$$\dot{E}(\underset{\sim}{X}_t\underset{\sim}{X}_t^T) = 0, \quad AE(\underset{\sim}{X}_t\underset{\sim}{X}_t^T) + E(\underset{\sim}{X}_t\underset{\sim}{X}_t^T)A^T = -B\Sigma B^T. \tag{12}$$

According to this result, we are able to calculate the covariances of the stationary system processes by means of analytical or numerical evaluations without the integration procedures of the classical spectral method. For systems with a small number of state processes, it is suitable to apply direct solution methods such as the Gauß-Seidel algorithm or the Chen-Shieh algorithm[4]. For larger scaled systems, an indirect method is preferable. Particularly, the Stein-Smith iteration[5] seems to be well suited to solve the Ljapunow equations (12), approximately.

3. WIENER PROCESSES AND ITO'S CALCULUS

The normed Wiener process W_t is defined in the time range $0 \leq t < \infty$ by the following properties:

(i) W_t is normally distributed and has the expectations

$$E(W_t) = 0, \qquad E(W_t^2) = t, \qquad (t \geq 0) . \tag{13}$$

(ii) The initial value of W_t at the time $t_o = 0$ is given by $W_{t_o} = 0$ with probability one.

(iii) The process W_t possesses statistically independent increments, i.e.

$$E([W_{t_o}-W_{t_1}][W_{t_3}-W_{t_2}]) = 0 \qquad \text{for} \quad t_1 < t_2 < t_3. \tag{14}$$

These properties are sufficient to calculate all statistical characteristics of the Wiener process, in particular

$$E(W_tW_s) = \min(t,s), \qquad E([W_t-W_s]^2) = |t-s| . \tag{15}$$

The first expectation in (15) is the correlation function of W_t. The second is the square mean of its increment dW_t if we pass to the limit $s \to t$. From this limiting case, it finally follows the important symbolic notation

$$(dW_t)^2 = dt \tag{16}$$

which will repeatedly be used in the following text. The exact mathematical meaning of (16) is more complicated[1].

Similar as white noise ξ_t, the time derivative $\dot{W}_t$ of the Wiener process does not exist in the classical sense of the probability theory. However, to get more physical understanding, $\dot{W}_t$ is usually identified by ξ_t or equivalently by the following two symbolic notations:

$$dW_t = \xi_t dt, \qquad W_t = \int_0^t \xi \, d\tau. \tag{17}$$

More precisely, both relations in (17) are valid in the sense that the associated correlation functions coincide. To verify this coincidence, we consider the notation $X_t = \int_0^t Z_\tau d\tau$ instead of (17) in which Z_t is a stationary process with a vanishing mean value and the exponential correlation (1). Therewith, we are able to evaluate the associated correlation function

$$E(X_t X_s) = \min(t,s) - \frac{1}{2\rho}[1+e^{-\rho|t-s|}-e^{-\rho t}-e^{-\rho s}]. \tag{18}$$

Passing to the limit $\rho \to \infty$, we obviously obtain the same properties as they are noted in (14) and (15). Correspondingly, the first symbolic notation in (17) is justified, as well. It will repeatedly be used in the following text. In conclusion, it is worth to note that the definition of the Wiener process can be extended to the nonnormed and vectorial case. Analogously to the white noise processes of section 2, the Wiener vector process $\underset{\sim}{W}_t = (W_{1,t}, W_{2,t}, \ldots W_{n,t})^T$ is characterized by

$$E(\underset{\sim}{W}_t) = 0, \qquad E(\underset{\sim}{W}_t \underset{\sim}{W}_s^T) = \Sigma \min(t,s). \tag{19}$$

The associated covariance matrix $\Sigma = (\sigma_{ij})$ is equal to the identity matrix I provided that $\underset{\sim}{W}_t$ is normed with uncorrelated components. The reduction to this special case is always possible by introducing the new vector process $\underset{\sim}{V}_t$ given by $\underset{\sim}{W}_t = \sqrt{\Sigma}\, \underset{\sim}{V}_t$.

Dynamic systems under white noise can be described by the following n-dimensional equations:

$$\dot{\underset{\sim}{X}}_t = \underset{\sim}{f}(t,\underset{\sim}{X}_t) + \underset{\sim}{g}(t,\underset{\sim}{X}_t)\xi_t, \quad \underset{\sim}{X}_t = (X_{1,t}, \ldots X_{n,t})^T,$$

$$\underset{\sim}{X}_t = \underset{\sim}{X}_o + \int_{t_o}^{t} \underset{\sim}{f}(\tau,\underset{\sim}{X}_\tau)d\tau + \int_{t_o}^{t} \underset{\sim}{g}(\tau,\underset{\sim}{X}\ )\xi_\tau d\tau. \tag{20}$$

The first form in (20) is an ordinary differential vector equation, the second is the equivalent integral equation with the initial state $\underset{\sim}{X}_o$ at the time t_o. ξ_t is a scalar white noise process characterized by the expectations (2). According to (17), $\xi_\tau d\tau$ is now formally eliminated by the increment dW_τ.

$$\underset{\sim}{X}_t = \underset{\sim}{X}_o + \int_{t_o}^{t} \underset{\sim}{f}(\tau,\underset{\sim}{X}_\tau)d\tau + \int_{t_o}^{t} \underset{\sim}{g}(\tau,\underset{\sim}{X}_\tau)dW_\tau . \tag{21}$$

In a short notation, this integral equation is written in the following differential form:

$$d\underset{\sim}{X}_t = \underset{\sim}{f}(t,\underset{\sim}{X}_t)dt + \underset{\sim}{g}(t,\underset{\sim}{X}_t)dW_t , \qquad (t \geq t_o). \tag{22}$$

Such a form is called a stochastic differential equation. It is provided that the initial state $\underset{\sim}{X}_o$ is independent of the difference $W_t - W_{t_o}$. Furthermore, $\underset{\sim}{g}$ and $\underset{\sim}{f}$ are assumed to be nonanticipating functions[5] satisfying the usual regularity conditions.

The formal elimination of white noise leads to so-called stochastic integrals of the form $\int_{t_o}^{t} \underset{\sim}{g}_\tau dW_\tau$ which are not explainable by the classical integration theory since their values are depending on the choice of the intersection points. Due to Itô, they are defined as follows.

$$\int_{t_o}^{t} \underset{\sim}{g}_\tau dW_\tau = \lim_{n\to\infty} \sum_{i=0}^{n-1} \underset{\sim}{g}_{t_i}(W_{t_{i+1}} - W_{t_i}) . \tag{23}$$

Herein, $\underset{\sim}{g}_t$ is approximated by a step function with the special decomposition $t_o < t_1 < \ldots < t_n = t$ and the limiting procedure is performed in the sense of the mean square convergence. The definition (23) has the advantage that the expectation operator can simply be applied leading to

$$E\left(\int_{t_o}^{t} \underset{\sim}{g}_\tau d\tau\right) = \lim_{n\to\infty} \sum_{i=0}^{n-1} E(\underset{\sim}{g}_{t_i}) E(W_{t_{i+1}} - W_{t_i}) = 0 \tag{24}$$

since $W_{t_{i+1}} - W_{t_i}$ is independent on the nonanticipating stochastic function $\underset{\sim}{g}_{t_i}$. Simularly, the covariance matrix of (23) is calculated to

$$E\left(\int_{t_o}^{t} \underset{\sim}{g}_\tau dW_\tau \int_{t_o}^{t} \underset{\sim}{g}_\sigma^T dW_\sigma\right) = E\left(\int_{t_o}^{t} \underset{\sim}{g}_\tau \underset{\sim}{g}_\tau^T d\tau\right) \tag{25}$$

i.e. the double integral is reduced to a single integral of a simple form.

Due to the integral definition (23), the handling of the associated stochastic differentials follows a special rule, the so-called Itô formula. To introduce such a calculus, we first consider the special integral $\int_0^t W_\tau dW_\tau$ and calculate its first two moments applying the definition (23). It turns out[1] that the following result is holding in the mean square:

$$\int_{t_o}^{t} W_\tau dW_\tau = \frac{1}{2} W_t^2 - \frac{1}{2} t, \rightarrow W_t dW_t = \frac{1}{2} d(W_t^2) - \frac{1}{2} dt. \tag{26}$$

The associated differential form is noted close by. In both forms, we recognize the additional terms $\frac{1}{2}$ t or $\frac{1}{2}$ dt, respectively, which are not obtainable by a formal application of the ordinary rule of differentiation or partial integration. However, applying the Taylor expansion to $d(W_t^2)$ up to the second term yields to

$$d(W_t^2) = 2W_t dW_t + (dW_t)^2 \longrightarrow d(W_t^2) = 2W_t dW_t + dt \qquad (27)$$

which coincides with (26) if we replace $(dW_t)^2$ by dt in correspondence to (16). This formal connectedness finds its general foundation by the following formula of Itô. Let $\underset{\sim}{X}_t = (X_{i,t})$ and $\underset{\sim}{Y}_t = (Y_{i,t})$ be n-dimensional vector processes given by

$$d\underset{\sim}{X}_t = \underset{\sim}{f}(t,\underset{\sim}{X}_t)dt + \underset{\sim}{g}(t,\underset{\sim}{X}_t)dW_t, \qquad \underset{\sim}{Y}_t = \underset{\sim}{u}(t,\underset{\sim}{X}_t) \qquad (28)$$

and $\underset{\sim}{f}$, $\underset{\sim}{g}$ and $\underset{\sim}{u}$ vector functions of the same dimension. Then, the stochastic differential of $\underset{\sim}{Y}_t$ is determined by

$$d\underset{\sim}{Y}_t = [\underset{\sim}{u}_t + \sum_{i=1}^{n} \underset{\sim}{u}_{x_i} f_i + \frac{1}{2}\sum_{i,j=1}^{n}\sum \underset{\sim}{u}_{x_i x_j} g_i g_j]dt + \sum_{i=1}^{n} u_{x_i} g_i dW_t \qquad (29)$$

provided that the partial derivatives $\underset{\sim}{u}_t$, $\underset{\sim}{u}_{x_i}$ and $\underset{\sim}{u}_{x_i x_j}$ exist. Again, the double sum in (29) can formally interpreted as the second term of a Taylor expansion[6].

To show this practical routine, we consider the special scalar case $u = X_{k,t}X_{l,t}$. Analogously to (27), the Taylor expansion is applied leading to

$$d(X_{k,t}X_{l,t}) = X_{k,t}(f_l dt+g_l dW_t)+X_{l,t}(f_k dt+g_k dW_t)+$$
$$+ (f_k dt+g_k dW_t)(f_l dt+g_l dW_t). \quad (30)$$

Subsequently, we replace $(dW_t)^2$ by dt and neglect the terms $dtdW_t$ which are of the higher order $(dt)^{3/2}$.

$$d(X_{k,t}X_{l,t}) = (X_{l,t}f_k+X_{k,t}f_l+g_kg_l)dt+(X_{l,t}g_k+X_{k,t}g_l)dW_t. \quad (31)$$

It is obvious that this result coincides with the corresponding evaluation of Itô's formula (29). In concluding, we apply the expectation operator to (31) taking into account that dW_t is vanishing with $E(dW_t) = 0$. The rest of the equation is devided by dt in arriving to the following covariance equations:

$$\frac{d}{dt} E(X_{k,t}X_{l,t}) = E(X_{l,t}f_k)+E(X_{k,t}f_l)+E(g_kg_l). \quad (32)$$

In the linear case, this result coincides with the corresponding Ljapunow equation already derived in section 2. However, there are differences between the ordinary differential equation (20) and the stochastic differential equation (22) if $\underset{\sim}{g}$ is dependent on the solution process $\underline{X}_t$ in a special manner. In those cases, we have to supplement the stochastic differential equation (22) by correction terms[7] in order to investigate dynamic systems which should be excited by white noise in the physical sense. But, it is also possible to avoid such discrepancies by applying coloured noise excitations generated from white noise by means of shaping filters[8].

4. APPLICATIONS IN THE ESTIMATION THEORY

In what follows, we show some applications of Itô' calculus. Problems of interest are the estimation and measurement of square means and power spectra of stationary random vibrations. For this goal, we apply the linear filter theory in order to show physical backgrounds and to give a good understanding of such basic estimations. They are explained at a one degree of freedom system under white noise $\dot{W}_t$ described by the following state equations of motion:

$$\dot{X}_{1,t} = X_{2,t}\,, \qquad E(\dot{W}_t\dot{W}_s) = \delta(t-s),$$
$$\dot{X}_{2,t} = -2D_1\omega_1 X_{2,t} - \omega_1^2 X_{1,t} + \sigma\dot{W}_t. \tag{33}$$

Herein, $X_{1,t}$ and $X_{2,t}$ are the state processes of the displacement and the velocity, respectively. The parameter D_1 denotes the damping ratio and ω_1 is the natural frequency of the system. According to (3), σ^2 is the spectral intensity of the excitation by white noise.

First, we are interested in the higher order moments $E(X^i_{1,t}X^j_{2,t}) = m_{1\ldots12\ldots2}(t)$ of the system processes $X_{1,t}$ and $X_{2,t}$. To set up the associated moments equations, we apply Itô's formula (29) and the expectation operator as it is shown in (31) and (32). This procedure results in the following second order moments equations:

$$\begin{aligned} \dot{m}_{11} &= 2m_{12}, \qquad m_{12} = E(X_{1,t}X_{2,t}), \\ \dot{m}_{12} &= -\omega_1^2 m_{11} - 2D_1\omega_1 m_{12} + m_{22}, \\ \dot{m}_{22} &= -2\omega_1^2 m_{12} - 4D_1\omega_1 m_{22} + \sigma^2. \end{aligned} \tag{34}$$

In an analogous manner, we set up the equations of the fourth order moments.

$$\begin{aligned} \dot{m}_{1111} &= 4m_{1112}, \qquad m_{1112} = E(X^3_{1,t}X_{2,t}), \\ \dot{m}_{1112} &= -\omega_1^2 m_{1111} - 2D_1\omega_1 m_{1112} + 3m_{1122}, \\ \dot{m}_{1122} &= -2\omega_1^2 m_{1112} - 4D_1\omega_1 m_{1122} + 2m_{1222} + \sigma^2 m_{11}, \\ \dot{m}_{1222} &= -3\omega_1^2 m_{1122} - 6D_1\omega_1 m_{1222} + m_{2222} + 3\sigma^2 m_{12}, \\ \dot{m}_{2222} &= -4\omega_1^2 m_{1222} - 8D_1\omega_1 m_{2222} + 6\sigma^2 m_{22}. \end{aligned} \tag{35}$$

In the stationary case, all moments are independent on the

time. Therewith, the equations (34) and (35) have the stationary solutions

$$m_{11} = \frac{\sigma^2}{4D_1\omega_1^3}, \quad m_{22} = \frac{\sigma^2}{4D_1\omega_1}, \quad m_{12} = 0,$$
$$m_{1111} = 3\,\frac{\sigma^4}{(4D_1\omega_1^3)^2}, \quad m_{2222} = 3\,\frac{\sigma^4}{(4D_1\omega_1)^2}, \tag{36}$$
$$m_{1122} = \frac{\sigma^4}{(4D_1\omega_1^2)^2}, \quad m_{1222} = m_{1112} = 0.$$

The moments of the first and third order are not of interest since they are vanishing in the stationary case.

The aim is now to measure the square mean m_{11} of the displacement process $X_{1,t}$. For this purpose, we apply a measuring device simply realized by a first order low-pass filter with the limiting frequency ω_g.

$$X_{3,t} + \omega_g X_{3,t} = \omega_g X_{1,t}^2, \qquad H(i\omega) = \frac{\omega_g}{i\omega + \omega_g}. \tag{37}$$

The associated transfer function $H(i\omega)$, noted in (37), leads to a power spectrum which has the amplification one and is concentrated around the zero frequency provided ω_g is small. From this it follows that the low-pass filter (37) gives a correct estimation of the square mean m_{11}. More precisely, the measuring output of the filter is a stochastic process fluctuating around the square mean m_{11} of interest with a dispersion influenced by the filter bandwidth ω_g as well as by the data D_1 and ω_1 of the given system. To explain such influences in detail, we have to investigate both statistical characteristics, the mean value $m_3 = E(X_{3,t})$ and the square mean $m_{33} = E(X_{3,t}^2)$ of the nongaussian filter process $X_{3,t}$. Starting with m_3, we apply Itô's formula to (37) and set up a moment equation as follows:

$$\dot{m}_3 = -\omega_g m_3 + \omega_g m_{11}, \quad \longrightarrow \quad m_3 = m_{11} = \frac{\sigma^2}{4D_1\omega_1^3} . \qquad (38)$$

In the stationary case $\dot{m}_3 = 0$, the mean value of the filter process is equal the square mean of interest, i.e. the applied amplification and bandwidth of the filter is without any influence on the estimation mean. Such an estimation is called unbiased.

Next, we investigate the stationary dispersion of the measuring process $X_{3,t}$. The associated moment equation is

$$\dot{m}_{33} = -2\omega_g m_{33} + 2\omega_g m_{113}, \rightarrow m_{33} = m_{113} = E(X_{1,t}^2 X_{3,t}). \quad (39)$$

Since m_{113} is still unknown, we once again have to apply (29) arriving to the following three equations of the third order:

$$\dot{m}_{113} = -\omega_g m_{113} + 2m_{123} + \omega_g m_{1111},$$

$$m_{123} = -\omega_1^2 m_{113} - (2D_1\omega_1+\omega_g)m_{123} + m_{223} + \omega_g m_{1112}, \qquad (40)$$
$$m_{223} = -2\omega_1^2 m_{123} - (4D_1\omega_1+\omega_g)m_{223} + \omega_g m_{1122} + \sigma^2 m_3.$$

With the results (36) and (38), we are able to calculate the stationary solutions of (40), in particular, the moment m_{113} which determines the square mean m_{33} and the associated variance σ_{33}^2, as well.

$$\sigma_{33}^2 = m_{33} - m_3^2 , \qquad \eta_g = \omega_g/\omega_1,$$
$$\frac{\sigma_{33}^2}{\sigma_{11}^2} = \frac{2\eta_g[2 + (2D_1 + \eta_g)(4D_1 + \eta_g)]}{8D_1 + \eta_g[4 + (2D_1 + \eta_g)(4D_1 + \eta_g)]} . \qquad (41)$$

Herein, the result of interest is appropriately related to the variance σ_{11}^2. The frequency parameter η_g denotes the ratio of the filter bandwidth and the natural frequency of the system.

The related dispersion σ_{33}/σ_{11} is evaluated in Fig. 2 in dependence on the damping ratio D_1 and the frequency parameter η_g. We recognize that both patameters have a significant influence on the dispersion of the mean square estimation. Particularly, it is as smaller as narrower the related bandwidth η_g will be. Provided a positive damping of the system, the dispersion of the measuring process is vanishing if η_g

tends to zero. Such an estimation is called consistent. In practical measurements, however, we are always restricted to a nonvanishing bandwidth η_g in order to get a finite duration of measurement times $T_m = 1/\omega_g$. Then, the influence of the system damping D_1 becomes obvious. It leads to a rapidly increasing dispersion of the measurement when D_1 is smaller than o.1. From this it follows inversly that the measurement time T_m has to be enlarged if the dynamic system is lightly damped.

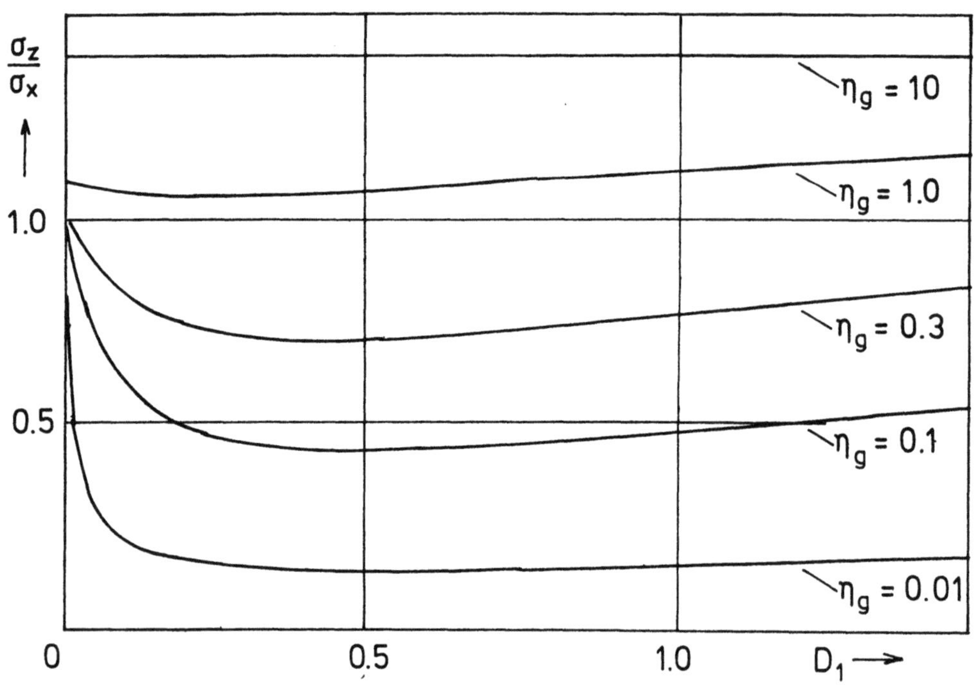

Fig. 2 Dispersion of the square mean estimation

For the identification problem, it is necessary to extend the above considerations to a frequency analysis of the response processes of vibrating structures. Similar as above, we are interested in simple estimation prescriptions in order to work out the basic influences on the quality and accuracy of measured power spectra. For this purpose, we first apply a spectral filter of the following form[9]:

$$\ddot{Z}_t + 2D_m\omega_m\dot{Z}_t + \omega_m^2 Z_t = 2\sqrt{D_m\omega_m}\; X_{1,t}, \qquad T_m = \frac{1}{D_m\omega_m}, \qquad (42)$$

Obviously, it can be realized by a second order circuit with the damping ratio D_m and the filter frequency ω_m. Its input is the response process of the linear oscillator (33) with the amplification factor $2\sqrt{D_m\omega_m}$ such that the spectrum $S_{zz}(\omega)$ of the stationary filter process $\dot{Z}_t$ is degenerating to a delta function for $D_m \to 0$.

$$\lim_{D_m\to 0} \frac{S_{\dot{z}\dot{z}}}{S_{11}} = \lim_{D_m\to 0} \frac{4D_m\omega_m\omega^2}{(\omega_m^2-\omega^2)^2+(2D_m\omega_m\omega)^2} = \delta(\omega\pm\omega_m). \qquad (43)$$

For small damping ratios, the filtering is concentrated around the frequency ω_m. Hence, the square mean $E(\dot{Z}_t^2)$ of the measuring process gives an estimation $\tilde{S}_{11}(\omega_m)$ of the power spectrum $S_{11}(\omega)$ at any arbitrary spectral frequency $\omega = \omega_m$. Naturally, this estimation is as better as longer the transient time duration T_m of the spectral filter (42) will be.

To investigate such an influence analytically, we again aplly Itô's formula (37) to the forth order system (33) and (42) and set up the associated moments equations. Then, we calculate their stationary solutions in arriving at the following result[9]:

$$E(\dot{Z}_t^2) = \tilde{S}_{11}(\omega_m), \quad \eta = \frac{\omega_m}{\omega_1}, \quad \beta = \frac{\omega_m D_m}{\omega_1 D_1},$$

$$\tilde{S}_{11}(\omega_m) = \frac{\sigma^2}{\omega_1^4} \frac{1 + \beta}{(1 - \eta^2)^2 + 4D_1^2(1 + \beta)(\eta^2 + \beta)} . \qquad (44)$$

Herein, the parameter β denotes the filter damping related to the system damping and η is the ratio of the spectral frequency of the analysed response process and the natural frequency of the system. Fig. 3 shows an evaluation of the result (44) in dependence on η and β for the system data $\omega_1 = 1\ s^{-1}$ and $D_1 = o.25$. We recognize that there is a systematic error in all measurement results which is only vanishing if the measurment time T_m goes to infinity. However, such a biased estimation is typical for spectral measurements and has to be taken into account since the measurement time T_m is finite in practical procedures.

In concluding, it is worth to note that the above spectral filter technique can be extended in order to measure the complete spectral matrix of general dynamic systems described

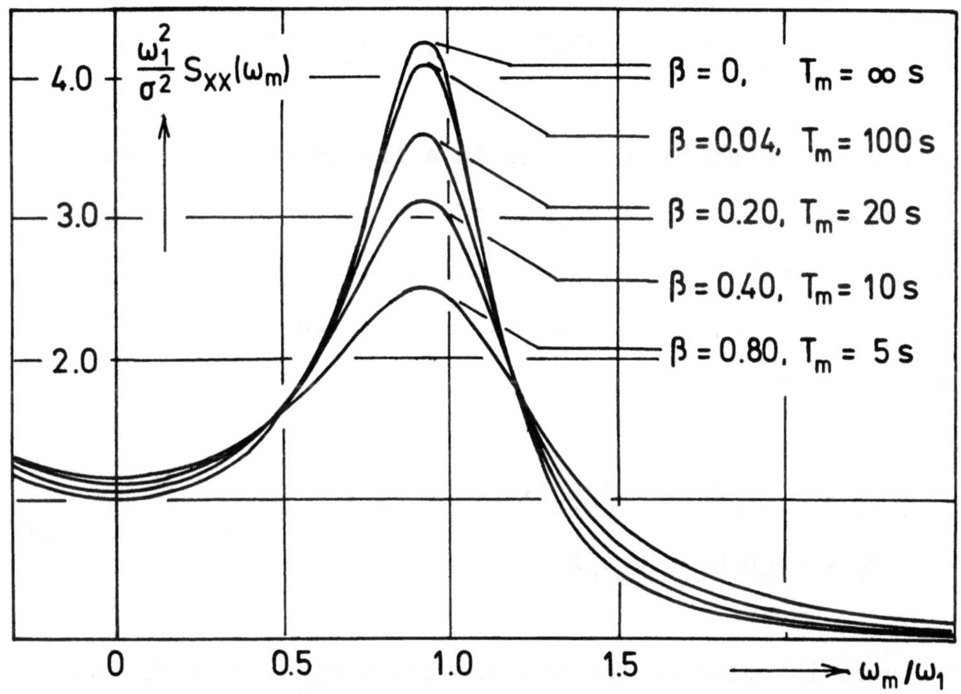

Fig. 3 Systematic errors in power spectra measurements

by the following n-dimensional first order form:

$$\dot{\underset{\sim}{X}}_t = A\underset{\sim}{X}_t + \dot{\underset{\sim}{W}}_t, \qquad E(\dot{\underset{\sim}{W}}_t \dot{\underset{\sim}{W}}_s^T) = \Sigma\,\delta(t-s). \qquad (45)$$

For this purpose, we consider a conjugate complex version[9] of (42) applied to each component $X_{i,t}$ of the state vector $\underset{\sim}{X}_t$ in (45).

$$\underset{\sim}{U}_t = -\omega_m(D_m+i\delta_m)\underset{\sim}{U}_t+\sqrt{2D_m\omega_m}\,\underset{\sim}{X}_t, \qquad i = \sqrt{-1},$$
$$\underset{\sim}{V}_t = -\omega_m(D_m-i\delta_m)\underset{\sim}{V}_t+\sqrt{2D_m\omega_m}\,\underset{\sim}{X}_t, \qquad \delta_m = \sqrt{1-D_m^2}. \qquad (46)$$

Herein, the parameter ω_m is again a filter frequency and D_m is the damping ratio of each pair of conjugate complex filter equations. Both have the same amplification $\sqrt{2D_m\omega_m}$ of the inputs $X_{i,t}$ $(i=1,2,\ldots n)$. Naturally, (46) can be reduced to the real-valued forms

$$\underset{\sim}{A}_t = -\omega_m D_m\underset{\sim}{A}_t + \omega_m\delta_m\underset{\sim}{B}_t + \sqrt{2D_m\omega_m}\,\underset{\sim}{X}_t,$$
$$\underset{\sim}{B}_t = -\omega_m D_m\underset{\sim}{B}_t - \omega_m\delta_m\underset{\sim}{A}_t \qquad (47)$$

simply by introducing the new processes $2\underset{\sim}{A}_t = \underset{\sim}{U}_t + \underset{\sim}{V}_t$ and $2i\underset{\sim}{B}_t = \underset{\sim}{U}_t - \underset{\sim}{V}_t$, respectively.

Associated to the uncoupled filter system (46), there is a covariance matrix $E(\underset{\sim}{U}_t\underset{\sim}{V}_t^T)$ which can be used as an estimation $\tilde{S}_{xx^T}(\omega_m)$ of the spectral matrix $S_{xx^T}(\omega)$ of the dynamic system (45). As already shown[9], the covariance analysis of both, the dynamic system (45) and the filter system (46), leads to the following final result:

$$E(\underset{\sim}{U}_t\underset{\sim}{V}_t^T) = \tilde{S}_{xx^T}(\omega_m) = [\omega_m(D_m-i\delta_m)I-A]^{-1}\times$$
$$\times[(2D_m\omega_m I-A)E(X_tX_t^T)-E(X_tX_t^T)A^T][\omega_m(D_m+i\delta_m)I-A]^{-1}. \qquad (48)$$

Passing now to the limiting case $D_m \to 0$ and $\delta_m \to 1$ of a filter concentrated around $\omega_m = \omega$, we obtain obviously the result of the classical spectral method

$$S_{xx^T}(\omega) = (-i\omega I - A)^{-1} \Sigma (i\omega I - A)^{-1}, \qquad (49)$$

$$\Sigma = -AE(X_t X_t^T) - E(X_t X_t^T)A$$

taking into account that the matrix Σ of the white noise excitation $\dot{\underset{\sim}{W}}_t$ is given by the Ljapunow equation (12) which is once more noted in (49) for the special case B(t) = I. Consequently, the estimation (48) is biased. Similar as in (44), it has systematic errors if finite measurement times T_m are provided. Evaluating the formula (48), however, we are able to check these times in order to get spectral measurements of a sufficiently good accuracy.

5. CONCLUSIONS

The contribution provides a comprehensive treatment of modern methods of covariance analysis. Starting with dynamic systems under white noise, we derive the associated Ljapunow equations of the covariances of the response processes. A generalization is achieved by the introduction of the Wiener processes leading to stochastic differential equations and the related Itô's calculus.

The most important advantage of this modern approach consists in the simple handling of the stochastic characteristics of the solution processes. This is illustrated by some applications in the estimation theory which are significant in the experimental identification procedures, particularly in measurements of square means and power spectra.

For sake of brevity, we have applied simple frequency filters to analyse dynamic systems under white noise. It is shown that the square mean measurement of the response processes is consistent. However, the spectral estimation has a systematic error which is only vanishing if the measurement times go to infinity. Therefore, it is always important to check the measurement times which can be performed via the damping ratios of the frequency filters, mentioned above.

REFERENCES

[1] Arnold, L., Stochastische Differentialgleichungen, Oldenbourg, München, 1973.

[2] Bunke, H., Gewöhnliche Differentialgleichungen mit zufälligen Parametern, Akademie-Verlag, Berlin 1972.

[3] Bucy, R.S. and Joseph, P.D., Filtering for Stochastic Processes with Applications to Guidance, Intersience Publ., New York, 1968.

[4] Chen, C.F. and Shieh, L.S., A note on expanding $PA + A^TP = Q$, IEEE Trans., AC-13, 1968, 122.

[5] Smith, R.A., Matrix equation $XA + BX = C$, SIAM J. Appl. Math., 16, 1968, 198.

[6] Sagirow, P., Introductory Course on Stochastic Processes in Dynamical Systems, CISM-Course, Udine, 1973 (nonpublished).

[7] Wong, E. and Zakai, M., On the relation between ordinary and stochastic differential equations and

applications to stochastic problems in control theory, Proc. Third Intern. Congress, IFAC, London, 1966.

[8] Wedig, W., Zufallsschwingungen, VDI-Berichte Nr. 221, VDI-Verlag, Düsseldorf, 1974, 195.

[9] Wedig, W., Anwendung einer Spektralfiltertechnik zur Identifikation linearer und nichtlinearer Systeme, Ing.-Archiv, 49, 5/6, 1980, 413.

INTEGRATION OF NONLINEAR STOCHASTIC SYSTEMS AND APPLICATIONS TO THE CRACK IDENTIFICATION

WALTER WEDIG
University of Karlsruhe

1. INTRODUCTION

The problem of the stochastic crack identification is explained using the nonlinear model of a crack oscillator under white noise. In a first investigation, we restrict our interest to one degree of freedom systems with a linear external damping. Its partially cracked restoring element is described by a bilinear spring characteristic.

We first give an introduction to the integration of such nonlinear systems in order to derive the square means of the stationary response processes. Then, we extend these investigations to an frequency analysis of such systems applying a spectral filter technique. The results of both are compared with experimental measurements of simulations.

It is shown that two different frequencies of the response processes can be analysed. The first one remains invariant with respect to the crack extension of the oscillator. The second is diminished with an increasing damage extension. Hence, the loss of stiffness of the oscillator is following from the difference measurement of both frequencies.

2. INTEGRATION OF NONLINEAR SYSTEMS

The following investigations are based on a nonlinear model of the cracked bending oscillator represented in Fig. 1. It has a linear external damping D and a spring characteristic F(x) which is asymmetric and piecewise linear with one salient point at the equilibrium position x = 0. Accordingly, the crack oscillator can be described by the two linear differential equations

$$\begin{aligned} \ddot{X}_t + 2D\omega_1\dot{X}_t + \omega_1^2 X_t &= \sigma\dot{W}_t, \qquad \text{for } X_t \geq 0, \\ \ddot{X}_t + 2D\omega_1\dot{X}_t + \omega_1^2(1-\gamma)X_t &= \sigma\dot{W}_t, \qquad \text{for } X_t \leq 0. \end{aligned} \tag{1}$$

The first equation holds for positive deflections X_t where the uncracked frequency ω_1 is valid according to the upper part of the spring characteristic. For $X_t \leq 0$, the cross section of the restoring element is opened leading to the diminished frequency $\omega_1\sqrt{1-\gamma}$ of the second equation in (1). Herein, γ denotes the crack parameter. The external excitation σW_t in (1) is assumed to be a stationary Gaussian white noise process with zero mean, a normed delta correlation and an intensity given by the parameter σ.

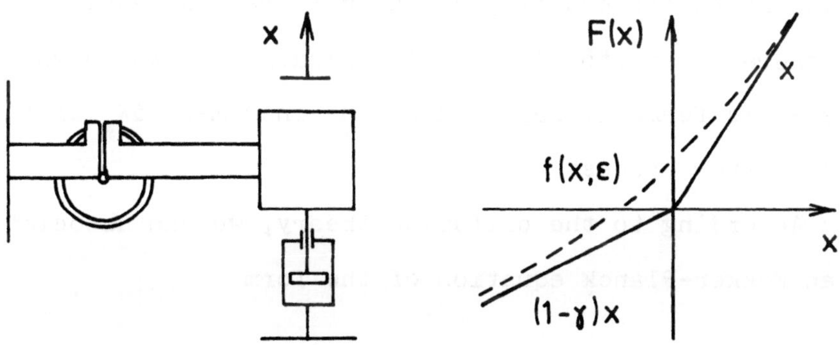

Fig.1 Mechanical model of the cracked oscillator

First investigations[1] of the stationary response processes of such nonlinear stochastic systems have been started by means of the wellknown diffusion theory. This theory is applicable when analytical characteristics are provided.

In our case, too, we are obviously able to construct such an analytical function $f(\varepsilon,x)$ which goes over to the piecewise linear characteristic $F(x)$ if ε tends to zero. Such an analytical representation can be used in order to show a first application of the diffusion theory[2]. For this purpose, we consider a scalar first order system of the form

$$\dot{X}_t + \omega_g f(\varepsilon,X_t) = \sigma\dot{W}_t, \qquad dX_t = -\omega_g f(\varepsilon,X_t)dt + \sigma dW_t. \quad (2)$$

Herein, ω_g denotes a frequency parameter of the system and $f(\varepsilon,X_t)$ is the nonlinear characteristic, mentioned above. Equivalent to the ordinary scalar differential equation in (2), there is an Itô differential equation in which white noise $\dot{W}_t$ is formally replaced by the increment dW_t of the normed Wiener process W_t.

According to the diffusion theory, we can associate to (2) an Fokker-Planck equation of the form

$$\frac{\partial p}{\partial t} + \frac{\partial}{\partial x}[-\omega_g f(\varepsilon,x)p] - \frac{1}{2}\frac{\partial^2}{\partial x^2}(\sigma^2 p) = 0. \quad (3)$$

It determines the density distribution function $p(x,t)$ of the nonlinear solution process X_t. In the stationary case, the density is obviously independent on the time t leading in (3) to an ordinary differential equation which can easily be integrated.

$$\frac{\partial p}{\partial t} = 0: \qquad p(x) = C \exp[-2\frac{\omega_g}{\sigma^2} \int f(\varepsilon, x)dx] . \tag{4}$$

The result noted in (4) is a simple exponential distribution valid for any arbitrary nonlinear spring characteristic. In particular, we are now able to pass to the limit $\varepsilon \to 0$ of the piecewise linear function F(x) in obtaining a special density distribution with the following two important properties. It is piecewise Gaussian and continuous at the salient point x=0.

From this it follows that such systems can be investigated by means of a piecewise calculation. For each semiaxis of x we set up a Gaussian distribution function

$$\begin{aligned} w_p(x) &= C_p \frac{2}{\sqrt{2\pi k_p}} \exp[-\frac{1}{2k_p}(x-m_p)^2], \quad \text{for } x \geq 0, \\ w_n(x) &= C_n \frac{2}{\sqrt{2\pi k_n}} \exp[-\frac{1}{2k_n}(x-m_n)^2], \quad \text{for } x \leq 0. \end{aligned} \tag{5}$$

Herein, the parameters m and k are mean values and variances of each semiaxis. Because of the submitted linearity, they are piecewise calculable by means of the linear theory of stochastic systems. The unknowns C_p and C_n in (5) are distribution constants following from the two conditions

$$w_p(x=+0)=w_n(x=-0), \quad \int_{-\infty}^{0} w_n(x)dx + \int_{0}^{\infty} w_p(x)dx = 1 \tag{6}$$

so that both distributions are continuous at the boundary x=0 and normed with the probability measure one.

3. SQUARE MEANS OF THE CRACK OSCILLATOR

In exactly the same manner, we now investigate the two-dimensional problem of the cracked bending oscillator. For this purpose, we rewrite the equations of motion (1) into the form of a first order system

$$\dot X_{1,t} = X_{2,t}, \qquad \dot X_{2,t} = -2D\omega_1 X_{2,t} - \omega_1^2 f(\varepsilon, X_{1,t}) + \sigma \dot W_t \tag{7}$$

introducing the state processes $X_{1,t}$ and $X_{2,t}$ of the displacement X_t and the velocity X_t, respectively. As already shown[3], we set up the associated two-dimensional Gaussian densities

$$w_{p,n}(x_1,x_2) = \frac{C_{p,n}}{\pi\sqrt{|K_{p,n}|}} \exp[-\tfrac{1}{2}(\mathbf{x}-\mathbf{m}_{p,n})^T K_{p,n}^{-1}(\mathbf{x}-\mathbf{m}_{p,n})] \tag{8}$$

and calculate the mean values $\mathbf{m} = (m_i)$ and the covariances $K = (k_{ij})$ for both, the positive and the negative subregion of the state space $\mathbf{x} = (x_i)$ for $i = 1,2$. Subsequently, we fit both density functions at $x_1 = 0$ in order to satisfy the continuity and normalization conditions (6). Integrating them over the entire state space, we finally obtain the following square means of the stationary response processes of (7):

$$E(X_{1,t}^2) = \frac{\sigma^2}{4D\omega_1^3}\left(1 + \frac{1}{1-\gamma} - \frac{1}{\sqrt{1-\gamma}}\right), \quad E(X_{2,t}^2) = \frac{\sigma^2}{4D\omega_1} \; . \quad (9)$$

Now, these results are valid in both, the positive and the negative semiregion of the state space.

In (9) we recognize that the square mean of the displacement process $X_{1,t}$ is increasing with the crack propagation. However, the square mean of the velocity process $X_{2,t}$ is basicly not influenced by any damage extension. This phenomenon can physically be explained by the fact that both processes are uncorrelated in the stationary case. Furthermore, it is wellknown from the analogous linear problem that the moments of the stationary velocity process are only influenced by the damping of the system and not by the restoring if a white noise excitation is provided. Nevertheless, we have performed some experiments[4] in order to verify such a phenomenon. The results are shown in Fig. 2. They are realized by means of an analogue simulation over a time duration of 75 minutes. The chosen data of the cracked bending oscillator are $\omega_1 = 1\ s^{-1}$, $D = 0.3$ and the γ variation given in Fig. 2. Herein, the straight lines represent the theoretical values calculated from (9). The other ones are measure-

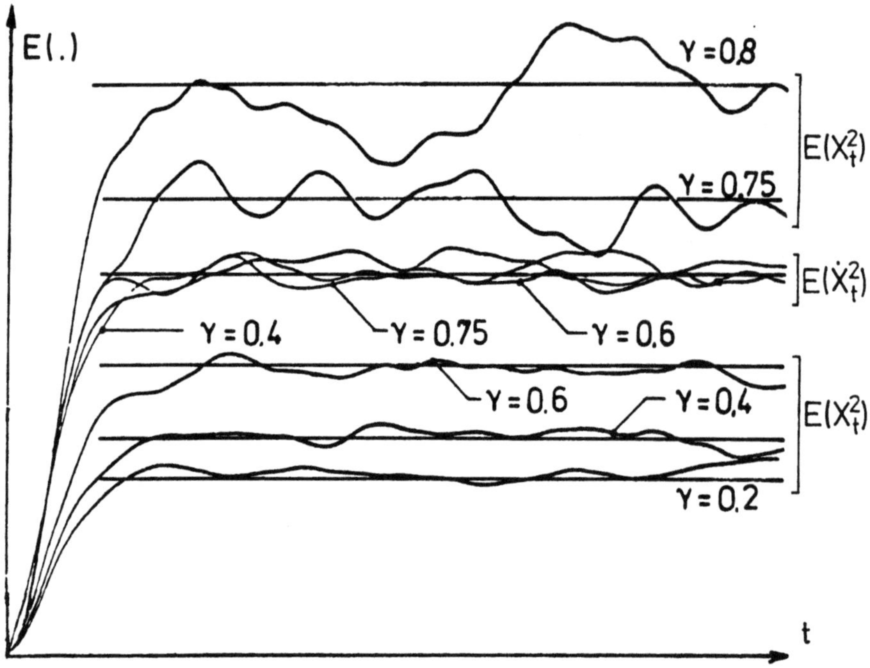

Fig.2 Experimental and theoretical results for the square means of the nonlinear response processes

ments or realizations of a low-pass filter applied in order to get consistent estimations for the square means of the nonlinear vibration processes. Obviously, there is a good coincidence between both, the theoretical calculation and the practical measurement. In particular, they are confirming the phenomenon derived above that the square means of the velocity response process is not influenced by any crack propagation. Indeed, such an invariant property gives a first possibility to measure the crack extension of a structure by the dimensionless square mean ratio

$$\omega_1^2 \frac{E(X_{1,t}^2)}{E(X_{2,t}^2)} = (1 + \frac{1}{1-\gamma} - \frac{1}{\sqrt{1-\gamma}}). \qquad (10)$$

Obviously, this crack identification can be performed without any knowledge where the crack is situated. Furthermore, it is independent on the external damping. Only, the knowledge of the uncracked natural frequency ω_1 of the structure is provided.

4. APPLICATION OF A SPECTRAL FILTER

To avoid the disadvantage mentioned above, we are looking for a specific frequency behaviour of the cracked bending oscillator. For this purpose, we make an extension of the analysis in order to calculate or estimate the correlation and spectral behaviour of the nonlinear solution processes by means of a spectral filter technique. As already described[3], the applied spectral filter has the form as follows:

$$\ddot{Z}_t + 2D_m\omega_m\dot{Z}_t + \omega_m^2 Z_t = 2\sqrt{D_m\omega_m}\, X_t \tag{11}$$

It can be realized by a linear second order circuit. Its spectral frequency distribution has a band-pass form. For a small filter damping D_m, it is concentrated around the frequency ω_m. If D_m tends to zero, the square means of the filter processes Z_t or $\dot{Z}_t$ lead to a consistent estimation for the power spectrum of X_t, provided[5] X_t is the response of a linear dynamic system.

If the input X_t of (11) is the nonlinear solution $X_{1,t}$ of (7), we have the advantage to use exactly the same calculation procedure, as before. Following it[3], we set up the associated four-dimensional Gaussian distribution functions. Then, we calculate the distribution parameters for each subspace $x_1 \gtrless 0$ by means of the linear stochastic theory and fit the density distributions at the boundary $x_1 = 0$, one-dimensionally. Finally, the square mean $E(\dot{Z}_t^2)$ leads to the following spectrum of the nonlinear solution process $X_{1,t}$ of the cracked bending oscillator.

$$S_{xx}(\eta) = \frac{\sigma^2/\omega_1^4}{1+\sqrt{1-\gamma}}\left[\frac{\sqrt{1-\gamma}}{(1-\eta^2)^2+(2D\eta)^2} + \frac{1}{(1-\gamma-\eta^2)^2+(2D\eta)^2}\right]. \tag{12}$$

Herein, the parameter η denotes the ratio of the spectral frequency ω and the uncracked natural frequency ω_1. Associated to (12), there is a correlation function $R_{xx}(\tau)$ of the form:

$$R_{xx}(\tau) = \frac{\sigma^2}{4D\omega_1^3}\frac{1}{1+\sqrt{1-\gamma}}\exp(-D\omega_1\tau)\,[\,\sqrt{1-\gamma}\,(\cos\nu_p\tau + (D\omega_1/\nu_p)\sin\nu_p\tau) + \frac{1}{1-\gamma}(\cos\nu_n\tau + \frac{D\omega_1}{\nu_n}\sin\nu_n\tau)],$$

$$\nu_p = \omega_1\sqrt{1-D^2}, \qquad \nu_n = \omega_1\sqrt{1-\gamma-D^2}. \tag{13}$$

It follows from the Fourier transform of (12). Both analytically derived results (12) and (13) are evaluated in Fig. 3 and marked by dotted lines therein. The frequency scale $0 \leq f$ $f \leq 3$[Hz] of the spectrum is above, the time scale $0 \leq \tau \leq 1.5$ [sec] is below. The data of the cracked bending oscillator are chosen with $D = 0.1$, $\gamma = 0.6$ and $\omega_1 = 10[\mathrm{sec}^{-1}]$ or $f_1 =$ 1.58 [Hz], respectively. Looking first at the frequency distribution, we observe the following invariant property. There exist two resonance peaks. The first one at the cracked frequency $\omega_1\sqrt{1-\gamma}$ and the second at ω_1. The first and main resonance is shifted to lower frequencies for an increasing crack parameter γ. The second is decreasing, however, it is basically not shifted to lower frequencies if the damage extension is increasing. Indeed, this phenomenon allows to measure both frequencies, separately. With the difference of both squared frequencies related to ω_1^2, we are then able to determine the actual crack parameter γ and therewith the loss of stiffness of the structure, as well.

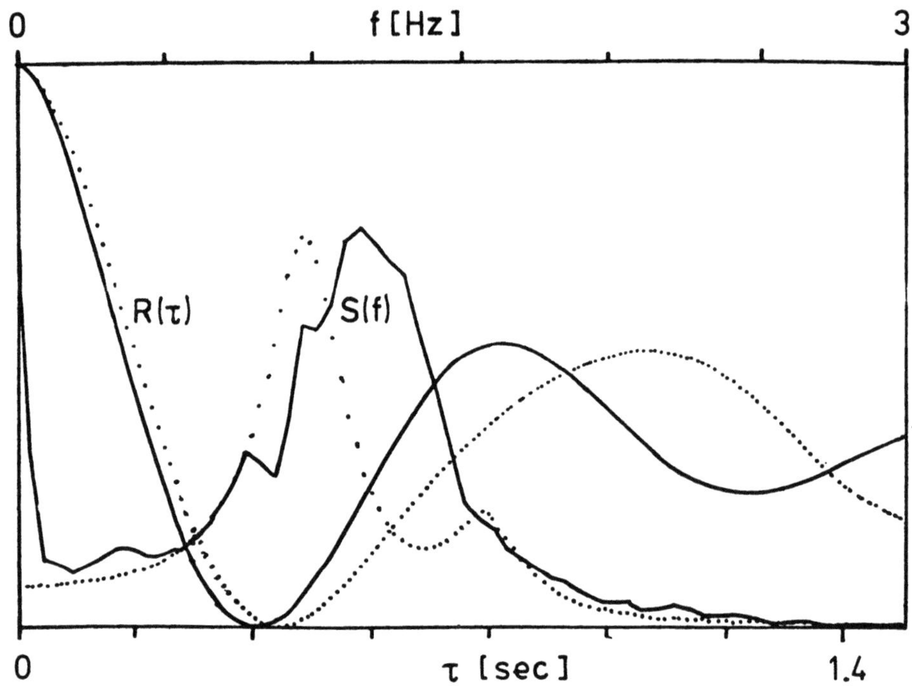

Fig.3 Measurements and analytical estimations of the correlation function and the power spectrum

Unfortunately, such a frequency phenomenon cannot be realized in a direct manner. This becomes obvious by the experimental results shown in Fig. 3. They, too, are realized via an analogue simulation for the same system data, mentioned above. The applied noise generator has had the bandwidth f_g = 5 kHz. The random signals of the simulated crack oscillator have been evaluated by means of a digital analyser performing 1,000 averages for one result. In a comparison of both, the measured correlation function and the calculated one, we observe in Fig. 3 a good coincidence only in its first half period. However, the deviations between both are increasing in the range of greater correlation times τ. Accordingly, there are significant deviations in the associated power spectra. In particular, the frequency separation, mentioned above, is not observable in such a global analysing of real experimental processes. But, it has to exist in a correlation analysis which is restricted to small correlation times.

5. PRACTICAL IDENTIFICATION PROCEDURE

With the above derived knowledge, we are now in the situation to improve the practical measurement procedure in order to verify the estimated phenomenon of the frequency separation which is most important for the identification problem. As already described[6], we propose the following procedure. Instead of a global analysing of the measured vibration processes, we first cut off the signals in a positive part P_t

and in a negative one N_t.

$$P_t = \begin{cases} X_t & \text{for } X_t \geq 0, \\ 0 & \text{for } X_t \leq 0, \end{cases} \qquad N_t = \begin{cases} 0 & \text{for } X_t \geq 0, \\ X_t & \text{for } X_t \leq 0. \end{cases} \tag{14}$$

This separation is performable by means of two comparators which are shown in Fig. 4 on the right side. The left side represents the analogue simulation of the cracked bending oscillator.

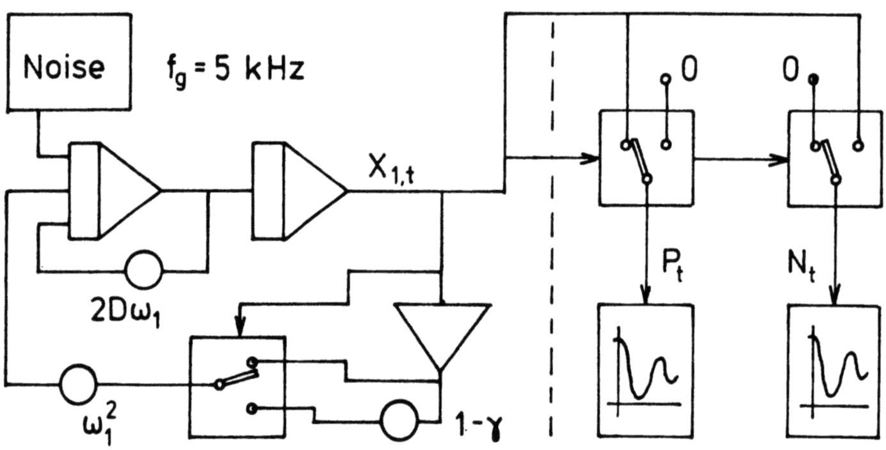

Fig.4 Analogue simulation of the frequency separation

After the cutting off (14), both different processes P_t and N_t are analysed in two correlators, separately. The results are given in Fig. 5. Above, we see the correlation

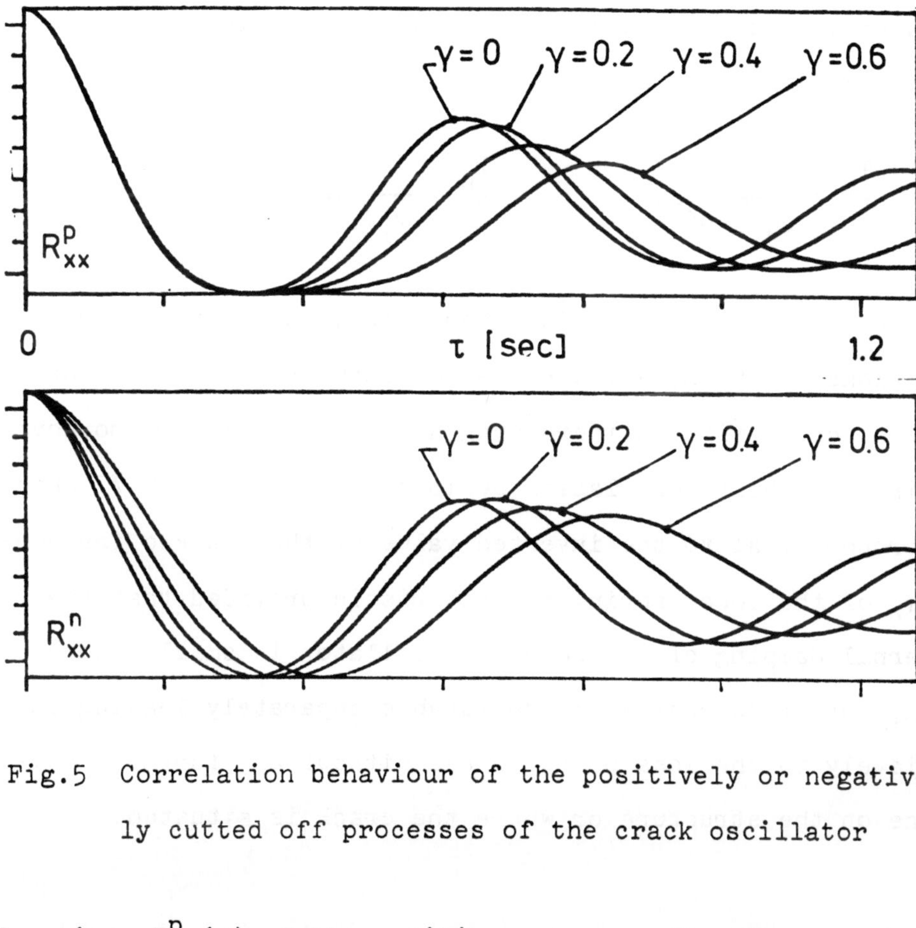

Fig.5 Correlation behaviour of the positively or negatively cutted off processes of the crack oscillator

functions $R_{xx}^p(\tau)$ of the positive processes P_t. The correlation functions $R_{xx}^n(\tau)$ of the negative part N_t are below. Both are digitally measured for the system data D = 0.1 and ω_1 = 10 [sec^{-1}]. The crack parameter is varied with γ= 0 (o.2) 0.6. In the positive correlation functions, we now observe an invariant first part up to half a period marked by the first minimum of the functions. Consequently, this correlation period can exactly be identified by the associated uncracked frequency ν_p of the upper spring characteristic of the crack oscillator.

$$\frac{T_p}{2} = \frac{\pi}{\omega_1\sqrt{1-D^2}}\,, \qquad \frac{T_n}{2} = \frac{\pi}{\omega_1\sqrt{1-\gamma-D^2}}\,. \tag{15}$$

In opposite to this invariant behaviour, the correlation functions of the negative processes show an increasing correlation period with an increasing damage extension γ. Moreover, their first half correlation periods coincide with $T_n/2$ which is proportional to the inverted value of the cracked frequency ν_n of the lower spring characteristic provided that the external damping of the bending oscillator is small. Therewith, both frequencies are measurable separately leading immediately to the loss of stiffness without any further knowledge on the structure or where the crack is situated.

6. FINAL REMARKS

To summarize briefly, we have to say that dynamic systems with piecewise linear characteristics can be investigated by means of a piecewise calculation if we are interested in mean values, covariances and stationary distribution functions of the nonlinear response processes. If we go further on to the correlation and spectral analysis, the applied filter technique is valid only for small correlation times.

Nevertheless, such an analytical investigation has pointed out the significant phenomenon of the frequency separation which is most important for the identification problem. The frequency separation is practically realizable in the correlation analysis by cutting off the measured signals in the time domain. Finally, it should be remarked that this cutting off procedure is a nonlinear filtering which has now much more applications[7] in completely different identification problems.

REFERENCES

[1] Wedig, W., Zur Integration stochastischer Systeme mit stückweise linearen Kennlinien, Ing.-Archiv, 49 201, 1980.

[2] Arnold, L., Stochastische Differentialgleichungen, Oldenbourg, München, 1973.

[3] Wedig, W., The integration of nonlinear stochastic systems with applications to the damage and ambiguity identification, ZAMM, 61, 7, 1981.

[4] Bräutigam, H. and Wedig, W., Schadensfrüherkennung beim angerissenen Biegeschwinger - Rißidentifikation durch Spektralanalyse zweier Resonanzen, VDI - Berichte Nr. 381, Düsseldorf, 45, 1980.

[5] Wedig, W., Anwendung einer Spektralfiltertechnik zur Identifikation linearer und nichtlinearer Systeme, Ing.-Archiv, 49, 413, 1980.

[6] Wedig, W., Schadensfrüherkennung - Theorie und Praxis der stochastischen Rißidentifikation, ZAMM, 62

T 7150, 1982.

[7] Wedig, W., Nichtlineare Identifikation linearer Systeme - Trennung benachbarter Eigenfrequenzen, VDI - Berichte Nr. 419, Düsseldorf, 7, 1981.

STOCHASTIC SYSTEMS WITH DISTRIBUTED PARAMETERS

WALTER WEDIG
University of Karlsruhe

1. INTRODUCTION

The contribution is devoted to multispanned panel systems consisting of one-dimensional structure elements under a stationary random excitation assumed to be uncorrelated in space and time. Up to now, the dynamic response of such distributed systems is investigated by means of discretization methods using a normal mode approach or local approximations by finite elements. Since these methods principally lead to infinte series of coupled equations of motion, there is an enormously increasing requirement in the numerical evaluation and the identification procedure.

Therefore, we propose mathematical models with distributed parameters which simplifies basically the problems, mentioned above. In the stochastic case, the variances of the stationary response is described by a finite set of ordinary differential equations in dependence on the space coordinate leading to a finite element method analogous to the wellknown situation of deterministic structures under a statical loading.

2. STRUCTURES UNDER DISTRIBUTED NOISE

As a typical example, we consider a multispanned panel system under a white field excitation. It consists of a uniform string preloaded by an axial force H_o and simply supported at its both ends. It has the length l, the mass μ and a transverse external damping β per unit length. It is excited by a boundary layer turbulence $Q_{x,t}$ in dependence on the axial coordinate x and the time variable t. As shown in Fig. 1, we assume an elastic bedding c(x) which may be distributed or concentrated at several positions a_i (i=1,..,n). In the last case, the elastic bedding can be given by a sum over the delta functions $\delta(x-a_i)$ as noted in (1). Provided

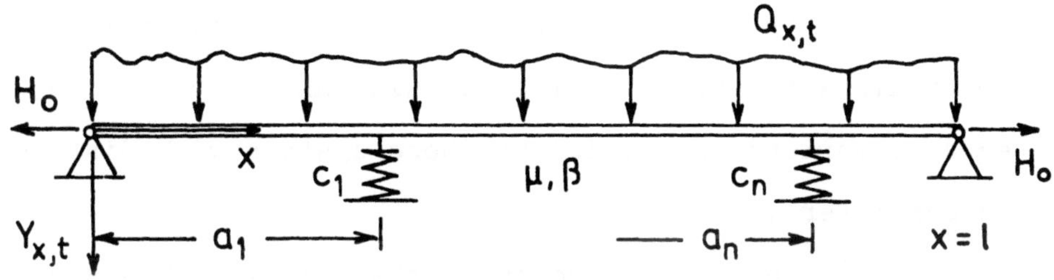

Fig.1 Multispanned panel system under distributed noise

small deflections, the distributed response process $Y_{x,t}$ is described by the following stochastic boundary value problem[1].

$$\mu \ddot{Y}_{x,t} + \beta \dot{Y}_{x,t} + c(x) Y_{x,t} - H_o Y''_{x,t} \triangleq Q_{x,t} \,,$$
$$c(x) = \sum_{i=1}^{n} c_i \, \delta(x-a_i), \qquad Y_{o,t} = Y_{l,t} = 0. \tag{1}$$

Herein, dots and dashes are denoting partial derivations with respect to time and space, respectively. The external loading $Q_{x,t}$ is a stationary field excitation with zero mean, an intensity σ and delta correlations in space and in time. Setting up

$$Q_{x,t} = \sigma \, W'_x \dot{W}_t, \qquad E(Q_{x,t} Q_{y,s}) = \sigma^2 \delta(x-y)\delta(t-s), \tag{2}$$

we represent it by the normed white field W'_x and the white noise process $\dot{W}_t$. By their increments dW_x and dW_t, both processes are related to the corresponding Wiener forms W_x and W_t which allow to apply the Itô calculus and therewith an integral covariance analysis[2] to these systems with distributed parameters.

For its application, we formulate the given boundary value problem (1) in its equivalent integral representation which is derived in two steps. First, we integrate the partial equation of motion twice with respect to the space coordinate and determine the arising constants of integration by means of the boundary conditions given in (1).

$$H_o Y_{x,t} = \int_0^l G(x,\xi)[\sigma W'_\xi \dot{W}_t - \mu \ddot{Y}_{\xi,t} - \beta \dot{Y}_{\xi,t} - c(\xi) Y_{\xi,t}] d\xi \,, \tag{3}$$

$$G(\xi,x) = \xi(1-x/l) \text{ for } \xi \leq x, \quad G(x,\xi) = x(1-\xi/l), \ (x \leq \xi).$$

Herein, $G(x,\xi)$ is the Green's function. It is symmetric with respect to the space coordinates x and ξ. In a second step, we rearrange the integral equation (3) in the time domain by introducing the new state field processes $Z_{x,t}$ and $X_{x,t}$.

$$Z_{x,t} = \int_0^l G(x,\xi) Y_{\xi,t} d\xi \,, \quad X_{x,t} = \int_0^l G(x,\xi) Y_{\xi,t} d\xi. \tag{4}$$

Obviously, they are defined via the corresponding displacement and velocity processes multiplied by the Green's function and integrated over the entire space range of the structure. Using the definitions (4), we are finally able to go over to a first order system which represents the equation (3) in the following form:

$$\dot{Z}_{x,t} = X_{x,t}, \qquad C_{x,t} = \int_0^1 G(x,\xi)c(\xi)Y_{\xi,t}d\xi\ ,$$
$$\mu\dot{X}_{x,t} = -\beta X_{x,t} - H_o Y_{x,t} - C_{x,t} + \sigma\dot{W}_t \int_0^1 G(x,\xi)dW_\xi\ . \tag{5}$$

It should be noted that the field process $C_{x,t}$ is introduced for sake of brevity. It is an abbreviation analogously defined to (4) but without the physical meaning of $Z_{x,t}$ and $X_{x,t}$.

More important is the fact that the state equations of $Z_{x,t}$ and $X_{x,t}$ have now the form of Itô's differential equations[3]. In (5) we observe that $W_\xi d\xi$ is replaced by the increment dW_ξ of the Wiener field. The corresponding term $W_t dt$ can be eliminated by dW_t when in the time domain too, we formally go over to the integral equations of (5). Therefore, we are now allowed to apply Itô's formula to both state processes $Z_{x,t}$ and $X_{x,t}$ in order to set up the associated covariance equations. This procedure can simply be performed in an analogous manner as we know it from the case of ordinary systems. It results in the following three equations of covariances:

$$\dot{E}(Z_{x,t}^2) = 2\,E(Z_{x,t}X_{x,t}),$$

$$\dot{E}(X_{x,t}^2) = -2\frac{\beta}{\mu}\,E(X_{x,t}^2) - 2\frac{H_o}{\mu}\,E(X_{x,t}Y_{x,t}) -$$

$$- \frac{2}{\mu}\,E(X_{x,t}C_{x,t}) + (\frac{\sigma}{\mu})^2 \int_0^1 G^2(x,\xi)d\xi, \tag{6}$$

$$\dot{E}(Z_{x,t}X_{x,t}) = E(X_{x,t}^2) - \frac{\beta}{\mu}\,E(Z_{x,t}X_{x,t}) -$$

$$- \frac{H_o}{\mu}\,E(Z_{x,t}Y_{x,t}) - \frac{1}{\mu}\,E(Z_{x,t}C_{x,t}).$$

To derive the equations (6), we set up the increments of the squared state processes. Then, we take the expectation of each increment using the Markow property

$$E(Y_{x,t}dW_t) = 0, \qquad E(\dot{Y}_{x,t}dW_t) = 0, \tag{7}$$

i.e. the time increment of the Wiener process is independent on the distributed displacement and velocity process of the structure. An analogous independence exists in the space domain such that the expectation of the double integral

$$E(\int_0^1 G(x,\xi)dW_\xi \int_0^1 G(x,\eta)dW_\eta) = \int_0^1 G^2(x,\xi)d\xi \tag{8}$$

is reduced to a single integral over the squared Green's function. This corresponds to the known formula[3] of the time domain where the space coordinates ξ and η are replaced by the time variables t_1 and t_2.

3. FINITE ELEMENTS OF STATIONARY COVARIANCES

The covariance equations (6) can exactly be solved if we are only interested in their stationary solutions. In this case, the covariances are time-independent, i.e. their time derivations $\dot{E}(.)$ are vanishing. Hence, the right hand sides of (6) can be setted equal to zero and simply be evaluated. Starting with the first equation of (6), we obtain

$$E(Z_{x,t}X_{x,t}) = 0 = \int_0^1 \int_0^1 G(x,\xi)G(x,\eta)E(Y_{\xi,t}\dot{Y}_{\eta,t})d\xi d\eta. \quad (9)$$

Inserting the definitions (4) and taking into account that the integrals in (9) have to be zero for all variations of the Green's function results in a vanishing correlation between the stationary state processes $Y_{\xi,t}$ and $\dot{Y}_{\eta,t}$ in correspondence to the situation of a one degree of freedom system. With this property the second equation in (6) can simply be evaluated.

$$E(X_{x,t}Y_{x,t}) = 0, \qquad E(X_{x,t}^2) = \frac{\sigma^2}{2\beta\mu}\int_0^1 G^2(x,\xi)d\xi. \qquad (10)$$

Subsequently, we insert (10) and the abbreviation $C_{x,t}$ into the third equation of (6) in obtaining the following final result:

$$H_oE(Y_{\xi,t}Y_{x,t})+\int_0^l G(x,\eta)c(\eta)E(Y_{\xi,t}Y_{\eta,t})d\eta = \frac{\sigma^2}{2\beta}G(x,\xi).(11)$$

It is worth to note that the field correlation $E(Y_{\xi,t}Y_{x,t})$ of the stationary displacement process is proportional to the Green's function provided a vanishing bedding of the structure. In case of a nonvanishing bedding, the result in (11) represents a linear Fredholm's integral equation which always can be reduced to an ordinary differential equation. For $c(x)$ = const., its solution is already derived in the paper[1].

In the present contribution, however, we are only interested in bedding concentrations given by the sum over the delta functions noted in (1). Inserting this special bedding, the integral in (11) can obviously be evaluated leading to the following basic relation[4]:

$$E(Y_{\xi,t}Y_{x,t})+\frac{1}{H_o}\sum_i c_iG(x,a_i)E(Y_{\xi,t}Y_{a_i,t}) = \frac{\sigma^2}{2\beta H_o}G(x,\xi). \quad (12)$$

This is a functional equation or more precisely, a simple algebraic relation between the general field correlation $E(Y_{\xi,t}Y_{x,t})$ and its special version $E(Y_{\xi,t}Y_{a_i,t})$ where $Y_{a_i,t}$ (i=1,2,...n) are the point processes of the structure's reponse acting at the positions $x = a_i$ of the bedding concentrations. To solve this functional equation, we first set $x = a_i$ for each position i resulting in a system of n conditional equations for the field point correlations. Solving them and inserting the results into (12), we are then able to determine the general correlation function at x and ξ by a simple summation procedure. In the special case of a structure with three panels or with two intersection points a_1 and a_2, we obtain for instance the following two conditional equations:

$$\begin{aligned}
&[1 + \frac{c_1}{H_o} G(a_1,a_1)]E(Y_{\xi,t}Y_{a_1,t}) + \frac{c_2}{H_o} G(a_1,a_2)\times \\
&\qquad \times E(Y_{\xi,t}Y_{a_2,t}) = \frac{\sigma^2}{2\beta H_o} G(a_1,\xi), \\
&\frac{c_1}{H_o} G(a_2,a_1)E(Y_{\xi,t}Y_{a_1,t}) + [1 + \frac{c_2}{H_o} G(a_2,a_2)]\times \\
&\qquad \times E(Y_{\xi,t}Y_{a_2,t}) = \frac{\sigma^2}{2\beta H_o} G(a_2,\xi).
\end{aligned} \tag{13}$$

They are linear algebraic equations with constant coefficients given by the spring parameters c_i, by the preload H_o

of the string and by the special values of the Green's function at the positions a_1 and a_2. The generalization of such a procedure is quite obvious. For a structure with three intersection points for instance, we obtain a finite element system of three equations. In case of other boundary conditions or higher order structural elements, the associated Green's functions are modified. But, the principle or the basic nature of the finite element construction in (13) is not altered.

Finally, we give an explicite solution for a two spanned panel system subdivided by one spring c at the position x = a. The associated solution has the form

$$E(Y_{\xi,t}Y_{x,t}) = \frac{\sigma^2}{2\beta H_o}[G(\xi,x) - \frac{G(\xi,a)G(a,x)}{G(a,a) + H_o/c}]. \qquad (14)$$

Obviously, it is valid in the entire range $0 \leq x,\xi \leq 1$ of the structure and completely determined by the global factor $\sigma^2/2\beta H_o$, by the stiffness ratio H_o/c and by the Green's function $G(\xi,x)$. From (14) it follows that the correlation function of the stationary response of the structure is a Green's function, as well. It is symmetric with respect to the space coordinates and satisfies vanishing boundary conditions at $x = 0,1$ and $\xi = 0,1$. Because of the properties (3) of the Green's function, the solution range of (13) is quadratic and subdivided in four parts. The first is situated in $0 \leq x, \xi \leq a$ and the second in $a \leq x, \xi \leq l$. The solutions of the two other ranges describe the statistical dependence of the two first ones. At each boundary line of the subranges, the correlation function (13) satisfies the geometrical and dynamic transition conditions. Finally, we pass over to the special case $x = \xi$ in order to determine the distributed square mean of the stationary displacement process $Y_{x,t}$. As shown in Fig. 2, we obtain simple quadratic functions.

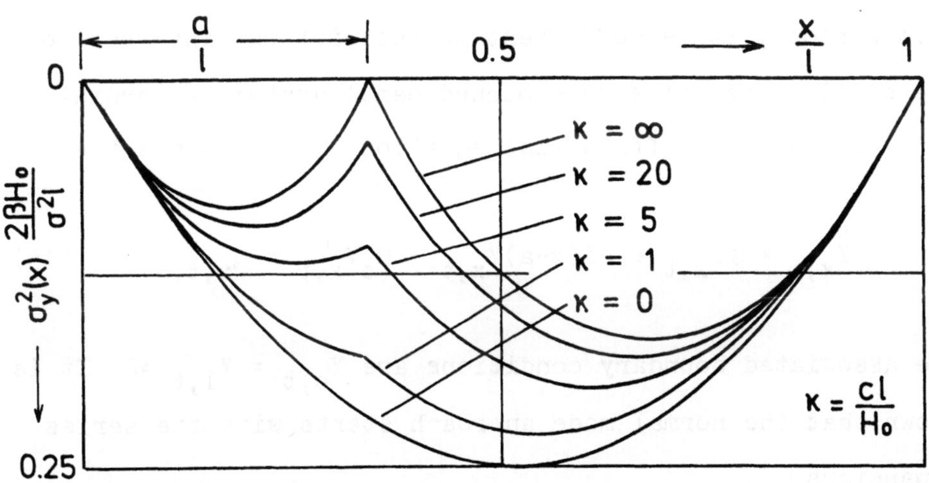

Fig. 2 Square means for several stiffness values

For an increasing stiffness parameter $\varkappa = cl/H_0$, the distributed square mean is devided into two different parabolic functions. Both solutions are interacting and coupled in such a manner that they are continous at x = a. If $\varkappa$ goes to infinity, they are completely separated.

4. APPLICATION OF A NORMAL MODE APPROACH

To get more confidence in the simple results (12), we now apply a normal mode approach to the same problems. For this purpose, it is sufficient to restrict our interest to the special case of a two-spanned panel system. According to (1), its partial differential equation of motion has the form

$$\mu \ddot{Y}_{x,t} + \beta \dot{Y}_{x,t} + c\delta(x-a)Y_{x,t} - H_o Y''_{x,t} = Q_{x,t} \qquad (15)$$

The associated boundary conditions are $Y_{o,t} = Y_{1,t} = 0$. It is known that the normal mode approach starts with the series expansions

$$Y_{x,t} = \sum_{n=1}^{\infty} Y_{n,t} V_n(x), \qquad Q_{x,t} = \sum_{n=1}^{\infty} Q_{n,t} V_n(x) \qquad (16)$$

Herein, $Y_{n,t}$ and $Q_{n,t}$ are stochastic processes while $V_n(x)$ are deterministic denoting the natural modes of the structure.

In the given boundary value problem, these modes are defined by an eigenvalue problem of the form:

$$V_n''(x) + \lambda_n^2 V_n(x) = 0, \qquad V_n(o) = V_n(1) = 0,$$
$$\int_o^1 V_n(x) V_m(x) dx = \gamma_n 1 \delta_{n,m}, \qquad n,m = 1, 2, \ldots \tag{17}$$

i.e. the eigenfunctions $V_n(x)$ are solutions of an ordinary second order differential equation. They have to satisfy the boundary conditions and the classical orthogonality condition, both noted in (17). The $\gamma_n 1$ are constants of normalization and $\delta_{n,m}$ denotes Kronecker's symbol.

We continue the analysis by introducing both series expansions (16) into the equation (15). Performing a comparison of all coefficients of $V_n(x)$ results in an infinite system of ordinary differential equations for each unknown random process $Y_{n,t}$.

$$\mu Y_{n,t} + \beta Y_{n,t} + H_o \lambda_n^2 Y_{n,t} + \frac{c}{\gamma_n 1} V_n(a) \sum_{m=1}^{\infty} V_m(a) Y_{m,t} = Q_{n,t}. \tag{18}$$

We recognize that this system contains uncoupled acceleration, damping, restoring and excitation terms. The bedding terms, however, possess a complete coupling arising from the infinite sum over all time functions $Y_{m,t}$. This is a consequence of the normal modes which satisfy only the geometric boundary conditions at the ends of the string but not the dynamic transition conditions at the concentrated spring suspension. Nevertheless, the covariance analysis of such infinite systems can be performed by means of the classical spectral method or more directly, by means of the Ljapunow equations[6]. To apply this direct analysis, we introduce the following infinite-dimensional matrices:

$$\underset{\sim}{S}_t = (Y_{n,t}), \quad \underset{\sim}{T}_t = (\dot{Y}_{n,t}), \quad \underset{\sim}{Q}_t = (Q_{n,t}), \quad n = 1,2..$$

$$\Delta = \mathrm{diag}(\lambda_n^2), \quad N = \mathrm{diag}(\frac{1}{\gamma_n 1}), \quad K = (V_n(a)V_m(a)). \qquad (19)$$

Using this notation, it is possible to go over to the first order system

$$\begin{pmatrix} \dot{\underset{\sim}{S}}_t \\ \hline \dot{\underset{\sim}{T}}_t \end{pmatrix} = \left(\begin{array}{c|c} 0 & I \\ \hline -\frac{1}{\mu}(H_o\Delta + cNK) & -\frac{1}{\mu}\beta I \end{array}\right) \begin{pmatrix} \underset{\sim}{S}_t \\ \hline \underset{\sim}{T}_t \end{pmatrix} + \begin{pmatrix} \underset{\sim}{0} \\ \hline \frac{1}{\mu}\underset{\sim}{Q}_t \end{pmatrix} \qquad (20)$$

or to the following compressed form:

$$\underset{\sim}{Y}_t = A\underset{\sim}{Y}_t + \underset{\sim}{R}_t, \quad \underset{\sim}{Y}_t = (\underset{\sim}{S}_t, \underset{\sim}{T}_t)^T, \quad \underset{\sim}{R}_t = (\underset{\sim}{0}, \frac{1}{\mu}\underset{\sim}{Q}_t)^T. \qquad (21)$$

Both are equivalent to (18).

It is known[6] that the stationary covariances of (21) are determined by the matrix equation

$$AE(\underset{\sim}{Y}_t\underset{\sim}{Y}_t^T) + E(\underset{\sim}{Y}_t\underset{\sim}{Y}_t^T)A^T = - E(\underset{\sim}{R}_t\underset{\sim}{R}_t^T). \tag{22}$$

Herein, the matrix A is given by (19) and $E(\underset{\sim}{R}_t\underset{\sim}{R}_t^T)$ can be calculated by utilizing the orthogonality of $V_n(\xi)$ and the delta correlations (2) of $Q_{x,t}$. Subsequently, the Ljapunow equation (22) is evaluated elementwise leading to the final result

$$E(Y_{n,t}Y_{m,t}) + \frac{c}{H_o}\frac{V_m(a)}{\gamma_m l\lambda_m^2}\sum_{k=1}^{\infty} E(Y_{n,t}Y_{k,t})V_k(a) = = \frac{\sigma^2}{2\beta H_o}\frac{\delta_{n,m}}{\gamma_n l\lambda_n^2}\,. \tag{23}$$

This result is an infinite system of coupled algebraic equations. Naturally, it can only approximately be solved by breaking off at a finite number of modes. For instance, if we break off with the tenth mode, we get a finite system with hundred unknown modal covariances $E(Y_{n,t}Y_{m,t})$ for n,m = 1, 2,...10 taking not into account the symmetry condition. This requires the inversion of a matrix with 10,000 elements. Applying the integral approach, however, we obtain instead of (23) the functional equation (12), here of the form

$$E(Y_{x,t}Y_{z,t}) + \frac{c}{H_o} E(Y_{x,t}Y_{a,t})G(x,a) = \frac{\sigma^2}{2\beta H_o} G(x,z). \tag{24}$$

As shown in the previous chapter, the evaluation of (24) requires to invert only a single equation in order to get results without any approximation or convergence problems. It is easy to show that the functional equation (24) coincides with the infinite system (23). To verify it, we just have to expand the equation (24) by means of the same natural modes $V_n(x)$. Using the series

$$E(Y_{x,t}Y_{z,t}) = \sum_{n,m=1}^{\infty}\sum E(Y_{n,t}Y_{m,t})V_n(x)V_m(z),$$
$$G(x,z) = \sum_{n,m=1}^{\infty}\sum \frac{1}{\gamma_n l \lambda_n^2}\,\delta_{n,m}V_n(x)V_m(z) \tag{25}$$

the coincidence follows in a direct manner. Consequently, there are no differences in the results of both methods, there are only different representations.

5. FINAL REMARKS

The present contribution is devoted to the classical problem of stationary vibrations of structures. As an example, we considerd multispanned panel systems consisting of a uniform string with elastic suspensions. The applied loading is given by the basic model of a stochastic field excitation which is uncorrelated in space and time.

The applied covariance analysis uses two different approaches. In the first one, the problem is formulated by its equivalent integral representation which allows to apply Itô's formula in order to investigate the covariances of the response processes. In the stationary case, this results in a finite element system simply evaluable by means of matrix calculus.

In the second approach, the solutions of interest are expanded in the natural modes of the structure leading to an infinte series of modal covariance equations. It is shown that they coincide with the correspondingly expanded equations of the integral analysis, mentioned above.

The advantage of the first method is quite obvious. Since it is based on mathematical models with distributed parameters, we obtain a concentrated description of the stationary response which simplifies enormously the numerical handling of such dynamic systems and therefore applications in the system identification, as well.

REFERENCES

[1] Wedig, W., Zufallsschwingungen von querangeströmten Saiten, Ing.-Archiv, 48, 325, 1979.

[2] Wedig, W., Moments and probability densities of parametrically excited systems and continuous systems, Proc. of ICNO - VII, Berlin, II,2, 1977.

[3] Arnold, L., Stochastische Differentialgleichungen, Oldenbourg, München, 1973.

[4] Wedig, W., Development of a new finite element method for stochastic systems with distributed parameters, Berichte zur Zuverlässigkeitstheorie der Bauwerke, University of München, 49, 79, 1980.

[5] Wedig, W., Stationäre Zufallsschwingungen von Balken - eine neue Methode zur Kovarianzanalyse, ZAMM 60, T 89, 1980.

[6] Wedig, W., Stochastic vibrations of slim masts induced by turbulent flows, Proc. of Int. Conference on Structural Dynamics, University of Southampton, 1, 131, 1980.

272 – H.G. NATKE : Identification of Vibrating Structures

ERRATA

page 66

equation (2.12) should read

$$y_M(k) = - \sum_{\nu=1}^{n} a_\nu y(k-\nu) + \sum_{\nu=0}^{n} b_\nu u(k-\nu)$$

page 73

line 3 from top should read

$$k \geq (n+1)$$

page 74

line 9 from top should read

$$(\nu=1,2,...,n)$$

page 99

before Table 3.1 insert the following line

able signals u and y are described by stochastic signals.

page 411, line 2 from above:

"guaranteed" instead of "quaranteed"

MIX
Papier aus verantwortungsvollen Quellen
Paper from responsible sources
FSC® C105338

If you have any concerns about our products,
you can contact us on
ProductSafety@springernature.com

In case Publisher is established outside the EU,
the EU authorized representative is:
Springer Nature Customer Service Center GmbH
Europaplatz 3, 69115 Heidelberg, Germany

Printed by Libri Plureos GmbH
in Hamburg, Germany